I0482758

Environmental Impact Statement on the Construction and Operation of a Proposed Mixed Oxide Fuel Fabrication Facility at the Savannah River Site, South Carolina

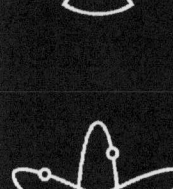

Chapters 1 through 8 and Appendices A through E

Final Report

U.S. Nuclear Regulatory Commission
Office of Nuclear Material Safety and Safeguards
Washington, DC 20555-0001

AVAILABILITY OF REFERENCE MATERIALS
IN NRC PUBLICATIONS

NUREG-1767, Vol. 1

Environmental Impact Statement on the Construction and Operation of a Proposed Mixed Oxide Fuel Fabrication Facility at the Savannah River Site, South Carolina

Chapters 1 through 8 and Appendices A through E

Final Report

Manuscript Completed: January 2005
Date Published: January 2005

Division of Waste Management and Environmental Protection
Office of Nuclear Material Safety and Safeguards
U.S. Nuclear Regulatory Commission
Washington, DC 20555-0001

ABSTRACT

The U.S. Department of Energy (DOE) has contracted with Duke Cogema Stone & Webster (DCS) to design, construct, and operate a proposed Mixed Oxide (MOX) Fuel Fabrication Facility that would convert depleted uranium and weapons-grade plutonium into MOX fuel. The proposed MOX facility would be located on the DOE's Savannah River Site in South Carolina. Use of the proposed facility to produce MOX fuel would be part of the DOE's surplus plutonium disposition program. The purpose of the DOE program is to ensure that plutonium produced for nuclear weapons and declared excess to national security is converted to proliferation-resistant forms.

This final environmental impact statement (FEIS) was prepared in compliance with the National Environmental Policy Act (NEPA), the U.S. Nuclear Regulatory Commission's (NRC's) regulations for implementing NEPA, and the guidance provided by the Council on Environmental Quality regulations implementing the procedural provisions of NEPA. This FEIS evaluates the potential environmental impacts of the proposed action. The document discusses the purpose and need for the proposed action, describes the proposed action and its reasonable alternatives, describes the environment potentially affected by the proposal, presents and compares the potential environmental impacts resulting from the proposed action and its alternatives, and identifies mitigation measures that could eliminate or lessen the potential environmental impacts. The document also includes comments received on the draft environmental impact statement and NRC's responses.

CONTENTS

CONTENTS (Cont.)

CONTENTS (Cont.)

CONTENTS (Cont.)

CONTENTS (Cont.)

CONTENTS (Cont.)

FIGURES

TABLES

TABLES (Cont.)

TABLES (Cont.)

TABLES (Cont.)

TABLES (Cont.)

EXECUTIVE SUMMARY

The consortium of Duke Project Services Group, Inc., COGEMA, Inc., and Stone & Webster, Inc., has formed a Limited Liability Company called Duke Cogema Stone & Webster (DCS). DCS has been hired by the U.S. Department of Energy (DOE) to design, construct, and operate a facility (the proposed MOX facility) that would convert depleted uranium and surplus weapons-grade plutonium into mixed oxide (MOX) fuel. The DOE is responsible for the surplus plutonium disposition program for the United States. Within this program, the U.S. Nuclear Regulatory Commission (NRC) has the independent responsibility of determining whether the proposed MOX facility can be built and operated in a safe and environmentally acceptable manner. The proposed action requiring the February 2003 draft environmental impact statement (DEIS) and this NRC final environmental impact statement (FEIS) involves a decision by the NRC whether to authorize DCS to construct and later operate the proposed MOX facility at DOE's Savannah River Site (SRS) in South Carolina. DCS has submitted to the NRC, among other documents, a revised Construction Authorization Request (CAR) and a revised environmental report (ER), in seeking authority to begin constructing the proposed MOX facility.

This FEIS was prepared by the staff of the NRC and its contractor, Argonne National Laboratory, and complies with the National Environmental Policy Act (NEPA), NRC regulations for implementing NEPA (Title 10, Part 51 of the *Code of Federal Regulations* [10 CFR Part 51]), and the applicable Council on Environmental Quality (CEQ) regulations.

The proposed MOX facility would convert 34 metric tons (MT) (37.5 tons) of surplus weapons-grade plutonium into MOX fuel. This facility would be built on 16.6 ha (41 acres) of land in the F-Area of the SRS. If the NRC approves the CAR, DCS plans to request a 10 CFR Part 70 license to possess and use special nuclear material at the proposed MOX facility. Such a license would allow DCS to operate the proposed MOX facility for 20 years. The facility would be designed for a maximum annual throughput of 3.5 MT (3.9 tons) of plutonium.

Feedstock (surplus plutonium dioxide and depleted uranium dioxide) would be required to be transported to the SRS to make the MOX fuel. The surplus plutonium is currently stored at seven DOE facilities at various locations in the United States. Additionally, depleted uranium hexafluoride would need to be transported from a DOE site (assumed to be the gaseous diffusion uranium enrichment facility in Portsmouth, Ohio) to a commercial fuel fabrication facility (assumed to be the Global Nuclear Fuel Americas, LLC, in Wilmington, North Carolina), where it would be converted to depleted uranium dioxide, which would then be transported to the SRS. Once manufactured, the MOX fuel would be transported to mission reactors, where it would be irradiated. For purposes of complying with NEPA's requirements, it is assumed that one or more reactors will later be authorized by the NRC to use MOX fuel, and the FEIS includes a generic evaluation of using MOX fuel in a reactor. In order for a specific commercial reactor to use MOX fuel, an amendment to its 10 CFR Part 50 NRC license would be required. The NRC would analyze the site-specific environmental impacts related to such an amendment if and when such a request was made to the NRC. Following irradiation and storage at reactor sites, the spent MOX fuel would be transported to a geologic repository (assuming one is later licensed by the NRC to operate) for final disposal, and the FEIS includes a discussion of spent MOX fuel transportation impacts.

In addition to presenting the potential environmental impacts of the proposed MOX facility and the related fuel cycle impacts, this FEIS discusses two proposed DOE facilities — the Pit Disassembly and Conversion Facility (PDCF) and the Waste Solidification Building (WSB) — which would also be located at the SRS, that would be required to support operation of the proposed MOX facility. The PDCF would be required to convert approximately 25.6 MT (28.2 tons) of surplus plutonium from a metallic form to plutonium dioxide powder. The remaining quantity of surplus plutonium, called "alternate feedstock," would be in a form that would be suitable to go directly to the proposed MOX facility. The proposed MOX facility would remove impurities from the plutonium dioxide and mix it with depleted uranium dioxide to make MOX fuel.

The WSB would process liquid waste streams from the PDCF and proposed MOX facility. The WSB may also be used for temporary storage and processing of other waste forms generated at the proposed MOX facility and the PDCF before such wastes are transferred to the SRS waste management system or shipped off-site for disposition. In addition, infrastructure upgrades would be needed to support the proposed MOX facility. These upgrades would include constructing waste transfer pipelines, realigning electric utility lines, and adding access roads.

A brief summary of FEIS Chapters 1-6 follows. Chapter 1 of the FEIS discusses the purpose and need for this action and its relationship to the DOE's surplus plutonium disposition program. The fundamental purpose of this DOE program is to ensure that surplus weapons-grade plutonium is converted to proliferation-resistant forms. The DOE's program is intended to lay the foundation for parallel disposition of excess Russian plutonium, thereby protecting against proliferation of materials capable of making weapons of mass destruction.

Chapter 2 of this FEIS describes the proposed action and alternatives to the proposed action, including the no-action alternative. The no-action alternative consists of the continued storage of surplus plutonium at various locations throughout the DOE complex, in the event the NRC does not approve the proposed MOX facility. This alternative is evaluated in detail in Chapter 4. Other alternatives to the proposed action discussed in Chapter 2 include alternate locations for the proposed MOX facility in the F-Area, alternate technology and design options, immobilizing surplus plutonium instead of producing MOX fuel, deliberately making off-specification MOX fuel, the "MIX MOX" alternative, and the Parallex Project (which involves irradiating the MOX fuel in Canadian deuterium uranium reactors).

Chapter 3 describes the environment that would be affected by the proposed action and includes discussions on soils, hydrology, air quality, local ecology, waste management, risks to human health, and socioeconomic issues.

Chapter 4 evaluates and compares the environmental effects of the proposed action and the no-action alternative. Significant or more important potential impacts are discussed in Chapter 4, which includes the following topics: (1) human health, (2) air quality, (3) hydrology,

(4) waste management, (5) accident impacts, (6) decommissioning, and (7) environmental justice. Indirect impacts of transportation of radioactive materials, conversion of depleted uranium, and reactor use are discussed in Chapter 4. The following potential impacts for the no-action alternative and proposed action are considered to be less significant and are discussed in Appendixes G and H: (1) geology, seismology, and soils; (2) noise; (3) ecology; (4) land use; (5) cultural and paleontological resources; (6) infrastructure; and (7) socioeconomics. A summary of the significant or more important potential impacts discussed in Chapter 4 is presented below.

The annual collective dose to members of the public (i.e., those living and working within 80 km [50 mi] of the SRS) produced by routine operation of the proposed MOX facility, the PDCF, and the WSB would be expected to result in a latent cancer fatality (LCF) rate of approximately 0.0009/yr or less. Routine operation of the proposed MOX facility, the PDCF, and the WSB is expected to produce small air quality impacts and would not cause exceedance of any ambient air quality standard level for criteria pollutants at the SRS.

Construction and routine operation of the proposed facilities would not be expected to cause any disproportionately high and adverse impacts to low-income or minority populations in the SRS vicinity. Of the accidents evaluated, a hypothetical PDCF tritium release accident had the highest estimated short-term impacts, approximately 3 LCFs among members of the off-site public. Such an accident also had the highest estimated 1-year exposure impact, including the ingestion dose, of up to 100 LCFs among members of the off-site public. However, it is regarded as highly unlikely that such an accident would occur, and the risk to any population, including low-income and minority communities, is considered to be low. Nevertheless, the communities most likely to be affected by a significant accident would be minority or low-income, given the demographics and prevailing wind direction. The extent to which low-income or minority population groups would be affected would depend on the amount of material released and the direction and speed of the wind.

Transportation of uranium and plutonium feedstock materials, transuranic waste, fresh MOX fuel, and spent MOX fuel would result in approximately 3,300,000 to 8,200,000 km (2,050,000 to 5,100,000 mi) traveled by 1,497 to 3,512 truck shipments over the operations period of the proposed MOX facility. Up to 1 LCF might be expected from the radioactive nature of the cargo. (Estimated LCFs for members of the public and the transportation crews were 0.2 to 0.4 and 0.1 to 0.3, respectively.) One to two latent fatalities from vehicle emissions were estimated, and no fatalities (0.078 to 0.20 fatality) from the physical trauma of potential vehicle accidents were estimated.

Chapter 4 of the FEIS also evaluates the use of MOX fuel in a generic reactor using a 40% MOX fuel core. For both normal operations and design-basis accidents, the impacts of using MOX fuel in a reactor would not be significantly different from the impacts of a reactor using 100% low-enriched uranium fuel. For highly unlikely beyond-design-basis accidents, the impacts for a reactor using a 40% MOX fuel core could be up to 14% greater than for a reactor using 100% low-enriched uranium fuel. Since no reactor licensee has yet sought the authority to use MOX fuel, the transportation of fresh MOX fuel is also evaluated on a generic basis, using a surrogate reactor located in the Midwest.

Chapter 4 also presents the costs and benefits of the proposed action. The primary benefit of operating the proposed MOX facility would be the resulting reduction in the supply of weapons-grade plutonium available for unauthorized use. Converting surplus plutonium in this manner is viewed as being a safer use/disposition strategy than the DOE's continued storage of surplus plutonium, as would occur under the no-action alternative, because it would reduce the number of locations where the various forms of plutonium are stored. Further, converting weapons-grade plutonium into MOX fuel in the United States — as opposed to immobilizing a portion of it as the DOE had previously planned to do — lays the foundation for parallel disposition of weapons-grade plutonium in Russia, which distrusts immobilization because of its failure to degrade the plutonium's isotopic composition. Converting surplus plutonium into MOX fuel is thus viewed as a better way of ensuring that weapons-usable material will not be obtained by rogue states and terrorist groups. Implementing the proposed action is expected to promote the above nonproliferation objectives.

In addition to the above primary benefits, there would be secondary economic benefits of the proposed action. Impacts of construction on the regional economic area (REA) and region of influence (ROI) would be beneficial with respect to jobs and income. During operations, the proposed MOX facility, PDCF, and WSB would be expected to generate 490 direct and 780 indirect jobs, producing a total income of $64 million a year in the REA. The economic cost benefit analysis for the proposed action shows an overall net benefit to the ROI and REA of $1,940 million. National economic costs for the proposed MOX facility, PDCF, and WSB are estimated to be $4,064 million (in 2003 dollars). The national economic benefits would include adding employment and income in various sectors of the economy through the purchase of goods and services required during construction and operation.

Chapter 5 of the FEIS identifies mitigation measures that could eliminate or lessen the potential environmental impacts of the proposed action. The NRC evaluated proposed mitigation measures identified by DCS and identified additional measures that could reduce or eliminate adverse environmental impacts of the proposed action. On the basis of its independent review, the NRC is making a preliminary conclusion that the potential significant impacts of the proposed action can be mitigated. However, any possession and use license issued to DCS should be conditioned on the commitments made by DCS and the various proposed NRC mitigation requirements discussed in Chapter 5.

Chapter 6 presents the many federal, state, and local environmental requirements that would be applicable to the proposed MOX facility.

After weighing the costs and benefits of the proposed action, comparing alternatives, and considering the comments received on the DEIS (see FEIS Appendix J), the NRC staff, in accordance with 10 CFR 51.91(d), includes in this FEIS its final NEPA recommendation regarding the proposed action. As discussed further in Chapter 2, the NRC staff continues to recommend that, unless safety issues mandate otherwise, the action called for is the issuance of the proposed license to DCS, with conditions to protect environmental values. As stated in Chapter 2, the NRC staff concludes that (1) the applicable environmental requirements presented in FEIS Chapter 6 and (2) the proposed mitigation measures discussed in FEIS

Chapter 5 would eliminate or substantially lessen any potential adverse environmental impacts associated with the proposed action.

Appendix J includes a summary of the comments and responses received on the DEIS. Ninety-four commenters submitted about 750 comments on the DEIS. Appendix J also identifies changes in the FEIS text based on the comments and revised accident analyses from new design information for the WSB provided by DCS since publication of the DEIS.

ACRONYMS AND ABBREVIATIONS

The following is a list of the acronyms, initialisms, abbreviations, and units of measure used in this document. Some acronyms and abbreviations used only in tables, figures, equations, or as reference callouts are defined in the respective tables, figures, equations, and reference lists.

Acronyms, Initialisms, and Abbreviations

7Q10	7-day low flow, 10-year recurrence flow
AADT	average annual daily traffic
ADU	ammonium diuranate
AEA	Atomic Energy Act
Ag	silver
AgNO$_3$	silver nitrate
ALARA	as low as reasonably achievable
ALI	annual limit on intake
ALOHA	Areal Locations of Hazardous Atmospheres (computer code)
Am	americium
ANL-W	Argonne National Laboratory-West
ANSI	American National Standards Institute
APA	aqueous polishing area
APSF	Actinide Packaging and Storage Facility
AQCR	Air Quality Control Region
BPIP	Building Profile Input Program
BRP	Reagents Processing Building
CAA	Clean Air Act
CANDU	Canadian Deuterium Uranium (reactor)
CAR	Construction Authorization Request
CAS	Chemical Abstract Services
CEDE	committed effective dose equivalent
CEQ	Council on Environmental Quality
CERCLA	Comprehensive Environmental Response, Compensation, and Liability Act
CFR	*Code of Federal Regulations*
CH-TRU	contact-handled transuranic (waste)
CIESIN	Center for International Earth Science Information Network
CIF	Consolidated Incineration Facility
CO	carbon monoxide
CO$_2$	carbon dioxide
CPT	cone-penetration test
CSWTF	Central Sanitary Wastewater Treatment Facility
CWA	Clean Water Act

D&D	deactivation and decommissioning
DCP	dry conversion process
DCS	Duke Cogema Stone & Webster
DDE	deep dose equivalent
DEIS	draft environmental impact statement
DOE	U.S. Department of Energy
DOT	U.S. Department of Transportation
DWPF	Defense Waste Processing Facility
EA	environmental assessment
EBR-II	Experimental Breeder Reactor-II
EDE	effective dose equivalent
EIS	environmental impact statement
EPA	U.S. Environmental Protection Agency
ER	Environmental Report
ERPG	Emergency Response Planning Guideline
ETF	Effluent Treatment Facility
FEIS	final environmental impact statement
FGR	Federal Guidance Report
FOF	F-Area Outside Facility
FONSI	Finding of No Significant Impact
FR	*Federal Register*
FSER	final safety evaluation report
FTE	full-time equivalent
FY	fiscal year
Ga	gallium
GE	General Electric
GENII	Generation II (computer code)
GRP	gross regional product
$H_2C_2O_4$	oxalic acid
HEPA	high-efficiency particulate air (filter)
HEU	highly enriched uranium
HF	hydrogen fluoride
HI	hazard index
HLW	high-level (radioactive) waste
HQ	hazard quotient
HRCQ	highway route controlled quantity
HSWA	Hazardous and Solid Waste Amendments
HVAC	heating, ventilation, and air conditioning
HYDOX	hydride-oxidation
ICRP	International Commission on Radiological Protection
IMPLAN	Intelligent Multi-Resource Planning (computer code)

INEEL	Idaho National Engineering and Environmental Laboratory
ISA	integrated safety analysis
ISCST3	Industrial Source Complex Short-Term (version 3) model
ISFSI	interim spent fuel storage installation
ITP	in-tank precipitation
KAMS	K-Area Material Storage (SRS)
LANL	Los Alamos National Laboratory
LCF	latent cancer fatality
L_{dn}	day-night average sound level
L_{eq}	equivalent sound pressure level
LEU	low-enriched uranium
LLC	Limited Liability Company
LLNL	Lawrence Livermore National Laboratory
LLW	low-level (radioactive) waste
LSA	low specific activity
LTA	lead test assembly
MAR	material at risk
MBTA	Migratory Bird Treaty Act
MC&A	material control and accounting
MEI	maximally exposed individual
MMI	Modified Mercalli Intensity (earthquake intensity scale)
MOX	mixed oxide (plutonium dioxide and uranium dioxide)
MPQAP	MOX Project Quality Assurance Plan
MSL	mean sea level
MWMF	Mixed Waste Management Facility
NAAQS	National Ambient Air Quality Standards
NEPA	National Environmental Policy Act
NERP	National Environmental Research Park
NESHAPs	National Emission Standards for Hazardous Air Pollutants
NMSS	Office of Nuclear Material Safety and Safeguards (NRC)
NNSA	National Nuclear Security Administration
NO_2	nitrogen dioxide
NOAA	National Oceanic and Atmospheric Administration
NOI	Notice of Intent
NO_x	nitrogen oxides
NPDES	National Pollutant Discharge Elimination System
NRC	U.S. Nuclear Regulatory Commission
NRHP	*National Register of Historic Places*
NSC	National Safety Council
NSPS	New Source Performance Standards

O_3	ozone
OAQPS	Office of Air Quality Planning and Standards (EPA)
OFASB	Old F-Area Seepage Basin
OHER	Office of Health and Environmental Research (DOE)
OML	oxalic mother liquor
ORR	Oak Ridge Reservation
OSHA	Occupational Health and Safety Administration
PAG	protective action guide
PAH	polycyclic aromatic hydrocarbon
Pb	lead
PDCF	Pit Disassembly and Conversion Facility
PEIS	programmatic environmental impact statement
PM	particulate matter
PM_{10}	particulate matter with a diameter less than or equal to 10 micrometers
$PM_{2.5}$	particulate matter with a diameter less than or equal to 2.5 micrometers
PMF	probable maximum flood
PSD	Prevention of Significant Deterioration
PSSCs	principal structures, systems, and components
Pu	plutonium
Pu (IV)	tetravalent plutonium
Pu (III)	trivalent plutonium
PuO_2	plutonium oxide
QA	quality assurance
RCRA	Resource Conservation and Recovery Act
REA	regional economic area
REG	mitigation measures instituted to ensure compliance with regulations, permits, and guidelines
RFETS	Rocky Flats Environmental Technology Site
ROD	Record of Decision
ROI	region of influence
S&D PEIS	Storage and Disposition Programmatic Environmental Impact Statement
SA	Supplement Analysis
SAAQS	State Ambient Air Quality Standard
SC	South Carolina; state route
SCAPA	Subcommittee on Consequence Assessment and Protective Action (DOE)
SCDHEC	South Carolina Department of Health and Environmental Control
SCDNR	South Carolina Department of Natural Resources
SCSHPO	South Carolina State Historic Preservation Officer
SER	safety evaluation report
SGT	Safeguards Transporter
SHPO	State Historic Preservation Office
SIP	state implementation plan

SNF	spent nuclear fuel
SNM	special nuclear material
SO_2	sulfur dioxide
SO_x	sulfur oxides
SPCC	spill prevention control and countermeasures
SPD	surplus plutonium disposition
SPD EIS	Surplus Plutonium Disposition Environmental Impact Statement
SPL	sound pressure level
SR	State Route
SRARP	Savannah River Archaeological Research Program
SREL	Savannah River Ecology Laboratory
SRS	Savannah River Site
SWB	standard waste box
TAP	toxic air pollutant
TCDD	tetrachlorodibenzo-para-dioxin
TEDE	total effective dose equivalent
TEEL	temporary emergency exposure limit
TI	transport index
TIGR	thermally induced gallium removal
TRAGIS	Transportation Routing Analysis Geographic Information System
TRU	transuranic (radioactive waste)
TRUPACT	transuranic package transporter
TSCA	Toxic Substances Control Act
TSD	Transportation Safeguards Division (DOE Albuquerque Operations Office)
TSP	total suspended particulates
U	uranium
UF_6	uranium hexafluoride
UO_2	uranium dioxide
U.S.C.	*United States Code*
VOC	volatile organic compound
VRM	visual resource management
WAC	waste acceptance criteria
WIPP	Waste Isolation Pilot Plant
WM PEIS	*Final Waste Management Programmatic Environmental Impact Statement for Managing Treatment, Storage, and Disposal of Radioactive and Hazardous Waste*
WMA	Wildlife Management Area
WSB	Waste Solidification Building

Units of Measure

Bq	becquerel(s)		km^2	square kilometer(s)
Btu	British thermal unit(s)		kV	kilovolt(s)
Ci	curie(s)		L	liter(s)
μCi	microcurie(s)		lb	pound(s)
cm	centimeter(s)		m	meter(s)
d	day(s)		m^2	square meter(s)
dB	decibel(s)		m^3	cubic meter(s)
dBA	A-weighted decibel(s)		μm	micrometer(s)
dps	disintegration(s) per second		mg	milligram(s)
°C	degree(s) Celsius		mi	mile(s)
°F	degree(s) Fahrenheit		mi^2	square mile(s)
ft	foot (feet)		min	minutes
ft^2	square foot (feet)		mm	millimeter(s)
ft^3	cubic foot (feet)		mo	month(s)
g	gram(s) or		mph	mile(s) per hour
	gravitational acceleration		mrem	millirem(s)
μg	microgram(s)		mSv	millisievert(s)
gal	gallon(s)		MT	metric ton(s)
gpm	gallon(s) per minute		MWh	megawatt-hour(s)
h	hour(s)		nCi	nanocurie(s)
ha	hectare(s)		Pa	Pascal(s)
hg	mercury		ppb	part(s) per billion
Hz	hertz		ppm	part(s) per million
in.	inch(es)		s	second(s)
K	kelvin degrees (temperature)		Sv	sievert(s)
kg	kilogram(s)		yd^3	cubic yard(s)
km	kilometer(s)		yr	year(s)

1 PURPOSE OF AND NEED FOR ACTION

1.1 Introduction

In 1992, at the end of the Cold War, the President commissioned the National Academy of Sciences to study management and disposition options for surplus weapons-usable plutonium. Several agreements were subsequently reached with Russia on the mutual reduction of plutonium stockpiles. The U.S. Department of Energy (DOE) is responsible for the surplus plutonium disposition program for the United States. Within this program, the U.S. Nuclear Regulatory Commission (NRC) has the independent responsibility of reviewing a proposal to design, construct, and operate a facility in the United States that would convert depleted uranium dioxide and weapons-grade plutonium dioxide into mixed oxide (MOX) fuel. A 1998 amendment to the Energy Reorganization Act of 1974 gave the NRC licensing and related regulatory authority over the proposed facility. In accordance with the National Environmental Policy Act (NEPA), 42 *United States Code* (U.S.C.) 4321 *et seq.*, the proposal to build and operate such a facility is being reviewed by the NRC in this final environmental impact statement (FEIS), to evaluate the potential environmental impacts that would result if the proposed action is taken.

The surplus plutonium disposition program is discussed in Section 1.1.1. The proposed action is described in Section 1.2, and the purpose and need for the proposed action are discussed in Section 1.3. Section 1.4 describes the process used by the NRC to determine the scope of this environmental impact statement (EIS), which identified the issues to be studied in detail and the issues that do not require detailed study.

1.1.1 Surplus Plutonium Disposition Program

Following the end of the Cold War, the United States and Russia took steps to mutually reduce their respective stockpiles of weapons-grade plutonium by declaring some of this plutonium excess to national security needs. The surplus plutonium disposition program involves making sure that this surplus plutonium cannot be used again to make nuclear weapons. The DOE evaluated a number of strategies to disposition the U.S. stockpile of surplus plutonium and has published two related EISs, a record of decision (ROD), and an amended ROD (DOE 1996, 1999, 2000, 2002). As part of this program, in 1999, the DOE selected a contractor, Duke Cogema Stone & Webster (DCS), to design, construct, and operate a facility that would convert uranium and weapons-grade plutonium into MOX fuel, as discussed further in Section 1.1.2.

To implement DOE's surplus plutonium disposition program, the DOE ROD in January 2000 set forth a "hybrid" approach, which involved immobilizing a portion of the surplus plutonium and converting the remaining portion into nuclear reactor fuel. Three new facilities were proposed for the DOE's Savannah River Site (SRS) in South Carolina to implement the hybrid approach. A Pit Disassembly and Conversion Facility (PDCF) would convert metallic weapons material, called pits, to plutonium dioxide powder. The proposed PDCF would be built and operated under the DOE's jurisdiction and authority. A plutonium immobilization plant was proposed to convert some of the plutonium dioxide powder from the PDCF and plutonium from other sources into ceramic cylinders to be encapsulated in vitrified high-level waste. The Mixed Oxide Fuel Fabrication Facility (hereafter referred to as "the proposed MOX facility") would convert the balance of the plutonium dioxide powder from the PDCF into MOX fuel for subsequent irradiation in U.S. commercial reactors authorized by the NRC to use such fuel.

Under its January 2000 ROD, the DOE planned to convert 33 metric tons (MT)[1] (36.4 tons) of surplus plutonium into MOX fuel and to immobilize 17 MT (19 tons) in the plutonium immobilization plant. Among the plutonium disposition program's purposes is to reduce over time the number of locations in the United States where the various forms of plutonium are stored, to better ensure that weapons-usable material does not fall into the hands of rogue states or terrorist groups. Irradiated MOX fuel would be highly radioactive, making it inaccessible for reuse as nuclear weapons material. In September 2000, Russia and the United States agreed to disposition 34 MT (37.5 tons) of surplus weapons-grade plutonium from their respective stockpiles (White House 2000). Under this agreement, disposition may be accomplished either by immobilization or by MOX fuel fabrication and subsequent irradiation.

However in April 2002, the DOE issued an amended ROD (DOE 2002), in which it decided not to pursue its hybrid approach due to budgetary constraints. The DOE determined that in order to make progress with available funds, only one approach could be supported. Russia does not consider immobilization alone to be an acceptable approach because immobilization, unlike the irradiation of MOX fuel, fails to degrade the isotopic composition of the plutonium. Russia further contends that the United States could easily retrieve plutonium from the immobilized waste at a later date and reuse that plutonium in nuclear weapons (DOE 2002). Because an immobilization-only approach would jeopardize Russia's continued involvement in the joint effort to reduce supplies of weapons-grade plutonium, the DOE decided that if only one disposition approach is to be pursued, the MOX fuel approach is the preferred one. The DOE concluded that implementation of the MOX-only approach is the key to successfully completing the September 2000 agreement between Russia and the United States (DOE 2002). Accordingly, the DOE decided to pursue a MOX-only approach, under which all 34 MT (37.5 tons) of surplus weapons-grade plutonium would be converted into MOX fuel, and the DOE canceled the plutonium immobilization plant. The DOE had earlier identified Duke Power Company's four reactors at the Catawba and McGuire stations (two at each station) as potential candidates to irradiate MOX fuel. The potential candidate reactors can accommodate up to 25.5 MT (28.2 tons) of surplus plutonium in MOX fuel. The DOE has not yet identified the additional candidate reactors necessary to accommodate the additional MOX fuel (8.5 MT [9.4 tons]) to be irradiated under the amended ROD.

The DOE also issued a supplemental NEPA analysis on April 24, 2003 (DOE 2003). The Supplement Analysis (SA) addressed the above-referenced changes in DOE's surplus plutonium disposition program, to determine whether the Surplus Plutonium Disposition Final Environmental Impact Statement (SPD EIS) (DOE 1999) should be supplemented. The SA

---

[1]   A metric ton (MT) equals 1,000 kilograms (kg) and is equivalent to 1.1 tons, or approximately 2,200 pounds (lb).

discussed how adoption of the MOX-only approach required additional aqueous processing steps at the proposed MOX facility to remove impurities — mainly chlorides — from the alternate feedstock material. Additional equipment at the proposed MOX facility to remove the chlorides includes two dissolution lines, an enlarged annular tank, and a chlorine gas wash column. The SA noted that the transuranic (TRU) waste generated by operation of the proposed MOX facility would, after processing at the Waste Solidification Building (WSB), be shipped from the SRS to the DOE's Waste Isolation Pilot Plant (WIPP). The DOE stated in its SA that prior to obtaining the necessary clearances for shipping TRU waste to WIPP, the amounts of such waste would be well within existing SRS storage capacity. The DOE further found that TRU waste generated by operation of the proposed MOX facility would meet the WIPP waste acceptance criteria, and that the impacts of packaging, transporting, and disposing of such waste would be bounded by prior DOE environmental analyses. The SA concluded that "the activities and potential environmental impacts associated with the proposed processing of 6.5 MT of surplus plutonium originally intended for immobilization and the increase in the total amount of surplus plutonium to be fabricated into MOX fuel from 33 MT to 34 MT are not different in kind, and only slightly in degree, from those described in the SPD EIS." Accordingly, the DOE found no requirements for supplementing the SPD EIS.

1.1.2 MOX Fuel Fabrication Facility

As referenced above, the DOE selected DCS to design, construct, and operate the proposed MOX facility. Because Congress gave the NRC licensing and related regulatory authority over the proposed MOX facility, its construction and operation will require NRC approvals, issued pursuant to the *Code of Federal Regulations*, Title 10, Part 70 (10 CFR Part 70), "Domestic Licensing of Special Nuclear Material." As part of its licensing review, the NRC has prepared this FEIS in accordance with the NRC's 10 CFR Part 51 regulations implementing NEPA and the generally applicable Council on Environmental Quality (CEQ) regulations in 40 CFR Part 1500. This FEIS addresses the direct, indirect, and cumulative impacts related to building, operating, and decommissioning the proposed MOX facility. Although the DOE has prepared previous EISs that cover impacts of the proposed MOX facility on a programmatic level, the NRC has prepared this EIS to incorporate additional site-specific information and design details in order to meet its NEPA requirements as stated in 10 CFR Part 51.

To obtain approval to construct the facility, DCS submitted a MOX Project Quality Assurance Plan (MPQAP) on June 22, 2000, an Environmental Report (ER) on December 19, 2000 (DCS 2000), a revised MPQAP on January 29, 2001, and a Construction Authorization Request (CAR) on February 28, 2001 (DCS 2001). The NRC then published its Notice of Intent to prepare an EIS for the proposed MOX

> **Categories of Impacts**
>
> Impacts of the proposed and connected actions include:
>
> - Direct effects — caused by the proposed action and occur at the same time and place,
>
> - Indirect effects — occur later in time or are farther removed in distance but are reasonably foreseeable, and
>
> - Cumulative impacts — potential impacts when the proposed action is added to other past, present, and reasonably foreseeable future actions.

facility (NRC 2001a). Because of design changes in the proposed MOX facility resulting from DOE's amended ROD, DCS submitted Revision 2 of the ER on July 12, 2002 (DCS 2002a), and an amended CAR on October 31, 2002 (DCS 2002b). DCS submitted Revision 3 of the ER on June 20, 2003 (DCS 2003a), which updated Revision 2 to incorporate responses to requests by the NRC for additional information and revised impacts from the WSB to include preliminary design details provided by the DOE. DCS submitted Revision 4 of the ER on August 14, 2003 (DCS 2003b), which updated impacts from the WSB based on recent revisions by the DOE. On June 10, 2004, DCS submitted Revision 5 to the ER (DCS 2004a). This revision incorporated changes in the facility design affecting waste volumes. In particular, the silver recovery process was removed from the design. Other changes included movement of the controlled area boundary to be colocated with the SRS site boundary, design refinements to the WSB, and the decision to route the liquid low-level waste (LLW) streams from the proposed MOX facility and the PDCF to the WSB rather than the Effluent Treatment Facility at the SRS. On the same date, DCS also submitted revisions to its CAR (DCS 2004b). If the amended CAR is approved, DCS plans to submit its application for a 10 CFR Part 70 operating license. The date for DCS filing such an application is not known at this time.

The NRC's decision-making process for the proposed MOX facility includes an environmental review and a safety review (see text box on the MOX licensing process). In addition to this EIS, which documents NRC's environmental review, the NRC will prepare two final safety evaluation reports (FSERs). The first FSER will evaluate the CAR and will address whether construction of the proposed MOX facility may be authorized pursuant to 10 CFR Part 70 and the Atomic Energy Act. In this regard, 10 CFR 70.23(b) states that the NRC will approve construction of a plutonium processing and fuel fabrication facility if it finds that the design bases of the principal structures, systems, and components (PSSCs) and the quality assurance (QA) program provide reasonable assurance of protection against natural phenomena and the consequences of potential accidents. The 10 CFR 70.23(b) safety findings on the CAR will be documented in the first FSER, now scheduled to be issued in February 2005. The NRC will use the safety findings in the first FSER and the environmental review in this EIS to decide whether or not to authorize construction of the proposed MOX facility.

If construction is authorized, a second FSER would address whether the proposed MOX facility, as built, may be authorized to operate under a 10 CFR Part 70 license. The second FSER would evaluate a DCS application for a license to possess and use special nuclear material (SNM) at the proposed MOX facility. DCS plans to submit such an application if the amended CAR is approved. The safety findings in the second FSER and the environmental review in this EIS would be used by the NRC to decide whether or not to issue an SNM possession and use license to DCS, which would authorize operation of the proposed MOX facility.

Under NEPA, the scope of this EIS is broader than that of the FSERs. This EIS addresses the environmental impacts of constructing, operating, and decommissioning the proposed MOX facility and the environmental impacts of the alternatives considered. This EIS does not address safety issues that are not considered to have potential environmental impacts. For example, the effects of a postulated criticality accident are presented here because such an accident could produce environmental impacts. However, the question of whether the criticality

MOX Licensing Process

DCS has chosen to request authorization to build and operate a mixed oxide (MOX) fuel fabrication facility in two steps. Step 1 was the Construction Authorization Request (CAR) initially filed by DCS in February 2001. The NRC staff is performing a safety review of the CAR and plans to issue a final safety evaluation report (FSER) on the CAR in February 2005. As reflected in this environmental impact statement (EIS), the NRC staff has also performed an environmental review evaluating the impacts of both the construction and operation of the proposed MOX fuel fabrication facility.

If the NRC staff grants the CAR, DCS plans as Step 2 of the process to apply for a license to possess and use special nuclear material (SNM) at the MOX fuel fabrication facility. If such an application is filed and accepted for docketing, the NRC staff would publish a notice of opportunity for hearing in the *Federal Register*. This notice would give individuals and organizations the opportunity to request the NRC to conduct an adjudicatory hearing regarding any DCS request for an SNM license. NRC hearings are governed by the requirements in 10 CFR Part 2. Regardless of whether or not an adjudicatory hearing is held, the NRC staff would perform a safety review of any DCS request for an SNM license, prepare a second FSER, and either issue DCS an operating license or deny the application. The MOX licensing process is further summarized in the chart below.

SAFETY REVIEWS	ENVIRONMENTAL REVIEW	ADJUDICATION
Construction Authorization	**Environmental Impact Statement**	**Adjudication Hearing**
• In a CAR, the applicant must identify principal structures, systems, and components (PSSCs) that reduce the risk of accidents and natural phenomena hazards.	• Pursuant to the *Code of Federal Regulations*, Title 10, Part 51 (10 CFR Part 51) implementing regulations for the National Environmental Policy Act (NEPA), the NRC staff prepares a single EIS.	• An adjudicatory hearing regarding the CAR is now being held.
• The applicant must also address baseline design criteria and quality assurance (QA) requirements. These include issues such as fire protection, criticality control, and quality standards and records.	• The NRC EIS includes impacts from both construction and operation of the proposed action and alternatives.	
• The NRC staff issues a construction-related FSER that documents its findings on the CAR and QA program description.		
License to Possess and Use SNM		
• DCS must also submit a license application for authorization to possess and use SNM.		
• The NRC staff would issue a second FSER that documents its findings relative to the license application.		

safety controls proposed by DCS would adequately prevent such an accident is part of the NRC's safety review and is not discussed in this EIS.

1.2 Description of the Proposed Action and Connected Actions

As described further in Section 1.2.1, the proposed action involves a decision by NRC whether or not to authorize DCS to construct and later operate the proposed MOX facility at the SRS to convert 34 MT (37.5 tons) of surplus weapons-grade plutonium to MOX fuel. Section 1.2.2 describes actions that are connected to the proposed action. Connected actions fall within the scope of the actions evaluated in an EIS (40 CFR 1508.25). More detailed technical information about the proposed action and connected actions is presented in Section 2.2.

1.2.1 Proposed Action

The proposed MOX facility would be built on 16.6 ha (41 acres) of land in the F-Area of the SRS (see Figures 1.1 and 1.2). DCS is expected to request a license for 20 years. The facility would be designed for maximum annual throughput of 3.5 MT (3.9 tons) of plutonium. Impacts in the ER are based on the maximum annual design capacity. This FEIS is based on a total of 34 MT (37.5 tons) of surplus plutonium. The rate at which DCS actually processes the plutonium would likely be less than the facility's design capacity. Therefore, actual annual impacts should be less than those presented in the ER. The period of operation would likely be less than the 20-year license period. The actual period of operation would vary depending on the annual throughput over time. The 20-year licensing period would allow deactivation and decommissioning to occur prior to license termination. For purposes of this FEIS, a period of operation of 10 years is

Proposed Action

- The proposed federal action is for the U.S. Nuclear Regulatory Commission to authorize Duke Cogema Stone & Webster (DCS) to build and operate a facility to fabricate mixed oxide (MOX) fuel.

- NEPA requires preparation of an EIS for major federal actions that could significantly affect the human environment.

- To operate the MOX facility, DCS would need an NRC license to possess and use special nuclear material (surplus plutonium from the U.S. nuclear weapons program).

- Under contract with the DOE, DCS would build and operate a facility to manufacture nuclear fuel using surplus plutonium.

- The NRC-licensed facility for fabricating nuclear fuel would be located on the DOE's Savannah River Site.

assumed to bound impacts. If the actual period of operation is longer than 10 years as a result of an actual throughput less than the maximum design capacity, the annual impacts would be less, even though they would occur over a longer period of time.

Direct effects of the proposed action include effects resulting from construction, operation, and decommissioning of the proposed MOX facility to convert 34 MT (37.5 tons) of surplus plutonium into MOX fuel. Plutonium dioxide powder would be processed at the proposed MOX

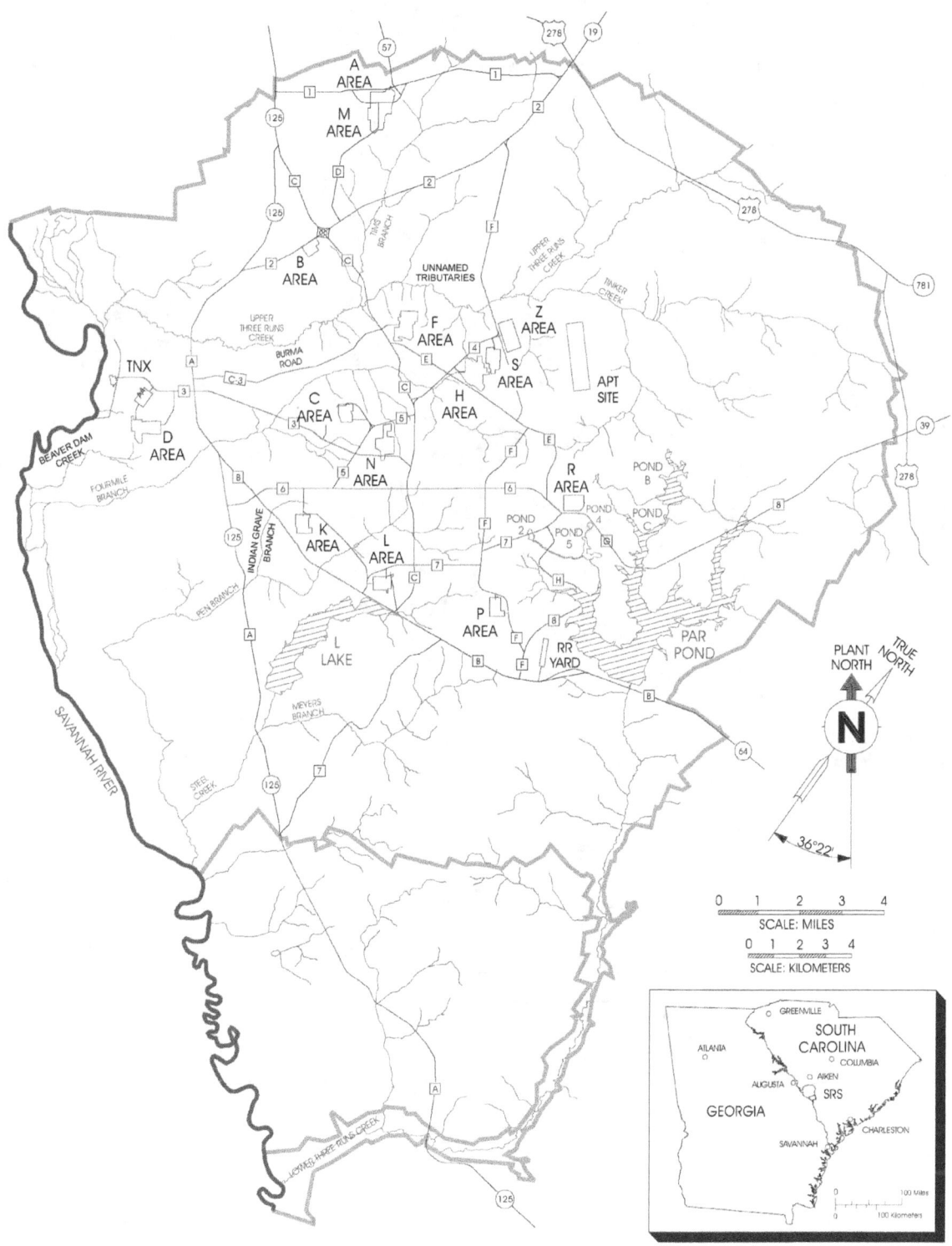

Figure 1.1. Location of the Savannah River Site and the F-Area (*Source*: DCS 2001).

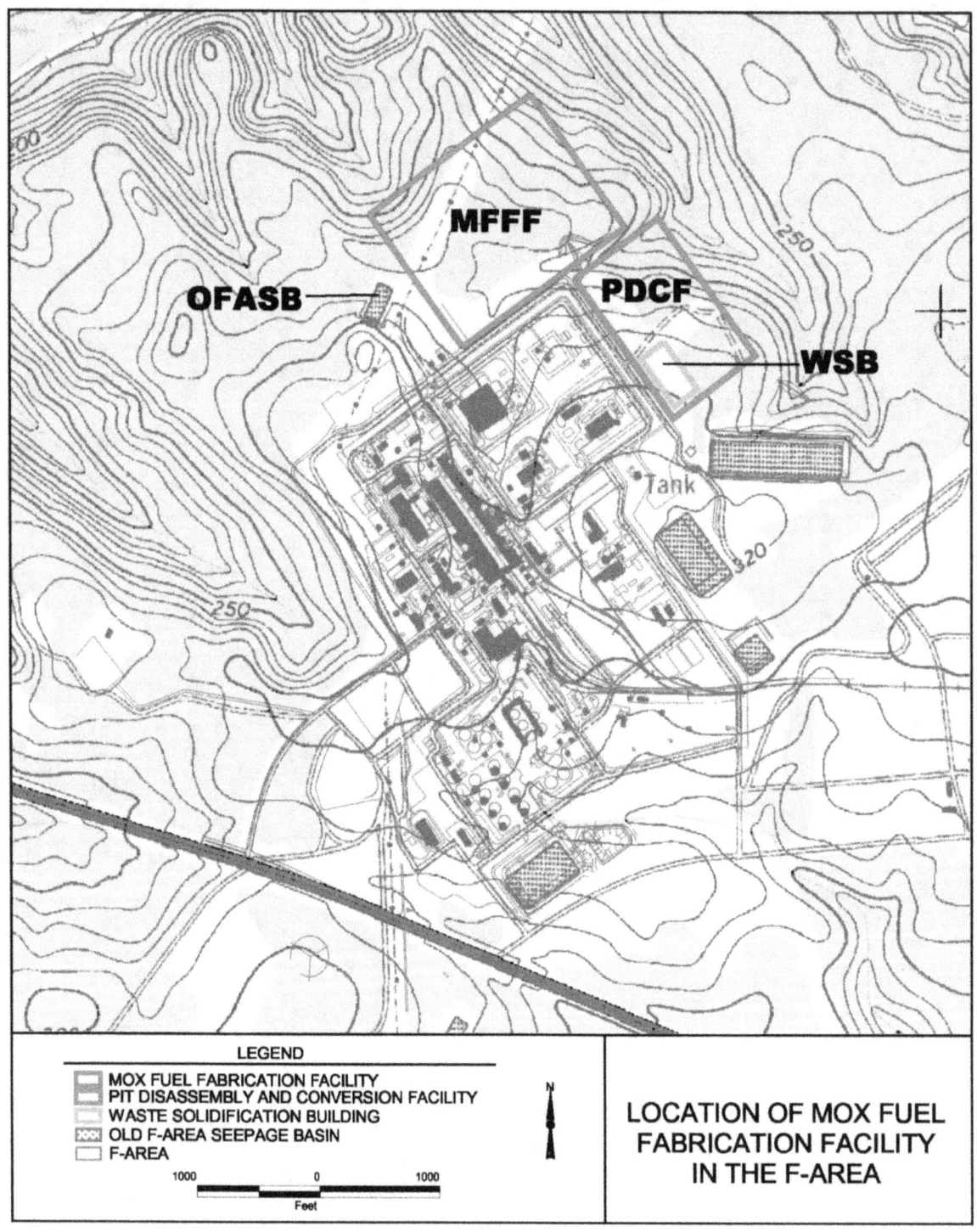

LEGEND

MOX FUEL FABRICATION FACILITY
PIT DISASSEMBLY AND CONVERSION FACILITY
WASTE SOLIDIFICATION BUILDING
OLD F-AREA SEEPAGE BASIN
F-AREA

**LOCATION OF MOX FUEL
FABRICATION FACILITY
IN THE F-AREA**

**Figure 1.2. Locations of the proposed MOX facility, the PDCF, and the WSB
in the F-Area on the SRS complex (*Source*: DCS 2002a).**

facility to remove impurities, such as americium and gallium, and would be mixed with the depleted uranium dioxide to form the MOX fuel. The final blend for MOX fuel would have a required plutonium content of 2.3% to 4.8% (percent by weight). The facility would be capable of producing MOX fuel with a plutonium content of up to 6% (DCS 2001).

1.2.2 Connected Actions

In order for the proposed MOX facility to fulfill its function, other "connected actions" would also occur. For example, the PDCF would be the source of some of the plutonium dioxide needed to make MOX fuel. Therefore, the PDCF must be constructed and authorized by the DOE to operate so that the proposed MOX facility would have the required material with which to make MOX fuel.

Connected Actions

Actions closely related to the proposed action that:

- Automatically trigger other actions which may require environmental impact statements,

- Cannot or will not proceed unless other actions are taken previously or simultaneously, or

- Are interdependent parts of a larger action and depend on the larger action for their justification.

Feedstock (surplus plutonium dioxide and depleted uranium dioxide) would be required to be transported to the SRS to make the MOX fuel. Because the surplus plutonium is currently stored at seven DOE facilities (see Figure 1.3 and Table 1.1), it would need to be transported to the SRS (DOE 2000). The depleted uranium hexafluoride would first be transported from a DOE site (assumed to be the gaseous diffusion uranium enrichment facility in Portsmouth, Ohio) to an existing commercial fuel fabrication facility (assumed to be the Global Nuclear Fuel-Americas, LLC, facility in Wilmington, North Carolina), where it would be converted to depleted uranium dioxide, which would then be transported to the SRS.

Two new DOE facilities (the PDCF and the WSB) are needed to support the proposed MOX facility. The PDCF would be required to convert approximately 25.6 MT (28.2 tons) of surplus plutonium metal to plutonium dioxide. The remaining quantity of surplus plutonium, called "alternate feedstock," would be in a form that would be suitable to go directly to the proposed MOX facility. The WSB would process liquid waste streams from the PDCF and the proposed MOX facility. Since the PDCF and WSB would not be under NRC's Atomic Energy Act jurisdiction, the safety issues pertaining to the PDCF and WSB will not be addressed by the NRC in the FSERs.

As discussed in Section 4.3.4, the wastes generated at the proposed MOX facility and the PDCF would be managed at the WSB, sent to the SRS waste management system, or sent to approved facilities off the SRS property for disposition. In addition, infrastructure upgrades would be needed to support the proposed MOX facility. These upgrades include waste transfer pipelines, electric utility line realignment, and addition of access roads.

The FEIS also evaluates transporting the fresh (unirradiated) MOX fuel made by the proposed MOX facility (assuming it is built and is authorized to operate) to mission reactors for irradiation.

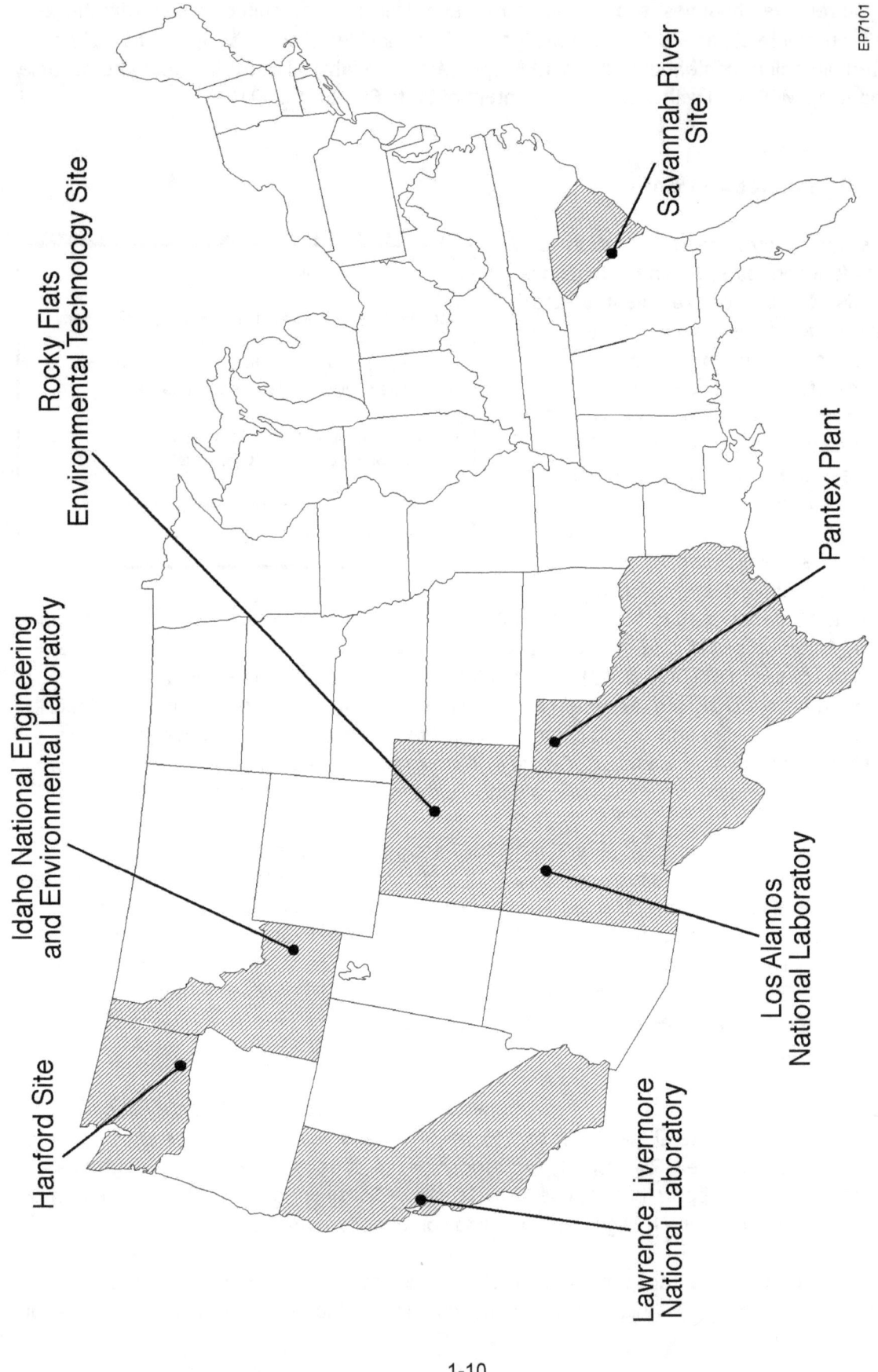

EP7101

Figure 1.3. Locations of DOE facilities containing surplus plutonium (*Source:* Adapted from DOE 1999).

This page is being withheld pursuant to 10 CFR 2.390(a).

to proliferation-resistant forms (DOE 1999). The purpose and need discussion establishes a range of reasonable alternatives to the proposed action that can satisfy this underlying purpose and need.

Following the subsequent September 2000 surplus plutonium disposition agreement between Russia and the United States (White House 2000), the DOE determined that a MOX-only approach best ensures the joint reduction of existing plutonium stockpiles held by the two nations, and concluded in its amended ROD that reliance on this approach is the key to successfully completing the agreement (DOE 2002). The result of this action would be to reduce over time the number of locations where the various forms of plutonium are stored and to ensure that this weapons-usable material does not fall into the hands of rogue states or terrorist groups.

1.4 Scope of the EIS

1.4.1 Scoping Process

On March 7, 2001, the NRC issued a Notice of Intent (NOI) in the *Federal Register* (66 FR 13794) to prepare an EIS for construction and operation of the proposed MOX facility at the SRS near Aiken, South Carolina. In the NOI, NRC announced plans for two scoping meetings: one in North Augusta, South Carolina, on April 17, 2001, and another in Savannah, Georgia, on April 18, 2001. In a second *Federal Register* notice on April 11, 2001 (66 FR 18223), the NRC announced that a third scoping meeting would be held in Charlotte, North Carolina, on May 8, 2001.

The three scoping meetings were held as planned. At each meeting, the NRC staff distributed background materials on the MOX

Proposed Action Elements
• Construction, operation, and decommissioning of proposed MOX facility, PDCF, and the WSB;
• Infrastructure upgrades;
• Shipment of surplus plutonium from the DOE sites to the SRS;
• Transport of depleted uranium hexafluoride from the DOE facility at Portsmouth, Ohio, to the commercial fuel fabrication facility in Wilmington, North Carolina;
• Transport of depleted uranium oxide from the Wilmington facility to the SRS;
• Transport of MOX fuel and fuel irradiation in surrogate reactors; and
• Spent MOX fuel transport to a geologic repository.

fuel program and NRC's plans for conducting licensing and environmental reviews for the facility. An open house held before each meeting provided attendees an opportunity to view informational materials and talk informally with NRC staff. During the meeting, the NRC staff presented an overview of the NRC's role in the facility licensing process and described the NRC's approach to meeting its obligations under NEPA. The presentations were followed by a question and answer period in which the NRC staff responded to questions from attendees. The majority of time at the meetings was devoted to allowing individuals to express their views on the scope of the EIS.

A total of about 300 individuals attended the three scoping meetings, and about 80 of them asked questions or provided oral comments at the meetings. In addition, approximately 60 individuals or organizations submitted written comments to the NRC by regular mail, fax transmittal, e-mail, or in person at the meetings. Some of the individuals who provided written comments also spoke at the meetings. Some individuals attended and offered comments at more than one meeting. Although issues raised during the scoping period were considered in the preparation of the draft environmental impact statement (DEIS), some of those issues were either analyzed in less detail or were not analyzed at all, depending on their relevance to the proposed action and the anticipated impacts. The full scoping summary report (NRC 2001b) is included as Appendix I.

The scoping process helped to determine the scope of the EIS and identify significant issues to be analyzed in depth. For instance, two technology options for the proposed action were identified during the scoping process. The first option is to substitute sand filters for the proposed high-efficiency particulate air (HEPA) filters to control air emissions from the facility. The second option is to substitute a dry process for the proposed wet process to remove impurities from plutonium dioxide powder. Cumulative impacts of the proposed action, in addition to other contaminant sources, were also identified as a relevant issue.

The no-action alternative, if NRC does not authorize construction or operation of the proposed MOX facility, was also refined through the scoping process. In addition to the no-action alternative of continued storage of all of the surplus weapons-grade plutonium at the present DOE sites in an unaltered form, the public suggested considering immobilizing all of the surplus weapons-grade plutonium at the SRS as a no-action alternative.

The scoping process identified several relevant areas of concern to the public.[2] Concerns were expressed about the existing groundwater contamination at the SRS and the potential for the proposed facility and waste disposal to further deteriorate groundwater quality. Existing deep boreholes at the SRS were identified as a possible conduit for contaminant migration. Concerns were also expressed about the existing contamination of the Savannah River and the potential for the proposed facility to affect surface water quality. The impacts of facility-induced surface water quality changes on the downstream fishing and marine economy and on the downstream tidal wetlands were also concerns raised at the scoping meetings. Similarly, concerns were expressed regarding air quality impacts from both chemical and radiological materials.

The potential for human health impacts to the public and workers was also a concern. This included workers at the proposed facility, at the SRS, at the proposed reactors, and at disposal facilities. It was also suggested that the impacts to groups other than the "Standard Man" be assessed, such as unborn fetuses, children, and elderly populations. Impacts from possible accidents at the proposed facility during transport of radioactive materials and at the proposed reactors also were a significant concern. It was suggested that the worst-case accidents should be evaluated, including natural disasters and terrorist acts.

---

[2] The Scoping Summary Report (Appendix I) contains a complete summary of all comments received.

Some issues identified during the scoping process were considered to be beyond the scope of the EIS. In general, these issues are not directly related to the assessment of potential impacts from the proposed action now under consideration. The lack of in-depth discussion in the EIS, however, does not imply that an issue or concern lacks value.

A number of commenters requested that the SPD EIS prepared by the DOE be supplemented and many of the decisions already made by the DOE be revisited. Because the scope of the EIS was limited to the action now under review by the NRC, issues pertaining to decisions already made by the DOE and not affected by new information were addressed by referencing the appropriate DOE analysis.

Comments that seek to alter international treaties or affect national, state, or local laws, statutes, or regulations (e.g., comments that asked to alter Price-Anderson Act[3] limits) were not addressed because they do not pertain to reasonably foreseeable impacts arising from the construction and operation of the proposed MOX facility.

Comments on the scope of assessing reactor use impacts in the EIS for the proposed MOX facility were varied. Considering that the environmental impact of reactor use of MOX fuel was a significant issue with many commenters, it is appropriate to consider those impacts in the EIS. However, the currently available information does not lend itself to performing new analyses. The DOE's SPD EIS (DOE 1999) analyzed impacts of MOX fuel use at the McGuire, Catawba, and North Anna reactors. Therefore, the FEIS refers to the SPD EIS, but does not reanalyze generic reactor use impacts of MOX fuel. The specific environmental impacts resulting from the use of 40% MOX fuel cores in any particular reactor would be addressed by the NRC in reviewing the requisite 10 CFR Part 50 license amendment application. Duke Energy has submitted a license amendment request to the NRC to place lead test assemblies in its reactors. As discussed in Section 4.4.3, impacts associated with the lead test assemblies are considered to be outside the scope of this EIS because these activities would occur regardless of any decision by the NRC on the proposed MOX facility.

A number of commenters requested that the EIS analyze the impacts of having to upgrade the emergency response equipment and retrain emergency responders in the communities around the SRS, at the reactors, and along transportation routes. Other commenters requested that the EIS identify capabilities of local, regional, and national medical facilities to manage the casualties resulting from potential accidental releases and assess the readiness of communities to evacuate certain areas along the transportation routes in case of an accident. These issues are discussed in the EIS to the extent that they are required as mitigation measures presented in Chapter 5.

Many commenters raised a number of different issues concerning terrorism. The Scoping Summary Report stated that the EIS would not address the impacts of terrorism because these impacts are not considered to be reasonably foreseeable as a result of the proposed action. However, following the events of September 11, 2001, the Commission decided to consider the

---

[3] The Price-Anderson Act limits the liability of the nuclear industry in the event of a nuclear accident in the United States.

question of whether NEPA requires the evaluation of such impacts. By order dated December 18, 2002 (CLI-02-24), the Commission ruled that NRC has no obligation under NEPA to consider intentional malevolent acts in conjunction with the licensing of the proposed MOX facility.

In response to the cancellation of the plutonium immobilization facility (DOE 2002), the NRC delayed the issuance of the DEIS. The NRC held three public meetings in North Augusta, South Carolina; Savannah, Georgia; and Charlotte, North Carolina, and solicited additional written comments on how the immobilization of surplus plutonium as a no-action alternative should be discussed (NRC 2002). The NRC also solicited views on other alternatives that should be considered in the DEIS. In response, most commenters said they still wanted immobilization considered as an alternative in the DEIS, while some urged the NRC to instead focus on the proposed action. As discussed further in Section 2.3, the NRC has determined that immobilization of plutonium did not require an in-depth evaluation in the DEIS, because it was not a reasonable alternative to the proposed action. In response to the NRC's solicitation on other alternatives that should be considered, the alternative of deliberately producing off-specification MOX fuel was identified. This alternative is discussed in Section 2.3.

With respect to the proposed PDCF, the DOE's change from a "hybrid" to a MOX-only approach resulted in a change in the scope of the DEIS from that described in the NRC's March 7, 2001, NOI. The NRC stated there that the PDCF would not be part of the NRC's NEPA review of the proposed MOX facility (NRC 2001a). Initially, the PDCF had independent utility apart from the MOX facility, since the DOE planned to build and operate the PDCF along with the plutonium immobilization plant regardless of whether MOX fuel was also produced (DOE 2000). Now, because of the DOE's subsequent decision in its amended ROD to cancel the plutonium immobilization plant and implement a MOX-only approach (DOE 2002), the PDCF no longer has independent utility apart from the proposed MOX facility. Thus, for NEPA purposes, the PDCF must be evaluated in the EIS to avoid an improper segmentation of the potential impacts discussion.

1.4.2 Issues Studied in Detail

As discussed in the Scoping Summary Report (Appendix I), the goal of this EIS is to set forth the impact analyses in a manner that is readily understandable by the public. Significant or more important impacts are discussed in Chapter 4 of this FEIS. On the basis of the NRC's analyses and consideration of comments received during the scoping process, the following topics are discussed in detail in Sections 4.2 and 4.3 for the no-action alternative and the proposed action, respectively: (1) human health, (2) air quality, (3) hydrology, (4) waste management, (5) accident impacts, (6) decommissioning, and (7) environmental justice. Transportation of radioactive materials, conversion of depleted uranium, and use of MOX fuel in reactors are discussed in Section 4.4. Cumulative impacts are discussed in Section 4.5. The cost-benefit analysis for the no-action and proposed action alternatives, which builds on the comparison of alternatives in Section 2.4, is provided in Section 4.6. Mitigation actions to address the potential impacts are discussed in Chapter 5.

1.4.3 Issues Eliminated from Detailed Study

Impacts found to be less significant are discussed in FEIS Appendixes G and H. These impacts include those pertaining to geology, seismology, soils, noise, ecology, land use, cultural and paleontological resources, infrastructure, socioeconomics, and aesthetics.

1.4.4 Preparation of the Final Environmental Impact Statement

The NRC made the DEIS available for public review and comment in February 2003 in accordance with 10 CFR 51.73, 10 CFR 51.74, and 40 CFR 1503.1. The NRC provided a 75-day public comment period (which ended May 14, 2003) on the DEIS. The length of the comment period exceeded the minimum of 45 days specified in 10 CFR 51.73.

During that period, the NRC held three public meetings to receive oral comments regarding the contents of the DEIS. These public meetings were held on March 25, 2003, in Savannah, Georgia; March 26, 2003, in North Augusta, South Carolina; and March 27, 2003, in Charlotte, North Carolina. The NRC published notice of these meetings in the *Federal Register* (68 FR 97208, February 28, 2003), on its Web site, and in local newspapers.

Approximately 45 people provided oral comments at the public meetings. A certified court reporter recorded the oral comments and prepared written transcripts. The transcripts of the public meetings are part of the public record for the proposed project and were used in developing the comment summaries contained in Appendix J. In addition to oral comments received at the public meetings, the NRC received written comments, letters, facsimile transmittals, and e-mails regarding the DEIS and associated issues. A summary of the comments and responses are included in Appendix J. The written comments and transcripts are reproduced in Appendix L.

The NRC has reviewed each comment letter and all transcripts of the public meetings and has grouped comments relating to similar issues and topics, as permitted by the Council on Environmental Quality's NEPA regulations and the NRC regulations at 10 CFR 51.91 and 40 CFR 1503.4(b). Because the comments were voluminous, Appendix J provides summaries of all substantive comments received on the DEIS. The NRC then prepared responses to each of the comments or summaries of comments. Commenters are identified in each summary with a commenter number. Appendix K contains an index of commenter names and commenter numbers.

1.4.5 Other National Environmental Policy Act Documents Related to This Action

In preparing the EIS, the following other NEPA documents were considered:

Storage and Disposition of Weapons-Usable Fissile Materials Final Programmatic Environmental Impact Statement, DOE/EIS-0229, U.S. Department of Energy, Office of Fissile Materials Disposition, Washington, D.C., December 1996.

Surplus Plutonium Disposition Final Environmental Impact Statement, DOE/EIS-0283, U.S. Department of Energy, Office of Fissile Materials Disposition, Washington, D.C., November 1999.

Record of Decision for the Surplus Plutonium Disposition Final Environmental Impact Statement, U.S. Department of Energy, Washington, D.C., January 11, 2000 (*65 Federal Register* [FR] 1608).

Final Environmental Impact Statement for a Geologic Repository for the Disposal of Spent Nuclear Fuel and High-Level Radioactive Waste at Yucca Mountain, Nye County, Nevada, DOE/EIS-0250, U.S. Department of Energy, Office of Civilian Radioactive Waste Management, Feb. 2002.

1.5 Cooperating Agencies

No cooperating agencies have been involved in preparation of the EIS.

1.6 Other State and Federal Agencies

Several federal, Native American, state, and local agencies and organizations were contacted to gather relevant information for this EIS. The scope of the analysis necessitated obtaining information from state agencies in both South Carolina and Georgia. The following is a list of all agencies contacted during early stages of the DEIS preparation:

Federal Agencies

> U.S. Department of Energy, Savannah River Operations Office
> U.S. Department of Energy, Office of Fissile Material Disposition
> U.S. Fish and Wildlife Service

Native American Organizations

> Catawba Indian Nation
> Pee Dee Indian Association
> Ma Chis Lower Alabama Creek Indian Tribe
> Muscogee (Creek) Nation
> Indian People's Muskogee Tribal Town Confederacy
> Yuchi Tribal Organization, Inc.
> United Keetowah Band of Cherokee Indians

State Agencies

> South Carolina State Historic Preservation Office, Department of Archives and History
> South Carolina Department of Natural Resources, Wildlife and Freshwater
>   Fisheries Division

South Carolina Department of Health and Environmental Control, Bureau of Air Quality
South Carolina Department of Transportation
Georgia Department of Natural Resources, Wildlife Resources Division
Georgia Department of Natural Resources, Environmental Protection Division,
   Air Protection Branch

Towns, Cities, and Counties

Columbia County, Georgia
Town of Grovetown, Georgia
Town of Harlem, Georgia
City of Augusta/Richmond County, Georgia
City of Blythe, Georgia
City of Hephzibah, Georgia
Aiken County, South Carolina
City of Aiken, South Carolina
Town of Jackson, South Carolina
Town of New Ellenton, South Carolina
City of North Augusta, South Carolina
Town of Wagener, South Carolina
Barnwell County, South Carolina
City of Barnwell, South Carolina
Town of Blackville, South Carolina
Town of Williston, South Carolina

School Districts

Columbia County Board of Education, Georgia
Richmond County Board of Education, Georgia
Aiken County Board of Education, South Carolina
Williston School District #19, South Carolina
Williston School District #29, South Carolina
Williston School District #45, South Carolina

1.7 References for Chapter 1

DCS (Duke Cogema Stone & Webster) 2000. *Mixed Oxide Fuel Fabrication Facility Environmental Report.* Docket Number 070-03098. Charlotte, NC. Dec.
DCS 2001. *Mixed Oxide Fuel Fabrication Facility Construction Authorization Request.* Docket Number 070-03098. Charlotte, NC. Feb.
DCS 2002a. *Mixed Oxide Fuel Fabrication Facility Environmental Report, Revision 1 & 2.* Docket Number 070-03098. Charlotte, NC.
DCS 2002b. *Amended Mixed Oxide Fuel Fabrication Facility Construction Authorization Request.* Docket Number 070-03098. Charlotte, NC.
DCS 2003a. *Mixed Oxide Fuel Fabrication Facility Environmental Report, Revision 3.* Docket Number 070-03098. Charlotte, NC. June.

DCS 2003b. *Mixed Oxide Fuel Fabrication Facility Environmental Report, Revision 4.* Docket Number 070-03098. Charlotte, NC.

DCS 2004a. *Mixed Oxide Fuel Fabrication Facility Environmental Report, Revision 5.* Docket Number 070-03098. Charlotte, NC. June 10.

DCS 2004b. *Mixed Oxide Fuel Fabrication Facility Construction Authorization Request, Revision 6/10/04.* Docket Number 070-03098. Charlotte, NC.

DOE (U.S. Department of Energy) 1996. *Storage and Disposition of Weapons-Usable Fissile Materials Final Programmatic Environmental Impact Statement.* DOE/EIS-0229. Office of Fissile Materials Disposition, Washington, DC. Dec.

DOE 1999. *Surplus Plutonium Disposition Final Environmental Impact Statement.* DOE/EIS-0283. Office of Fissile Materials Disposition, Washington, DC. Nov.

DOE 2000. "Record of Decision for the Surplus Plutonium Disposition Final Environmental Impact Statement." *Federal Register* 65:1608, Jan. 11.

DOE 2002. "Surplus Plutonium Disposition Program." Amended Record of Decision. *Federal Register* 67(76):19432-19435, April 19.

DOE 2003. *Changes Needed to the Surplus Plutonium Disposition Program, Supplement Analysis and Amended Record of Decision.* DOE/EIS-0283-SA1. Office of Fissile Materials Disposition, Washington, DC, April.

NRC (U.S. Nuclear Regulatory Commission) 2001a. "Notice of Intent to Prepare an Environmental Impact Statement for the Mixed Oxide Fuel Fabrication Facility." *Federal Register* 66:13794, March 7.

NRC 2001b. *Environmental Impact Statement Scoping Process Scoping Summary Report, Mixed Oxide Fuel Fabrication Facility Savannah River Site.* U.S. Nuclear Regulatory Commission, Aug. [Reproduced in Appendix I of this EIS.]

NRC 2002. "Notice of Delay in Issuance of the Draft and Final Environmental Impact Statements for the Mixed Oxide Fuel Fabrication Facility." *Federal Register* 67: 20183-20185, April 24.

Tuckinan, M.S., 2003. "Proposed Amendments to the Facility Operating License and Technical Specifications to Allow Insertion of Mixed Oxide (MOX) Fuel Lead Assemblies and Request for Exemption from Certain Regulations in 10 CFR Part 50," personal communication from Tuckinan (Duke Power, Charlotte, NC), to U.S. Nuclear Regulatory Agency (Washington, DC). February 27.

White House 2000. *Agreement between the Government of the United States of America and the Government of the Russian Federation Concerning the Management and Disposition of Plutonium Designated as No Longer Required for Defense Purposes and Related Cooperation.* White House, Washington, DC. Sept.

2 ALTERNATIVES, INCLUDING THE PROPOSED ACTION

This chapter presents details of the alternatives considered in this environmental impact statement (EIS). The no-action alternative, which is discussed in Section 2.1, considers the continued storage of surplus plutonium in various locations throughout the U.S. Department of Energy (DOE) complex in the event the U.S. Nuclear Regulatory Commission (NRC) either denies Duke Cogema Stone & Webster's (DCS's) construction authorization request for the Mixed Oxide Fuel Fabrication Facility (the proposed MOX facility) or, later, denies DCS's subsequent request for a Title 10, Part 70 of the *Code of Federal Regulations* (10 CFR Part 70) license to possess and use special nuclear material. Section 2.2 presents the technical details of the proposed action and the connected actions.

Section 2.3 considers several alternatives to the proposed action and explains why they are not analyzed further in Chapter 4. These alternatives include alternate locations for the proposed MOX facility in the F-Area, alternative technology and design options, immobilization of surplus plutonium, deliberately making off-specification MOX fuel, the MIX MOX alternative, and the Parallex Project.

The NRC recognizes that under the provisions of 10 CFR 70, the Commission may approve construction of the proposed MOX facility and subsequently deny the DCS application for a 10 CFR Part 70 license to possess and use special nuclear material. Although this is a possible outcome relative to the proposed action, the NRC is not considering construction alone as a separate alternative because the NRC would not knowingly select an alternative involving construction of a facility that cannot be used for its intended purpose.

Section 2.4 compares the potential impacts related to the proposed action with those of the no-action alternative. Section 2.5 presents the NRC staff's final environmental recommendation on the action to be taken.

2.1 No-Action Alternative — Continued Storage of Surplus Plutonium

The no-action alternative would be a decision by the NRC not to approve the proposed MOX facility. It is reasonable to assume that if the NRC does not approve the proposed MOX facility, the DOE's surplus plutonium would remain in storage at DOE facilities. The surplus plutonium inventory is now stored at seven DOE sites. If this storage were to continue, it is possible that limited new construction would be required at one or more of these sites to upgrade storage conditions. However, the impacts of such construction, if required, would be addressed under a separate site-specific environmental review by DOE. For purposes of this EIS, the impacts of continued storage of surplus plutonium are assumed to be essentially the same as those analyzed by DOE in the Surplus Plutonium Disposition Environmental Impact Statement (SPD EIS) (DOE 1999a). However, the analysis in this EIS also considers the DOE's action to consolidate the storage of 6 MT (6.6 tons) of non-pit surplus plutonium from the Rocky Flats Environmental Technology Site to the Savannah River Site's K-Area Material Storage (KAMS) facility (DOE 2002b). The impacts of the no-action alternative are presented in Section 4.2 and Appendix G.

2.2 Proposed Action — Description of Mixed Oxide Fuel Fabrication Facilities and Connected Actions[1]

2.2.1 Introduction

The proposed MOX facility is designed to convert surplus weapons-grade plutonium and depleted uranium dioxide (UO_2) into MOX fuel that could be used at commercial nuclear power plants authorized to use such fuel. If the construction authorization for the proposed MOX facility is granted, the facility would be built on the north-northwest side of the F-Area at the SRS (see Figure 1.2 in Section 1.2). The Pit Disassembly and Conversion Facility (PDCF) would be built by DOE on the north-northeast side of the F-Area. The PDCF would be used to recover the plutonium metal from the pits[2] of disassembled weapons and would convert the weapons-grade plutonium to plutonium dioxide powder, which would subsequently be transferred to the proposed MOX facility as feedstock.

Within the boundaries of the PDCF, the DOE would also construct the Waste Solidification Building (WSB) (see Figure 1.2). The WSB would be used to process several liquid waste streams from the proposed MOX facility and the PDCF and convert them to solid transuranic (TRU) waste or low-level waste (LLW). This section describes the general layout of the proposed MOX facility, the processes to be used to manufacture MOX fuel, and the systems that would be used to handle the waste streams from the facility. As discussed in Section 1.2.2, since the PDCF and WSB are connected actions, these proposed DOE facilities are also discussed in Sections 2.2.2 and 2.2.4, respectively. Other elements of the proposed action as described in Section 1.2 that are not discussed in Chapter 2 are discussed in Chapter 4. Direct and indirect impacts of the proposed action and connected actions are presented in Sections 4.3 and 4.4, and Appendix H.

As discussed in Section 1.4.1, the technology option of substituting a sand filter for the proposed high-efficiency particulate air (HEPA) filters to control air emissions from the proposed MOX facility was identified during the scoping process. This technology option is described in Section 2.2.5 and is analyzed in Section 4.3.8.

---

[1] Except as noted, the descriptions provided in this section are based on information from DCS (2000, 2001, 2002, and 2004) and DOE (1999a).

[2] Pits are weapon components with a spherical metal core made of plutonium metal and several outer layers.

2.2.2 Pit Disassembly and Conversion Facility

2.2.2.1 Description of the Pit Disassembly and Conversion Facility

The PDCF would be built by the DOE and would not be subject to NRC licensing. The facility would be used to recover plutonium metal from weapon components, and convert it to an unclassified (i.e., no longer exhibiting any characteristics that are protected for reasons of national security) plutonium dioxide. The plutonium dioxide would be transferred to the proposed MOX facility. In addition to excess weapon components, the PDCF would be able to receive excess plutonium metal in other forms and be capable of converting it to plutonium dioxide.

The PDCF would be designed to process up to 3.5 MT (3.9 tons) of plutonium metal into plutonium dioxide annually. Facility operations would require a staff of about 400 personnel. The facility would be built in a hardened space of thick-walled concrete that meets all applicable standards for processing special nuclear material. One or possibly both levels of the two-story building would be below grade. Areas of the facility in which plutonium would be processed or stored would be designed to survive natural phenomena such as earthquakes, floods, and tornadoes, as well as potential accidents associated with fissile and radioactive materials. Ancillary buildings would be required for support activities.

Activities involving radioactive materials or externally contaminated containers of radioactive materials would be conducted in gloveboxes. The gloveboxes would be interconnected by a contained conveyor system to move materials from one process step to the next. Gloveboxes would remain completely sealed and operate independently, except during material transfer operations. Built-in safety features would limit the temperature and pressure inside the gloveboxes and ensure that operations remain within criticality safety limits. When dictated by process needs or safety concerns, an inert atmosphere would be maintained in gloveboxes. The exhaust from the gloveboxes would be continuously monitored for radioactive contamination. The atmosphere in the gloveboxes would be kept at a lower pressure than that of the surrounding areas so that any leaks of gaseous or suspended particulate matter would be contained and filtered appropriately. The building ventilation system would include HEPA filters and would be designed to maintain confinement, thus precluding the spread of airborne radioactive particulates or hazardous chemicals within the facility or to the outside environment. Both intake and exhaust air would be filtered, and exhaust gases would be monitored for radioactivity.

Beryllium may be a constituent of some of the pits that would be disassembled in the PDCF. Because inhalation of beryllium dust and particles has been proven to cause a chronic and sometimes fatal lung disease, beryllium is of special interest from a health effects perspective. However, the process operations in the PDCF are expected to generate only larger, nonrespirable turnings and pieces of the metal, and all work would be performed in gloveboxes. No grinding would be done that could cause small pieces of beryllium to become airborne.

The PDCF would accommodate the following surplus plutonium-processing activities: pit receipt, storage, and preparation; pit disassembly; plutonium conversion; oxide blending and sampling; nondestructive assay; product canning; product storage; product inspection and sampling for international inspection; product shipping; declassification of parts not made from special nuclear materials; highly enriched uranium (HEU) decontamination, packaging, storage, and shipping; tritium capture, packaging, and storage; and waste packaging, sampling, and certification. Additional areas for support activities would be needed, including office space, change rooms, a central control room, a laboratory, mechanical equipment rooms, mechanical shops, an emergency generator to supply power to critical safety systems in the event of a power outage, a warehouse, shipping and receiving areas, waste storage, guard stations, entry portals, and parking.

2.2.2.2 Processes Occurring in the PDCF

At the PDCF, the storage containers in which the plutonium is received would be removed from their overpacks (outer shipping containers), the contents verified, and the information regarding the material entered into the PDCF's material accountability system. Pits and plutonium metal would be placed in a short-term receiving vault, checked for radiological contamination, and transferred to the pit storage vault until processing. Before being processed in the pit disassembly line, the pits would be segregated on the basis of the potential presence of tritium.[3] Pits without tritium would go into the pit bisector glovebox, and those containing tritium would start in the Special Recovery Line glovebox.

In the pit bisector glovebox, external structures would be cut away from the pit, and the pit would be cut in half. Nonbonded pits (pits whose components separate easily) would be separated into plutonium metal, HEU, classified metal shapes, and classified nuclear material parts. The plutonium parts would be assayed as part of the material accountability program. HEU would be sent to the HEU-processing station for material accountability, electrolytic decontamination, and packaging; the classified metal shapes and metal shavings would go to the declassification furnaces; the nuclear material parts to the storage at the pit conversion facility; and the plutonium to the hydride-oxidation (HYDOX) station for the next step of the process. Bonded pits, which cannot be separated prior to processing, would be sent to the HYDOX station intact. For these pits, HEU, classified metal shapes, and classified nuclear material parts would be separated from the plutonium metal during the HYDOX process, then sent to the HEU-processing station, declassification furnaces, and storage at the pit conversion facility, respectively. Recovered HEU would be stored in a vault at the pit conversion facility until shipped to the Y-12 Facility at the Oak Ridge Reservation (ORR) for declassification, storage, and eventual disposition. The HEU would meet Y-12 acceptance criteria prior to shipment to the ORR.

---

[3] Tritium can be used as a boosting fuel in high-energy atomic weapons. Although the operators of the pit conversion facility would know which pits contain tritium, the pit types and the number of surplus pits that contain tritium are classified.

Pits with tritium would also be bisected, and the HEU, classified metal shapes, and classified nuclear material parts would be separated from the plutonium; this would occur in the Special Recovery Line glovebox. Under normal circumstances, all of the tritium associated with a given pit would be captured and recovered during the tritium removal process in the Special Recovery Line. It is expected that the tritium in a small number of pits will have absorbed into the plutonium. For these pits, an additional step would occur in the Special Recovery Line glovebox: the plutonium would be heated in a vacuum furnace to drive off the tritium as a gas. The tritium would then be captured on a catalyst bed and packaged as LLW for treatment and disposal. HEU and classified metal shapes would be decontaminated and sent to the HEU-processing station and declassification furnaces, respectively; classified nuclear material parts would be placed in storage at the pit conversion facility. After confirmation that the plutonium metal was free of tritium, the plutonium would be assayed as part of the special nuclear material accountability program and transferred to the HYDOX station. Recovered HEU would be stored in a vault at the pit conversion facility until shipped to the ORR for declassification, storage, and eventual disposition. The HEU would meet Y-12 acceptance criteria prior to shipment to the ORR.

In the HYDOX module, plutonium metal would react with hydrogen, nitrogen, and oxygen at controlled temperatures and pressures in a pressure vessel to produce plutonium dioxide. The plutonium metal would first be reacted with hydrogen gas to form a hydride. Then the vessel would be purged of the hydrogen and the hydride reacted with nitrogen gas to form a nitride. The nitrogen would then be purged and replaced with oxygen for the final reaction forming plutonium dioxide. The plutonium dioxide product would be collected and assayed for the material accountability program to confirm that all of the plutonium metal entering the HYDOX process left as an oxide.

In the primary canning module, the cans of plutonium dioxide would be placed into a primary storage can made of stainless steel. This can would then be welded shut and leak tested to ensure that the weld was sound. If the can were to fail the leak test, it would be reopened and rewelded. After passing the leak test, the primary can would be sent to the electrolytic decontamination module. After decontamination, each can would be rinsed, dried, and surveyed to verify decontamination, then sent to the secondary canning module.

In the secondary canning module, primary cans would be placed into secondary stainless steel storage cans meeting the DOE's long-term storage requirements. Also in this module, secondary storage cans would be welded shut and leak tested. After leak testing, each can would be marked with a laser to identify the can and its contents, and passed to the nondestructive assay module.

In the nondestructive assay module, each can would be assayed to confirm its contents. Following assay, the cans would be moved into the main storage vault and would be available for international inspection. After inspection, the cans would be transferred to another vault that would also be subject to international inspection. The cans would subsequently be transferred to the proposed MOX facility.

2.2.2.3 Radioactive Effluents and Wastes at the PDCF

Potential effluents and wastes from the PDCF are described in a Los Alamos National Laboratory report (LANL 1998) and the SPD EIS (DOE 1999a). The facility would be designed to minimize the quantities of both the effluents and wastes. Preliminary estimates indicate that small quantities of various plutonium isotopes and americium-241 and tritium gas would be emitted to the air from the facility. No releases to surface water would be expected directly from the PDCF. The facility would be expected to generate small quantities of TRU waste, LLW, mixed waste, and nonradioactive hazardous waste. All liquid radioactive wastes generated in the PDCF would be sent to the WSB for treatment. The treated waste would either be sent to an approved disposal facility or discharged to a permitted outfall on the SRS. Radioactive solid wastes generated at the facility would be packaged in accordance with the acceptance criteria of the receiving disposal site and sent to the WSB for temporary storage and final processing before being shipped to an approved disposal facility. Mixed waste and hazardous waste generated at the facility would be sent to the SRS waste management system or to an off-site permitted facility for disposition. Nonradioactive/nonhazardous solid waste would be sent to an approved landfill. An evaluation of waste management impacts for this EIS is presented in Section 4.3.4.

2.2.3 MOX Fuel Fabrication Facility

2.2.3.1 Description of the MOX Fuel Fabrication Facility

As designed, the project site would occupy an area of about 16.6 ha (41 acres). Approximately 6.9 ha (17 acres) of the site would be developed with buildings, other facilities, and paving. The remaining 9.7 ha (24 acres) would be landscaped with either grass or gravel.

No highways, railroads, or waterways traverse the proposed MOX facility site, and material and personnel would be moved to and from the site on existing SRS roads. The proposed MOX facility would consist of the following buildings:

- MOX Fuel Fabrication Building
- Emergency Diesel Generator Building
- Standby Diesel Generator Building
- Secured Warehouse Building
- Administration Building
- Technical Support Building
- Reagents Processing Building
- Receiving Warehouse Building

All of these buildings except the Administration Building and the Receiving Warehouse Building would be enclosed within a double fence perimeter intrusion detection and assessment system. The area within this system would total about 5.7 ha (14 acres) and would be designated as the "Protected Area" (10 CFR Part 73).

The Technical Support Building, located between the Administration Building and the MOX Fuel Fabrication Building, would house the main support facilities for MOX Fuel Fabrication Building personnel and would contain the access facilities for the Protected Area and the MOX Fuel Fabrication Building. The building would not be directly involved in the principal processing functions of the facility. Supporting activities and facilities located in this building would include health physics, an electronics maintenance laboratory, a mechanical maintenance shop, personnel locker rooms, and a first aid station.

The MOX Fuel Fabrication Building would have three major functional areas: the MOX Processing Area, the Aqueous Polishing Area, and the Shipping and Receiving Area. The MOX Processing Area would include the blending and milling area, pelletizing area, sintering area, grinding area, fuel rod fabrication area, fuel bundle assembly area, a laboratory area, and storage areas for feed material, pellets, and fuel assemblies. Space would also be provided in the MOX Fuel Fabrication Building for support equipment, such as temporary waste storage; heating, ventilation, and air conditioning (HVAC) equipment; HEPA filter plenums; inverters; switchgear; and pumps. The Aqueous Polishing Area would be used to remove impurities from the feed plutonium coming from the PDCF as well as from the plutonium in the alternate feedstock for use in the MOX Processing Area. The aqueous polishing process would extract impurities from the weapons-grade plutonium dioxide. The Shipping and Receiving Area would contain the equipment and facilities used to handle incoming and outgoing materials to and from the MOX Processing Area and Aqueous Polishing Area.

The Emergency Diesel Generator Building would contain the emergency diesel generator to provide the emergency on-site electrical power supply for safety related structures, systems, or components. The Standby Diesel Generator Building would contain the diesel generators that would provide the on-site electrical power source in the event of loss of off-site power. The Secured Warehouse Building would include the Material Receipt Area, the Storage Area, the MOX Fresh Fuel Package Storage Area, the Parts Washing Facility, the Vehicle Access Portal, and the Vehicle Gatehouse. The Material Receipt Area would serve as the receiving facility for most of the materials (including depleted uranium dioxide), supplies, and equipment necessary for facility operations. The Administration Building, located outside of the Protected Area of the complex, would provide administrative support to the facility and its operations. Space would be provided in the building for facility management, facility operations, finance and administration, health and safety, quality assurance, and management personnel.

The Reagents Processing Building, located adjacent to the Aqueous Polishing Area of the MOX Fuel Fabrication Building, would provide storage for pure reagent-grade chemicals and facilities for preparation of chemical solutions used in the Aqueous Polishing Area. The Reagents Processing Building would consist of several separate rooms or areas for the various chemicals. Concrete curbs around the chemical storage areas would provide for spill containment. Chemicals would be transferred to the Aqueous Polishing Area from the Reagents Processing Building via piping located in a below-grade concrete trench between the two buildings.

The Receiving Warehouse Building would be a single-story, pre-engineered metal building located outside of the perimeter intrusion detection and assessment system. The building would consist of the Unloading Dock, the Materials Receiving Area, the Inspected Warehouse Holding Area, the Material Transfer Dock, offices, vestibule, and the Inspection Guard Station.

2.2.3.2 Processes Occurring in the Proposed MOX Facility

The proposed MOX facility is being designed to convert plutonium dioxide and depleted uranium dioxide to MOX fuel. Operations at the facility would begin with the receipt of the plutonium dioxide and depleted uranium dioxide feed materials. The plutonium dioxide would then be purified in the aqueous polishing process before being blended with the depleted uranium dioxide. The blended material would then be formed into pellets, the pellets incorporated into fuel rods, the fuel rods placed in fuel assemblies, and the assemblies loaded into transport casks for shipment to the nuclear power plants authorized to use MOX fuel. The technology used in the fuel fabrication process includes recycling of waste and scrap streams. The major steps in the aqueous polishing and fuel fabrication processes are shown in Figures 2.1 and 2.2, respectively.

2.2.3.2.1 Feed Materials

The plutonium dioxide feed material from the PDCF, transported in approved shipping containers, would be received in the shipping and receiving area of the MOX Fuel Fabrication Building. The feed material would be offloaded, the packaging would then be removed, and control would be transferred to the responsible facility manager. Material control and accounting (MC&A) and radiation protection functions would then be performed, and the feed material would be moved to the MOX Processing Area.

Alternate feedstock (feed material not coming from the PDCF) would be received as plutonium dioxide. Some of this material might contain higher than normal salt contaminants, some would contain chloride contaminants, and some would contain trace amounts of enriched uranium. All alternate feedstock would be milled to a uniform particle size to facilitate dissolution. The alternative feedstock would be analyzed for contaminants.

If chloride contaminant concentrations were found to be above feedstock specifications, they would be removed by conversion to chlorine gas. The chlorine gas would be passed through a scrubber to convert the chlorine to a sodium chloride solution. If the chloride contaminants were within feedstock specifications, the feedstock would be processed as described in Section 2.2.3.2.2.

For uranium-rich alternate feedstock, an additional scrubbing column would be used to remove uranium to levels that meet the specification for purified plutonium.

Depleted uranium dioxide feed material, packaged in drums and shipped by truck, would be received at the Material Receipt Area of the Secured Warehouse Building. Conventional materials and supplies would be received at the Secured Warehouse Building. The materials

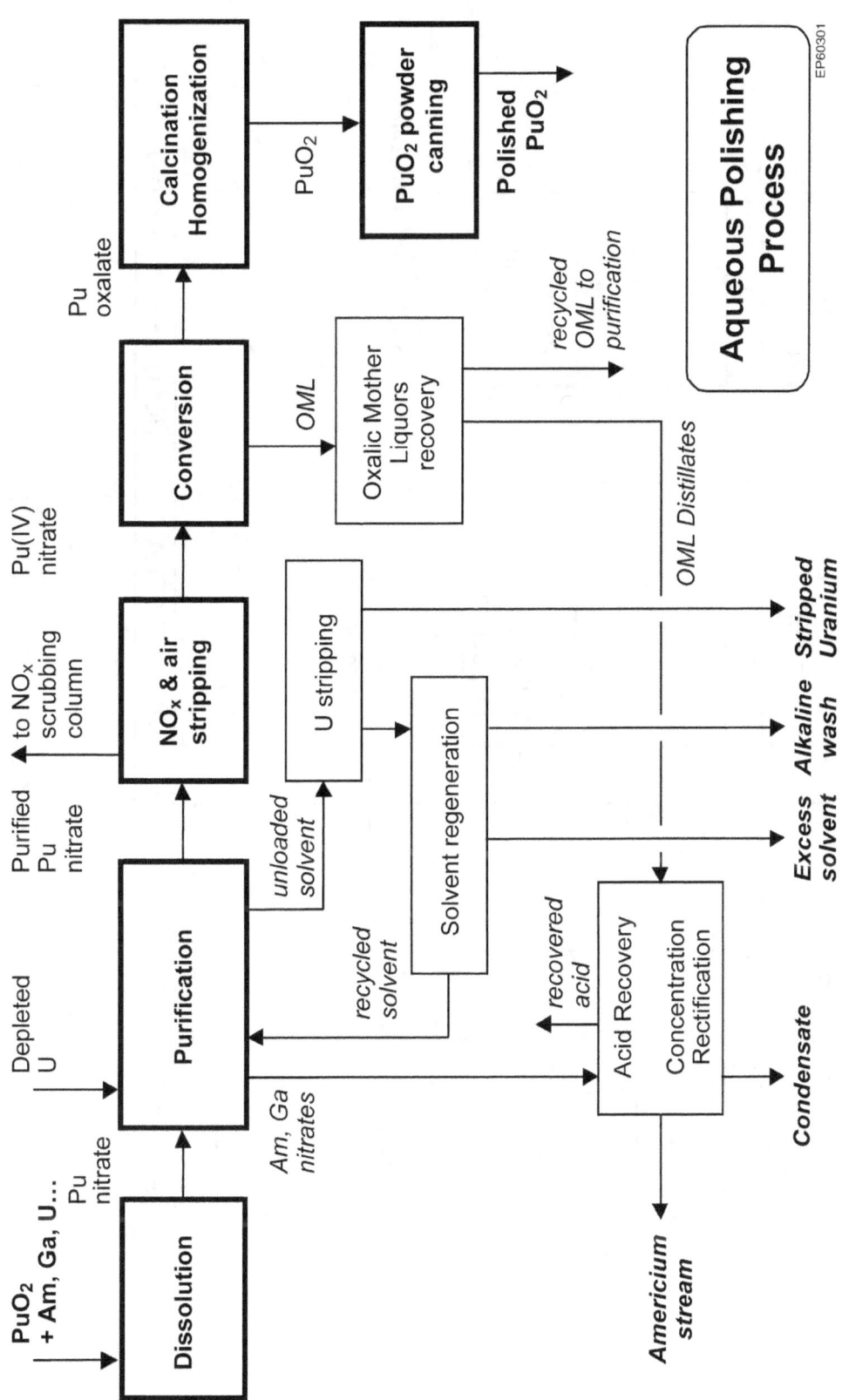

Figure 2.1. Principal steps in the aqueous polishing process (*Source:* DCS 2004).

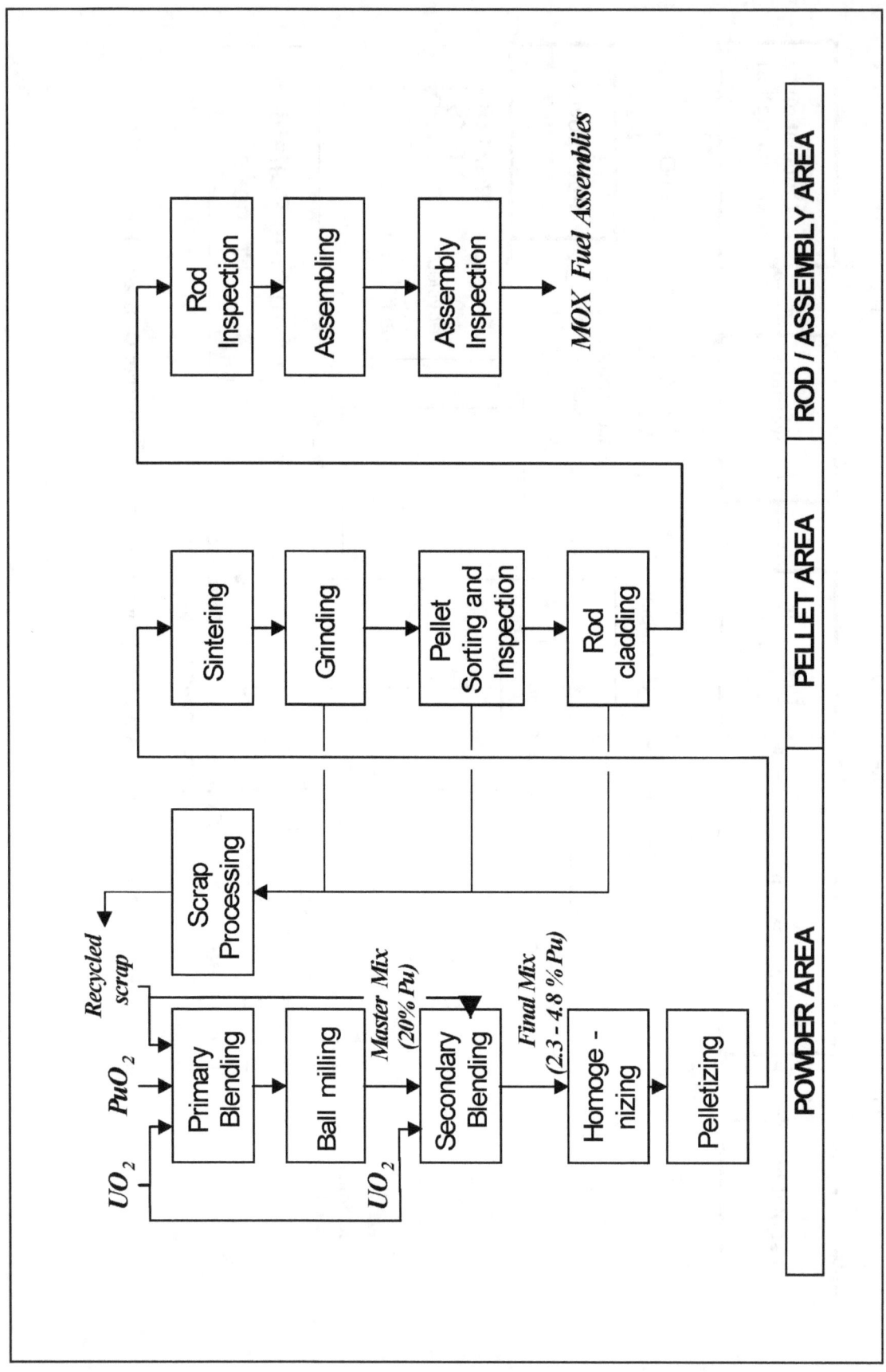

Figure 2.2. Principal steps in the fuel fabrication process (*Source:* DCS 2002).

would be inventoried, sorted, and removed to storage in the Secured Warehouse Building or delivered via on-site vehicles to the proper processing area.

2.2.3.2.2 Aqueous Polishing Process

The plutonium dioxide received at the facility would contain small amounts of impurities, mainly gallium, americium, highly enriched uranium, and, in the case of alternate feedstock, additional impurities. These impurities would have to be removed before the plutonium could be used in reactor fuel. The chloride contaminants would be removed from alternate feedstock before further aqueous polishing (see Section 2.2.3.2.1). The aqueous polishing process would remove remaining impurities in three major steps: dissolution, purification, and conversion.

The *dissolution step* would involve dissolving the plutonium dioxide powder in a water-based (aqueous) solution of silver (Ag^{2+}) and nitric acid at nearly room temperature. An electrical current would be passed through the solution to help dissolve the powder.

In the *purification step*, the plutonium in the aqueous solution would be separated from uranium, americium, gallium, and other impurities by solvent extraction. In this process, the aqueous solution and an organic solvent solution are mixed. The organic solvent does not readily mix with or dissolve in water, and the two solutions will separate if they are allowed to settle. However, by forcibly mixing the two solutions and adjusting chemical parameters in the aqueous solution, individual metals like plutonium can be selectively extracted from the aqueous solution into the organic solvent. In the process proposed by DCS, the solvent extraction process would involve mixing the aqueous solution with an organic solvent composed of 30% tri-butyl phosphate in dodecane. The mixing would occur in the middle of tall and narrow process vessels called columns. During mixing, the solvent would selectively extract the plutonium and uranium from the aqueous solution. The less dense solvent containing uranium and plutonium would then separate from the aqueous solution at the top of the columns. The impurities would remain in the denser aqueous solution and would be removed at the bottom of the column.

The solvent solution containing the uranium and plutonium would be washed with a nitric acid solution. This wash solution would be returned to the acid recovery unit for recycling of the acid. The plutonium and uranium in the organic solvent would then be mixed with an aqueous solution containing hydroxylamine nitrate. This process would reduce the tetravalent plutonium [Pu(IV)] to trivalent plutonium [Pu(III)], which would allow the plutonium to be removed from the organic solvent in an aqueous solution of nitric acid, hydrazine nitrate, and hydroxylamine nitrate. The organic solvent, which would then contain only high-enriched uranium and residual amounts of plutonium, would be mixed with another aqueous "wash" solution to remove the residual plutonium. The washed solvent would be routed to the uranium stripping process. High-enriched uranium would be stripped from the solvent by mixing the solvent with dilute nitric acid in another separation column. The stripped uranium solution would be diluted with depleted uranium before being transferred to the WSB for further treatment. The solvent, which would no longer contain significant amounts of uranium, plutonium, or impurities, would be routed to the solvent recovery mixer-settlers to be recycled.

The Pu(III) solution would be converted back to a solution of Pu(IV) by driving nitrous fumes (dinitrogen tetroxide [N_2O_4] and nitrogen dioxide [NO_2]) through the plutonium solution in a packed column. The offgas would be routed through an offgas treatment system before being discharged to the atmosphere.

The conversion step would be a continuous oxalate conversion process. The oxidized Pu (IV) would be reacted with excess oxalic acid ($H_2C_2O_4$) to precipitate plutonium oxalate. Plutonium oxalate would be collected on a filter, then dried in a screw calciner to produce purified plutonium dioxide powder. The purified plutonium dioxide powder would be blended and stored in cans.

Offgas from the screw calciner would be routed through the process offgas treatment unit and HEPA filters prior to discharge to the atmosphere through the exhaust stack. The filtered oxalic mother liquors would be concentrated, reacted with manganese to destroy the oxalic acid, and recycled to the beginning of the extraction cycle, to minimize losses of plutonium to waste.

A liquid americium waste stream would be generated by the aqueous polishing process described above. DCS estimates that approximately 24.5 kg (54.0 lb) of americium-241 would annually become part of this waste stream, an amount that would contain 84,000 Ci of radioactivity (DCS 2002). This liquid waste stream — together with an excess acid stream and an alkaline wash stream — would be combined into the high-alpha activity waste to be piped from the proposed MOX facility to the WSB, where it would be solidified through the use of the WSB's planned evaporation, neutralization, and cementation methods. (The WSB is discussed further in EIS Section 2.2.4). The maximum annual volume of these streams from the proposed MOX facility is estimated to be 44,200 L (11,700 gal) (DCS 2004).

2.2.3.2.3 MOX Fuel Fabrication Process

The MOX fuel fabrication process would consist of four major steps: (1) powder master blend and final blend production, (2) pellet production, (3) rod production, and (4) fuel rod assembly.

The first operation would be the production of the powder master blend. The purified plutonium dioxide from the aqueous polishing process would be mixed with depleted uranium dioxide and recycled scraps to produce an initial mixture that would be approximately 20% plutonium. This mixture would be ground in a ball mill and mixed with additional depleted uranium dioxide and recycled scraps to produce a final blend with the required plutonium content (typically from 2.3 to 4.8%). This final blend would be further homogenized to meet stringent plutonium distribution requirements. During the final homogenizing, lubricants and pore-formers would be added to control the density of the final mixture.

The final homogenized powder blend would be pressed to form green pellets. The green fuel pellets would be sintered to obtain the required ceramic qualities. Sintering is the process of heating the green pellets in a furnace at temperatures of up to 1,700°C (3,100°F). The sintering step would remove organic products dispersed in the pellets and remove the pore-formers that were added during powder homogenization. The sintered pellets would be ground to a specified diameter and sorted. Recovered powder from grinding and discarded pellets would be recycled through a ball mill and reused in the powder processing.

Fuel rods would be loaded to an adjusted pellet length column, welded, pressurized with helium, and then decontaminated in gloveboxes. The decontaminated rods would be removed from the gloveboxes and placed on racks for inspection and assembly. Fuel rods would be inserted into the fuel assembly frame, and the fuel assembly construction would be completed. The fuel assembly would be subjected to a final inspection before shipment to reactors.

2.2.3.3 Radioactive Effluents and Wastes at the Proposed MOX Facility

2.2.3.3.1 Airborne

DCS has proposed to treat exhausts from the Fuel Fabrication Building and remove airborne radioactive materials with (at a minimum) a two-stage HEPA filter system before exhaust air is discharged to the environment. The exhaust streams would include those from building ventilation; gloveboxes; the process vents of tanks, vessels, and other equipment in the Aqueous Polishing Area; and the sintering furnaces in the Processing Area.

The filtered exhausts would be discharged through a common stack (MOX vent stack) on the roof of the Fuel Fabrication Building. Stack effluents would be continuously monitored. The stack would be 37 m (120 ft) above grade.

2.2.3.3.2 Liquids

After sampling and characterization, liquid waste streams containing radioactive materials would be transferred to the WSB for processing and treatment. Thus, no radioactive liquids would be released directly from the facility to the environment. Within the Aqueous Polishing Area, recycling would be used extensively to reduce liquid waste volumes and impurities before transfer to the WSB.

The liquid waste streams from the Aqueous Polishing Area would include the following:

- Chloride removal waste
- Liquid americium stream
- Excess acid stream
- Excess low-level radioactive solvent waste
- Stripped uranium stream
- Rinsing water
- Contaminated drains

2.2.3.3.3 Solids

Solid radioactive wastes would be placed in appropriate containers (typically 55-gal drums), assayed, and transferred to the WSB for processing and disposal. Whenever practical, the solid wastes would be compacted to reduce volume and disposal costs.

The solid radioactive wastes generated in the Fuel Fabrication Building would include TRU solid wastes and LLW (which would include uranium and/or plutonium contamination). Other potentially radioactive, mixed, or nonradioactive hazardous wastes that might be generated by the facility would be transferred to the WSB, SRS waste management system, or an off-site permitted facility for disposition. Impacts associated with management of wastes from the proposed MOX facility are presented in Section 4.3.4.

2.2.4 Waste Solidification Building

2.2.4.1 Description of the Waste Solidification Building

The WSB, which is not subject to NRC licensing, would be constructed by the DOE on the PDCF site south of the PDCF to process the following liquid waste streams from the PDCF and the proposed MOX facility:

- MOX facility high-alpha-activity waste stream
- MOX facility stripped uranium stream
- PDCF laboratory liquid stream
- PDCF low-level liquid waste streams
- MOX facility low-level liquid waste streams

In addition, space would be provided in the WSB for temporary storage and minimal processing (e.g., sorting, packaging) of other waste streams, including solid LLW and TRU waste.

The WSB would occupy approximately 6,970 m^2 (75,000 ft^2) of land and would be a combination concrete and steel-frame structure (DCS 2003a,b, 2004). Concrete would be utilized as necessary to protect against the potential impacts of natural phenomena hazards events. In addition, a concrete-cell configuration would be used in areas where the proposed MOX facility high-alpha stream is processed. Process enclosures adjacent to the cells would provide worker protection to accommodate operations and maintenance activities. The shielding and confinement would also serve as fire isolation barriers. Secondary confinement features, such as dikes, sumps, and leak detection, would be provided for those areas with liquid spill potential. The major pieces of process equipment would be tanks, evaporators, and cementation equipment. Other equipment may include reverse osmosis, filtration, and activated carbon and ion exchange columns.

The processed liquid would be mixed in the WSB with concrete and poured into containers to produce solid waste. Cold chemical processing rooms, waste container storage, and truck loading/unloading areas may also be contained in hardened structures. The waste container storage area would be at grade. The waste receipt area would have tanks to separately receive high-alpha waste, stripped uranium waste, and the PDCF laboratory liquid stream waste. The tank volumes would be sufficient to receive and store waste from six weeks of processing the high-alpha-activity and stripped uranium waste streams by the proposed MOX facility and eight weeks of processing the laboratory liquid stream by the PDCF. Additional receipt storage would be available for low-level liquid waste streams from the proposed MOX facility, PDCF, and WSB internal sources.

The proposed MOX facility would transfer a liquid high-alpha-activity waste and liquid LLW streams to the WSB. The PDCF would transfer LLW streams. Within the WSB, these waste streams would be treated separately. The WSB would process the liquid wastes into TRU waste and LLW solid waste forms acceptable for shipment and disposal at their respective disposal locations. Treated effluents from liquid LLW streams would be discharged to a permitted outfall. The TRU waste form would be stored until cleared for shipment to the Waste Isolation Pilot Plant (WIPP) (DOE 2003). The LLW form would be sent to a suitable disposal site.

Within the WSB, the waste streams would be collected into receipt tanks, chemically adjusted, evaporated, neutralized, solidified in containers, stored, and shipped. These processes would be located inside a hardened (reinforced concrete) structure. Emissions from the process areas would pass through a HEPA filtration confinement system before release through an exhaust stack.

2.2.4.2 Processes Occurring in the WSB

The WSB would be designed to process and solidify three waste streams from the proposed MOX facility and two waste streams from the PDCF. The processes that would be conducted for each waste stream are described below.

2.2.4.2.1 Proposed MOX Facility High-Alpha-Activity Waste Stream

The proposed MOX facility high-alpha-activity waste stream, consisting of the liquid americium waste stream and two other liquid waste streams from the proposed MOX facility, namely the excess acid stream and the alkaline waste stream, would be pumped approximately 610 m (2,000 ft) from the proposed MOX facility to the WSB in a double-walled stainless steel pipe. The maximum volume received would be anticipated to be approximately 33,300 L (8,800 gal) per year, which would be received in approximately 25 transfers, at a frequency of about once every two weeks.

The WSB receipt tanks would be sized to hold three transfers (six weeks capacity in two 9,500-L [2,500-gal] tanks). The MOX facility high-alpha-activity stream collection tanks are sized for three months capacity. This arrangement would provide continued MOX facility processing capacity in the event of a shutdown of WSB operations because of maintenance or other disruptions. The tanks would be agitated or recirculated to mix the contents.

In the WSB, the proposed MOX facility high-alpha-activity waste stream would be sent to an evaporator to reduce its water content. The acidic bottoms collected in the evaporator would be neutralized with sodium hydroxide, mixed with cement, and poured into approved containers. The TRU waste collected in the containers would meet the WIPP waste acceptance criteria and would eventually be shipped to the WIPP for disposal (DOE 2003). The overheads from the evaporation step would be condensed, collected, sampled, and subjected to further evaporation or chemical treatment as necessary and finally would be sent to the Clean Water Treatment System for final treatment and discharge to a permitted outfall (see Section 2.2.4.2.4).

2.2.4.2.2 MOX Facility Stripped Uranium Stream

The proposed MOX facility stripped uranium stream would be pumped approximately 610 m (2,000 ft) from the proposed MOX facility to the WSB in a double-walled stainless steel pipe. The nominal waste volume of this stream would be 174,000 L (46,000 gal) per year, received in approximately 42 transfers at a frequency of about one every week.

The WSB receipt tanks would be sized to hold six transfers (six weeks of MOX facility capacity). The proposed MOX facility tanks would be sized to hold three months of MOX facility waste. The tanks would be agitated or recirculated to mix the waste.

In the WSB, the proposed MOX facility stripped uranium stream would be evaporated, the bottoms neutralized with sodium hydroxide, and the resulting waste mixed with cement and deposited into approved containers. The waste in the containers would be classified as LLW and would be shipped to a LLW disposal facility. The overheads from the evaporation step would be condensed, collected, sampled, and subjected to further evaporation or chemical treatment as necessary and finally would be sent to the Clean Water Treatment System for final treatment and discharge to a permitted outfall (see Section 2.2.4.2.4).

2.2.4.2.3 PDCF Laboratory Liquid Stream

The PDCF laboratory liquid stream would be pumped approximately 240 m (800 ft) to the WSB from the PDCF in a welded-jacketed stainless steel pipe, which would be direct buried. The volume of this waste stream is anticipated to be a nominal 41,600 L (11,000 gal) per year (DCS 2004), and would be received in approximately 12 transfers (3,400 L [900 gal] each) at a frequency of about one transfer every month.

The WSB receipt tank would be sized to hold two transfers (eight weeks of PDCF laboratory liquid stream capacity) in one 11,400-L (3,000-gal) tank. The PDCF tank is sized to provide up to 8 weeks of PDCF processing capacity in the event of a shutdown of WSB operations for maintenance or processing anomalies. The tank would be agitated or recirculated to mix the waste.

In the WSB, the PDCF laboratory liquid stream would be evaporated, the bottoms neutralized with sodium hydroxide, and the resulting waste would be mixed with cement and deposited into approved containers. The waste in the containers would be classified as LLW and would be shipped to a LLW disposal facility. The overheads from the evaporation step would be condensed, collected, sampled, and subjected to further evaporation or chemical treatment as necessary and finally would be sent to the Clean Water Treatment System for final treatment and discharge to a permitted outfall (see Section 2.2.4.2.4).

2.2.4.2.4 MOX Facility and PDCF Low-Level Liquid Streams

The proposed MOX facility and the PDCF would generate various aqueous liquid streams with either very low radioactive contamination or the potential for radioactive contamination due to their origin. These streams would be transferred, through double-walled transfer lines, to a receipt tank or tanks at the WSB. In addition, low-level liquid waste streams would be generated in the WSB from the evaporator overhead associated with the treatment of other liquid waste streams sent to the WSB from the proposed MOX facility and the PDCF (see Sections 2.2.4.2.1, 2.2.4.2.2, and 2.2.4.2.3). All of these waste streams would be transferred to the Clean Water Treatment System in the WSB. The Clean Water Treatment System would be designed using standard wastewater treatment technologies to meet U.S. Environmental Protection Agency (EPA), South Carolina Department of Health and Environmental Control (SCDHEC), and DOE discharge limits for the SRS. The discharges would be to a permitted outfall.

2.2.4.3 Radioactive Effluents and Wastes at the WSB

The WSB would be designed to minimize effluents to the air. The facility would also be designed to minimize effluents to surface water, as discussed in Section 2.2.4.2.4.

As discussed in Section 2.2.4.2, the WSB would receive five liquid waste streams, three from the proposed MOX facility and two from the PDCF, and convert those waste streams to solid TRU waste or solid LLW. An evaluation of waste management impacts for this EIS is presented in Section 4.3.4. The solidified TRU waste would eventually be shipped to WIPP for disposal (DOE 2003). LLW would be disposed of at a suitable disposal site.

2.2.5 Sand Filter Technology Option

This section describes the technology option of using a sand filter for air filtration instead of high-efficiency particulate air (HEPA) filters. Although DCS has selected the use of HEPA filters as its preferred option for removal of particulate contaminants before exhaust air is released to the atmosphere, this EIS also evaluates the use of a sand filter (Orr 2001). The differences in impacts are discussed in Section 4.3.8.

It is useful to understand the physical differences in the two types of filters. HEPA filters are designed to remove extremely fine particles suspended in the air. HEPA filters are enclosed in rigid casing with full-depth pleated filter medium. The filter medium is normally fibrous borosilicate glass, which is formed into a sheet folded into a series of accordion pleats. The standard HEPA filter measures 61 cm × 61 cm × 29.2 cm deep (24 in. × 24 in. × 11.5 in. deep). The filter edge will be a high-temperature silicon gasket to prevent bypass leakage, and improper installation or damage to the sealing surface can dramatically reduce the filter's efficiency and performance. HEPA filters function and are used in the HVAC system similarly to standard home air filters. DCS proposes to use HEPA filters in multiple stages. The proposed MOX facility would have many HEPA filters (Orr 2001).

Sand filters have a long history of use in DOE facilities at the SRS and at the Hanford Site near Richland, Washington. At the SRS, DOE currently uses sand filters in the F-Area, H-Area, and the Savannah River Laboratory. Unlike the case for HEPA filters, a facility would typically use only a single sand filter. A sand filter designed for the proposed MOX facility would be rectangular and would require a surface area of about 313 m^2 (33,650 ft^2). The filter would be about 3 m (10 ft) deep and would consist of gravel layers overlaid with sand layers arranged in order of decreasing particle size (Orr 2001). A coarse sand layer would be placed at the top of the filter to maintain integrity of the lower sand layers during filter operations. Air enters through a supply tunnel at the bottom of the structure and is collected at the top of the sand filter. Large fans or blowers are used to draw the air through the sand filter media. Suspended particles in the air are trapped by the sand filter. No routine maintenance is required during operation of sand filters (Orr 2001).

It is also useful to understand the performance differences in the two types of filters. Both filter types have approximately the same efficiencies for collection of particulates. Neither filter type is designed to trap gases. The filters would perform differently during some accidents. As discussed below, the selection of filter type can affect the facility design.

Several commenters during the public scoping meetings urged the NRC to evaluate the use of sand filters instead of HEPA filters, claiming sand filters would be better from a safety standpoint, particularly in case of a fire at the facility. Fires often generate large volumes of smoke that threaten the effective functioning of the filtration system by rapidly loading the filters with smoke particles. The resulting pressure drop across the filter could cause a break in the filter, followed by the release of contamination to the environment. This situation would not occur with sand filters in case of a fire because they have a much larger surface area that could trap smoke particles (Orr 2001). The integrity of HEPA filters could also be compromised during explosion accidents.

Given the potential vulnerability of HEPA filters during fire accidents, the proposed MOX facility is designed to mitigate the effects of an internal fire. The facility is designed into numerous fire areas to limit the amount of combustibles involved in a single fire; this reduces the amount of soot reaching individual banks of HEPA filters and ensures that the HEPA filters will not fail because of excessive plugging. If a sand filter was used, fewer fire areas could be used because sand filters are more resistant to smoke and sudden pressure changes. However, in the evaluation of the impacts of using sand filters instead of HEPA filters, changes in facility design are not considered.

2.3 Alternatives Considered But Not Analyzed in Detail

This section discusses some of the more significant alternatives identified during the scoping process and alternatives identified by DCS, but that are not subjected to in-depth evaluations in Chapter 4. Such alternatives include alternate locations for the proposed MOX facility in the F-Area, technology and design options, immobilization of surplus plutonium, off-specification MOX fuel, and the Parallex Project.

2.3.1 MOX Facility Location in F-Area

The DOE previously selected the SRS as the location of the proposed MOX facility, after evaluating several alternative sites across the country (DOE 2000). In its subsequent Environmental Report, DCS described the process the DOE used in choosing the specific site for the proposed MOX facility within the SRS F-Area (DCS 2000). The currently proposed location of the MOX facility was selected from five proposed sites within the F-Area. Final site selection was based on three siting qualification criteria that the site must meet and nine siting evaluation criteria that were more qualitative in nature (DCS 2002). The currently proposed location of the facility, as identified in Figure 1.2, was the only location that met all of the qualification criteria and scored the highest when all of evaluation criteria were considered. The criteria used by DCS in the selection process were as follows (DCS 2002):

Siting Qualification Criteria

1. Free from subsurface contamination,
2. Adequate terrain and area, and
3. Free from Resource Conservation and Recovery Act/Comprehensive Environmental Response, Compensation, and Liability Act (RCRA/CERCLA) features.

Siting Evaluation Criteria

1. No known or protected plant or animal species,
2. Water table significantly below the facility substructure,
3. Relatively level area in a higher location for increased security and so as not to block drainage,
4. Proximity to existing roads and the PDCF site,
5. Location with respect to subsurface soft zones,
6. Availability of utilities,
7. Location with respect to wetland areas,
8. Proximity to archaeological features, and
9. Interference with existing site operations.

Based on the above, this EIS does not consider alternatives to the SRS in which to locate the proposed MOX facility, nor does it further consider alternative locations within SRS F-Area.

2.3.2 Technology and Design Options

The general design of the proposed MOX facility was provided in DOE's SPD EIS (DOE 1999a). In developing the detailed proposed MOX facility design, DCS used the technology at Cogema's MELOX and La Hague facilities, with modifications to meet U.S. regulations, codes, and standards. A general description of the proposed MOX facility design is provided in Section 2.2.3. In its Environmental Report, DCS (2002) considered a number of technology and design alternatives. The technology and design alternatives considered by DCS were discussed if they had a possibility of having some potential impact or significance from an environmental perspective. These technology and design alternatives are summarized below. In evaluating these technology and design alternatives, NRC concluded that, with the exception of sand filters compared to HEPA filters, further detailed analysis was not warranted in Chapter 4. This technology option is also summarized in Section 2.2.5.

2.3.2.1 Dry Compared to Wet Impurity Removal

A polishing process is used to remove gallium and other impurities from the plutonium dioxide feedstock before pellet production. These impurities affect the performance of the MOX fuel in a reactor. Although the proposed aqueous (wet) polishing process meets the criteria for controlling the gallium content to less than 120 parts per billion (ppb) (Framatome ANP 2001), it also generates liquid radioactive and mixed wastes. An alternate technology for purifying the plutonium dioxide is the dry process. The dry process generates significantly less liquid waste and involves thermally induced gallium removal (TIGR). However, in an experimental setting, the TIGR process only reduced the gallium content to 25,000 ppb. The DOE considered the dry process in the SPD EIS (DOE 1999a) and concluded that the dry process would not meet the technical requirements for MOX fuel. The best reported gallium removal (Kolman et al. 2000) results in impurity contents are over 100 times the required criteria. Thus, the dry process was not further evaluated because it could not meet the technical specifications set for MOX fuel. In addition, TIGR remains an experimental process requiring further testing to scale the process to production while ensuring uniform pellet feedstock (DCS 2002; Kolman et al. 2000).

2.3.2.2 Reagent Storage

DCS considered two options for locating reagent storage and solution preparation for the aqueous polishing process. The options were to locate the storage and solution preparation process in the same area as the Aqueous Polishing Area or to locate them in a separate building and to pump reagents to the aqueous polishing process. Because of the potential explosion hazards of the chemical reagents, DCS decided to use a separate building to reduce this hazard. Because the design alternative to this approach involves potentially larger environmental impacts, namely, an increase in the explosion accident consequences, consideration of colocating the aqueous polishing and reagent storage is not evaluated in Chapter 4.

DCS also considered whether to store the chemical reagents in aboveground or belowground tanks. Belowground tanks have the advantage of limiting immediate human exposure to spills. However, there is increased environmental risk associated with leaking belowground tanks. DCS decided to use aboveground tanks with concrete curbs to contain potential spills and overflows. The NRC considered the design alternative of belowground storage tanks. However, this alternative would likely pose a greater risk of groundwater contamination. For this reason, consideration of belowground tanks is not evaluated in Chapter 4.

2.3.2.3 Acid Recovery Process

DCS added an evaporator to the acid recovery process. This evaporator reduces the activity of the distillates and recycles approximately half of the volume of distillates in lieu of using fresh demineralized water. This also results in a volume reduction of liquid wastes that would be processed and treated by the WSB. Because the design alternative to this approach involves larger environmental impacts, namely, a demand for more process chemical shipments and handling and larger waste volumes, further consideration of the aqueous polishing process without the acid recovery process as an alternative is not evaluated in Chapter 4.

2.3.2.4 Glovebox Cooling

In the MELOX design, gloveboxes are cooled at a higher air flow rate to remove heat generated from the reactor-grade plutonium. Because weapons-grade plutonium has a lower heat release, gloveboxes at the proposed MOX facility can be cooled using natural convective cooling. This results in a reduced airflow and permits a smaller HEPA filter size. The smaller filter size reduces the volume of solid TRU waste generated by filter replacement. Because the alternative to this design consideration (i.e., higher glovebox air flow) is unnecessary to meet any conceivable alternative relative to the proposed MOX facility's purpose and need to disposition weapons-grade plutonium, use of higher glovebox flows and larger HEPA filter banks is not evaluated as an alternative in Chapter 4.

2.3.2.5 Treatment of Aqueous Laboratory Waste

Aqueous laboratory wastes at the MELOX facility are precipitated and solidified, resulting in TRU wastes. DCS decided to remove the plutonium from the laboratory waste and recycle this plutonium into the aqueous polishing process. This step reduces the classification of the laboratory waste from TRU waste to LLW. Because the alternative laboratory waste management design would involve generation of more TRU waste and, therefore, have larger environmental impacts, inclusion of plutonium in laboratory waste streams is not further evaluated as an alternative in Chapter 4.

2.3.2.6 Pellet Grinding Process

In the facility design, DCS replaced the two-stage cyclone separator in the MOX powder processing operation with a decloggable metallic filter. This filter would reduce the TRU waste volume that would result from the periodic replacement of other filters downstream of the pellet grinding process. Therefore, the use of a two-stage cyclone separator instead of a decloggable filter would result in the generation of additional TRU wastes. Since additional TRU waste poses a larger environmental impact, use of a two-stage cyclone separator is not evaluated further as an alternative in Chapter 4.

2.3.2.7 Facility Heat Exchangers

DCS considered two options to remove heat from the facility. The options were to use water-cooled or air-cooled heat exchangers. Water-cooled exchangers can have impacts associated with cooling tower drift or blowdown. To reduce these potential impacts, DCS decided to use air-cooled heat exchangers. Because the water-cooled exchangers would involve generation of cooling tower drift or blowdown and, therefore, larger environmental impacts, using this type of exchanger is not further evaluated as an alternative in Chapter 4.

2.3.2.8 Physical Security Barriers

DCS considered several options to provide a physical security barrier around the proposed MOX facility. One of these was the construction of an earthen berm. Because this method would have resulted in a larger disturbed area for the site, DCS decided to use physical security barriers that resulted in less land disturbance. Because the earthen berms would involve a larger disturbed area and, therefore, larger environmental impacts, use of berms is not further evaluated as an alternative in Chapter 4.

2.3.2.9 Material Transfer from the PDCF to the Proposed MOX Facility

As discussed in Section 1.2.1, the PDCF would produce plutonium dioxide feedstock for the proposed MOX facility. The material would need to be transferred to the proposed MOX facility. DCS considered three transfer options: (1) tunnel, (2) closed transfer trench, and (3) vehicle transfer. Because the first two options would result in greater land disturbance, DCS decided to use vehicles to transfer the plutonium dioxide feedstock. Because the tunnel or closed transfer trenches would involve a larger disturbed area and, therefore, larger environmental impacts and because vehicle-related impacts would be small for the short distance between the facilities, use of tunnels or trenches is not further evaluated as an alternative in Chapter 4.

2.3.3 Immobilization of Surplus Plutonium

As discussed below, the NRC has concluded that immobilizing surplus plutonium is not a reasonable alternative to the proposed action, and, therefore, this alternative does not require detailed analysis in Chapter 4.

Before the DOE's January 2002 decision to cancel the plutonium immobilization plant, plutonium immobilization was available as a no-action disposition alternative to the proposed action. The DOE had already evaluated the environmental impacts of this alternative as alternative 12a in the SPD EIS (DOE 1999a), so that a new NRC analysis of this alternative was not required. However, as discussed in Section 1.4.1, following the DOE's January 2002 decision, the NRC solicited views on whether the immobilization alternative should still be evaluated in this EIS. The comments solicited did not identify any persuasive reasons to further consider the immobilization alternative.

The NRC has now determined for two reasons that immobilization is no longer a reasonable alternative to the proposed action. First, immobilization of the 34 MT (37.5 tons) of surplus plutonium would not meet a key element of the purpose and need for the proposed action, as described in Section 1.3. Due to budgetary constraints, the DOE decided to cancel the immobilization portion of the surplus plutonium disposition program and adopt a MOX-only approach. The DOE determined that in order to make progress with available funds, only one approach could be supported. The DOE stated that after evaluating the feasibility of implementing two disposition approaches, it believed that the best way to make the most progress with available funds while maintaining Russian interest in and commitment to surplus plutonium disposition was to pursue a MOX-only disposition strategy (DOE 2002a). The DOE further stated that Russia does not consider immobilization alone to be an acceptable approach. In the DOE's judgment, reliance by the United States on immobilization would therefore cause Russia to abandon its plutonium disposition efforts. Because immobilization fails to degrade the isotopic composition of the plutonium, Russia distrusts the immobilization alternative, as it would leave open the possibility of future retrieval and reuse of the plutonium in nuclear weapons (DOE 2002a). As discussed further in Section 1.1.1, the DOE therefore concluded that reliance on a MOX-only approach is the key to successfully completing the September 2000 agreement between Russia and the United States.

The second reason that immobilization is no longer a reasonable alternative to the proposed action is its connection with the conduct of United States foreign policy. Evaluating the immobilization alternative now would involve the NRC in foreign policy matters that the DOE has been conducting on behalf of the United States. In the NRC's view, an alternative that would block the implementation of an agreement with another country involves foreign policy matters that are outside NEPA's scope. Therefore, the NRC concludes that immobilization is not a reasonable alternative requiring detailed analysis in this FEIS.

2.3.4 Off-Specification MOX Fuel

During public information meetings in September 2002, NRC was asked to consider an alternative in which MOX fuel would be manufactured but not irradiated in commercial nuclear power plants. Under this alternative, as understood by the NRC, off-specification fuel rods would be manufactured in the proposed MOX facility and transported to spent fuel pools. These spent fuel pools could be located at either commercial nuclear power plants or interim spent fuel storage installations (ISFSIs). Once at the pool, the rods would be commingled with spent fuel rods, and possibly even incorporated into vacant positions in existing spent fuel assemblies. The final configuration would be a proliferation-resistant form that would be a candidate for the National Academy of Sciences' (NAS') spent fuel standard for surplus plutonium disposition (NAS 2000).

Since the demands for fuel quality and specifications would be lower, the fuel rods could be manufactured "off-specification." The so-called "off-specification" fuel rods would offer both environmental costs and benefits, as described below. Therefore, the NRC gave some consideration to this alternative based on the information provided by principal proponents of this approach (Macfarlane et al. 2001).

The alternative would involve a modified approach to manufacturing MOX fuel. The final powder blend would still have to be homogenized, pressed into pellets, and the pellets sintered in order to manufacture off-specification fuel rods. However, most impurities, including gallium and americium, could remain in the finished rods. This could significantly reduce liquid radioactive waste volumes associated with polishing the feedstock plutonium. As a result, the demand on the WIPP to accommodate solidified high-alpha-activity waste derived from the aqueous polishing process would be reduced.

Since the off-specification rods would not be used in a reactor, any risks of reactor accidents involving MOX fuel would not occur. In addition, the cause of some accidents in the proposed MOX facility would be prevented. For example, if aqueous polishing could be eliminated, then the risks of inadvertent nuclear criticality, solution spills, electrolyzer fires, and explosions would be considerably lower.

Since the concentration of plutonium dioxide in each off-specification rod would not be constrained by reactor fuel specifications, the mass of plutonium dioxide in each rod could be higher. This would result in lower numbers of manufactured rods and correspondingly lower vehicle-related transportation risks associated with transporting rods to any reactor sites. Fewer rods would also reduce the time required to operate the proposed MOX facility, which could result in lower operational costs. Criticality issues arising from the higher concentration of plutonium could be avoided by mixing neutron-absorbing gadolinium and hafnium with the plutonium.

However, there would be environmental costs associated with this alternative. Americium-241 would not be removed by the aqueous polishing process. Since americium-241 is a high-specific-activity alpha-emitter and poses a direct radiation hazard, radiation exposures to facility workers, site workers, and the public would be higher during MOX facility operations, off-specification MOX rod transportation, and handling of the off-specification rods at the reactor site or ISFSI.

The costs of manufacturing off-specification MOX fuel rods would also be affected by the elimination of the "fuel credit." The fuel credit is a project cost offsetting factor that accounts for the price a reactor licensee would pay for completed MOX fuel that meets its specifications. The estimated additional project costs would be $1.0 billion, thereby raising the total project costs from $3.8 billion to $4.8 billion.

The benefit of producing electricity from the use of MOX fuel would also be eliminated by the manufacture of off-specification MOX fuel.

Having qualitatively weighed the costs and benefits of this alternative, the staff find that this alternative likely involves a net increase in environmental costs. Therefore, no compelling reason exists to pursue this alternative in further detail. In addition, it is uncertain that this proposal would meet the National Academy of Sciences' spent fuel standard for surplus plutonium disposition (NAS 2000). The off-specification rods would not be irretrievably configured in irradiated spent fuel, and the isotopic distribution of the plutonium in off-specification rods would not be altered. As a result, this form is unlikely to meet with approval from the Russian Federation, whose parallel progress on plutonium disposition under formal bilateral agreements is integral to the purpose of and need for the proposed action. As discussed above for the immobilization of plutonium alternative, because this alternative does not meet the purpose of and need for the proposed action, the off-specification alternative is not further analyzed in detail in the EIS.

2.3.5 Parallex Project Alternative

Another suggested alternative to the proposed action was to transfer the surplus plutonium to Canada under the Parallex Project. The Parallex Project was identified by DOE in its ROD for the Storage and Disposition of Weapons-Usable Fissile Materials Programmatic Environmental Impact Statement (DOE 1997) as a possible option for dispositioning some of the surplus plutonium. The Parallex Project is a joint Canadian, Russian, and U.S. demonstration effort to evaluate the feasibility of burning MOX fuel in heavy-water-moderated reactors. The Parallex Project is still ongoing. It is a limited scale test of approximately 27 kg (59 lb) of MOX fuel that was manufactured at the DOE's Los Alamos National Laboratory (LANL) and at the Bochvar Institute in Moscow, Russia. This MOX fuel was shipped to Canada and is currently being tested in a Canadian Deuterium Uranium (CANDU) reactor. Following irradiation, additional analyses will be required to evaluate the usefulness of this approach. The DOE prepared an environmental assessment (EA) for this action and issued a Finding of No Significant Impact (FONSI) (DOE 1999b).

The suggested alternative of considering the Parallex Project would mean that the PDCF, the WSB, and the proposed MOX facility would be constructed and operated, but that the MOX fuel would be transferred to Canada for irradiation in heavy-water-moderated reactors there. This suggested alternative would be similar to the proposed action, except that the surplus plutonium would be irradiated in Canada. Implementing this alternative would require a change in national policy regarding the disposition of surplus weapons plutonium that is the responsibility of the DOE. Therefore, this alternative is not considered further in this EIS.

2.3.6 MIX MOX Alternative

During the public comment meetings on the DEIS in March 2003, NRC was asked to consider an alternative in which surplus weapons-grade plutonium would be mixed with reactor-grade plutonium. This alternative was named "MIX MOX" by the proponent and is described further below.

Weapons-grade plutonium has a lower percentage of plutonium-240 than does reactor-grade plutonium. One concern with the immobilization alternative was that it would not isotopically degrade the plutonium. The MOX fuel alternative does isotopically degrade the plutonium. The depleted uranium (uranium-238) and plutonium-239 in MOX fuel would be converted to plutonium-240 when subjected to irradiation in a nuclear reactor. The MIX MOX alternative would change the overall percentage of plutonium-240 by adding/mixing surplus weapons-grade plutonium with reactor-grade plutonium. The source of reactor-grade plutonium would be European stockpiles. For example, Britain has approximately 60 MT of surplus reactor-grade plutonium that was generated from reprocessing spent nuclear fuel. The MIX MOX proponent stated that after the materials were mixed, they could be disposed of in a geologic repository.

Several details of the MIX MOX alternative have not been fully developed. For example, it is not clear if new facilities would be required to perform the mixing and whether any processing of portions of the surplus plutonium would be required prior to mixing. In addition, the percentages of the two plutonium materials required to achieve suitable isotopic degradation have not been determined. The legality, availability, and cost of purchasing the reactor-grade plutonium is uncertain. As such, the environmental impacts cannot be determined. Assuming that existing DOE facilities could be used, it is conceivable that the costs of the MIX MOX alternative could be slightly lower than the proposed action; however, the benefit of producing electricity from the use of MOX fuel would be eliminated by the MIX MOX alternative.

The MIX MOX alternative appears to satisfy one element of the purpose of and need for the proposed action. It appears to result in material that is proliferation resistant and would therefore reduce the threat of nuclear weapons proliferation. However, the MIX MOX alternative does not satisfy the second element of the purpose of and need for the proposed action. The current United States - Russia agreement does not allow for disposition of surplus plutonium using the MIX MOX alternative. Moreover, given that the environmental costs of the proposed action are considered to be small, the MIX MOX alternative is not a clearly superior alternative. Therefore, the NRC concludes that MIX MOX is not a reasonable alternative requiring detailed analysis in this EIS.

2.4 Comparison of Alternatives

In weighing the environmental, economic, and other benefits of the proposed action against its environmental, economic, and other costs, the NRC must also consider and compare reasonable alternatives to the proposed action. These evaluations will be factored into the ultimate decision of whether the action called for is the issuance of the proposed license, with any appropriate conditions to protect environmental values. The proposed action and the no-action (continued storage) alternative are compared in the text below and in Table 2.1. The terms used in impact categorization are defined in the text box to the right.

Determination of the Significance of Potential Environmental Impacts

For purposes of describing impacts in this EIS, each impact was assigned one of the following three significance levels:

- *Small:* The environmental effects are not detectable or are so minor that they will neither destabilize nor noticeably alter any important attribute of the environment.
- *Moderate:* The environmental effects are sufficient to alter noticeably, but not to destabilize, important attributes of the environment.
- *Large:* The environmental effects are clearly noticeable and are sufficient to destabilize important attributes of the environment.

The impacts of the no-action alternative and the proposed action are compared for each technical area considered in this EIS. The level of impacts associated with the no-action alternative evaluated includes those impacts incurred by continued storage of surplus plutonium at DOE sites if the proposed MOX facility is not approved by the NRC. As stated previously, projected impacts for the no-action alternative were based on the analysis presented in the DOE SPD EIS (DOE 1999a) and were not reevaluated for this EIS.

The proposed action was evaluated for impacts from the following activities:

- Construction, operation, and deactivation and decommissioning of the proposed MOX facility, PDCF, and WSB at the SRS;

- Transport of depleted uranium hexafluoride from a DOE site at Portsmouth, Ohio, to a commercial fuel fabrication plant at Wilmington, North Carolina, to produce uranium dioxide needed as feedstock for the MOX fuel fabrication process;

- Conversion of depleted uranium hexafluoride to uranium dioxide;

- Transport of the uranium dioxide from Wilmington to the SRS;

- Transport of fresh MOX fuel from the SRS to a surrogate reactor site;

- Reactor use of MOX fuel; and

- Transport of spent MOX fuel to a geologic repository.

The continued storage (i.e., the no-action) alternative would result in no new construction at the DOE locations currently storing surplus plutonium, with the possible exception of minor

Table 2.1. Comparison of alternatives[a]

Impact area	Continued storage (no action)	Proposed action
Human Health Risk		
Construction	Not applicable	Human health impacts would be small.
Radiological		Same exposure as SRS employees from existing SRS operations.
Chemical		No adverse impacts from inhalation of construction-related emissions.
Physical hazards		<1 fatality, 122 injuries annually over 3 to 5 years.
Normal Operations	Under current operating conditions, human health impacts would be small.	Human health impacts would be small.
Radiological (annual impacts)		
Collective public dose (person-Sv/yr)	0.029	0.016
Annual LCFs	0.002	0.0009
Public MEI dose (mSv/yr)	0.065	6.1×10^{-5}
Risk of LCF	4×10^{-6}	4×10^{-9}
Facility workers collective dose (person-Sv/yr)	1.4	2.6
Annual LCFs	0.08	0.2
Average facility worker dose (mSv/yr)	$\leq$3.2	<5
Risk of LCF	$\leq$0.0002	<0.0003
Chemical	Insufficient data	No adverse impacts from chemical exposures.
Physical hazards	Insufficient data	<1 fatality, 41 injuries annually over 10 or more years.
Accidents	If an accident occurred, human health impacts would be small to moderate, depending on the type of the accident. Risks would be small.	If an accident occurred, human health impacts would be small, moderate, or large depending on the type of the accident. Risks would be small.
Radiological		
Event	Beyond design basis earthquake	PDCF tritium release (short-term exposure).
Dose to collective public (person-Sv)	6.6	42
LCFs	0.4	3

Table 2.1. Continued

Impact area	Continued storage (no action)	Proposed action
Chemical	No data	Large accidental releases of chlorine or nitrogen tetroxide could have adverse impacts on SRS employees and would require rapid emergency response actions.

Air Quality

Impact area	Continued storage (no action)	Proposed action
Construction	Continued storage of surplus plutonium at the DOE sites would not require new construction, thus no impacts to air quality would occur.	Air emissions impacts would be small.
Annual standard level for $PM_{2.5}$		<0.1% of standard level.
24-h standard level for $PM_{2.5}$		4.3% of standard level.
CO, SO_2, NO_2 emissions from construction equipment		<0.29% of ambient standard level.
Operations	No violation of air quality standards at DOE sites from continued storage of surplus plutonium.	Air emission impacts would be small.
24-h standard level for $PM_{2.5}$		1.9% of standard level.
$PM_{2.5}$ annual standard level		0.01% of standard level.
Toxic air pollutants and PAHs		<0.04% of South Carolina standard levels.
Prevention of significant deterioration of air quality		<6.0% of PSD Class II Area increment for SO_2 emissions.
		<6.0% PM_{10} increments to Class II Areas.
		<1% of Class I increment of PM_{10} standard at Cape Romain National Wildlife Refuge 160 km (100 mi) from proposed facilities.

Hydrology

Impact area	Continued storage (no action)	Proposed action
Construction	Not applicable	Hydrological impacts would be small.
Surface water		No surface water use or discharges to surface waters during construction.
Groundwater		139 million L/yr (37 million gal/yr). Total use for construction would be 10% of A-Area loop water demand and 3% of excess capacity.

Table 2.1. Continued

Impact area	Continued storage (no action)	Proposed action
Operations	No impacts on water use from continued surplus plutonium storage at DOE sites.	Hydrological impacts would be small.
Surface water		No significant impacts from discharges to an NPDES outfall and discharge of treated sanitary waste effluents.
Groundwater		76 million L/yr (20 million gal/yr). Total use by proposed facilities would be 5% of A-Area loop water demand in 2000 and 2% of excess capacity.

Waste Management

Impact area	Continued storage (no action)	Proposed action
Construction	No impacts to waste management systems from continued storage of surplus plutonium at DOE sites.	No TRU, LLW, or mixed LLW generation; small impacts to SRS treatment capacity for nonhazardous liquid waste.
Waste volumes generated during a 3-5-yr construction period:		
Hazardous [m^3 (yd^3)]		710 (929)
Nonhazardous liquid [m^3 (million gal)]		300,900 (79.5)
Nonhazardous solid [m^3 (yd^3)]		53,410 (69,858)
Operations	Small impacts on waste management systems from continued storage of surplus plutonium at DOE sites.	Small to moderate impacts on waste management systems at SRS and WIPP.
Waste volumes generated during 10-yr operation period:		
TRU [m^3 (yd^3)]		4,431 (5,796). TRU waste volume would be 13% of SRS storage capacity; 2.6% of WIPP disposal capacity.
Liquid LLW [m^3 (million gal)]		22,786 (6.0). The liquid LLW constitutes 4% of the discharge capacity of SRS.
Solid LLW [m^3 (yd^3)]		6,052 (7,916). Estimated volumes for solid LLW would represent about 21% of the SRS disposal capacity (if disposed of entirely at SRS).
Hazardous/mixed [m^3 (yd^3)]		120 (157). Estimated volume of hazardous waste would represent less than 2% of SRS storage capacity.

Table 2.1. Continued

Impact area	Continued storage (no action)	Proposed action
Nonhazardous liquid [m^3 (million gal)]		602,000 (159). Nonhazardous liquid waste would be 6% of SRS treatment capacity.
Nonhazardous solid [m^3 (yd^3)]		41,400 (54,149). Nonhazardous solid waste would be disposed off-site.
Environmental Justice		
Construction	No impacts would occur since no new construction would be needed for continued storage of surplus plutonium at DOE sites.	No exposure to radiological emissions and no adverse impacts from inhalation of construction-related chemical emissions, regardless of population group or income status.
Normal Operations	Radiological and nonradiological risks from continued storage of surplus plutonium would be small. No disproportionately high and adverse effects would occur.	No disproportionately high and adverse effects would occur from routine operations.
Accidents		An environmental justice impact is possible from a severe accident.
Aesthetics		
Construction and Operation	No impacts would occur because no new construction is needed for continued storage of surplus plutonium at the DOE sites.	Small impacts on visual resources from construction and operation of the proposed facilities.
Cultural and Paleontological Resources		
Construction	No impacts would occur because no new construction is needed for continued storage of surplus plutonium at the DOE sites.	Two archaeological sites, 38 AK 546/547 and 38 AK 757, would be directly affected by construction of the proposed MOX facility. The South Carolina State Historic Preservation Office accepted a data recovery plan for the sites, and data recovery was completed for both sites in 2002. Five additional eligible sites could experience indirect impacts by the construction workforce unless proper mitigation is used.
Operations	No impacts on cultural or paleontological resources are expected from continued storage of surplus plutonium at the DOE sites.	Routine operations would not impact archaeological sites near the proposed facilities.

Table 2.1. Continued

Impact area	Continued storage (no action)	Proposed action
Ecology		
Construction	No impacts would occur since no new construction is anticipated for continued storage of surplus plutonium at the DOE sites.	Impacts from habitat loss or noise generation during construction of the proposed facilities would be small.
Habitat loss		Impacts to wetlands and endangered/ threatened species would be small.
		Up to 14.7 ha (36.4 acres) of woodlands would be cleared for facilities, representing <1% of annual timber harvest at SRS, and trees would be small.
Noise impacts		Construction noise levels as high as 80 dBA could impact wildlife within 122 m (400 ft) of the project area.
Operations	Ecological impacts would be small.	Ecological impacts would be small.
Geology, Seismology, and Soils		
Construction	Continued storage of surplus plutonium at the DOE sites would not impact soils and geology since no new construction is expected.	Impacts to soils and geology would be small. Up to 50 ha (123 acres) would be disturbed in F-Area; some soil erosion and compaction.
Operations	Continued storage of surplus plutonium at the DOE sites would not impact soils and geology.	Impacts to soils and geology from routine operations would be small.
Infrastructure		
Construction	No new construction is expected, thus there would be no impacts to existing DOE infrastructure.	Impacts to existing infrastructure would be small.
Roads		An additional 4.8 to 6.4 km (3 to 4 mi) of roadways would be needed in the F-Area to support construction.
Electrical power		17,700 MWh/yr representing about 3.7% of SRS capacity would be needed during the 5-yr construction period.

Table 2.1. Continued

Impact area	Continued storage (no action)	Proposed action
Water		139 million L/yr (37 million gal/yr) representing about 3.3% of A-Area loop groundwater capacity.
Operations	Impacts occurring at DOE facilities during continued storage of surplus plutonium would be small.	Impacts to existing infrastructure would be small.
Electrical power		Use of about 186,000 MWh/yr, representing 36.4% of F-Area capacity, would occur during normal operations.
Water		76 million L/yr (20.1 million gal/yr) or about 5% of A-Area loop water demand in 2000 and 2% of excess capacity would be used.
Land Use		
Construction	No impacts would occur since no new construction of storage facilities for surplus plutonium is needed at the DOE facilities.	Small impacts to designated land use at SRS would occur for construction of the proposed facilities.
Normal Operations	No impacts to land use would occur at DOE facilities during continued storage of surplus plutonium.	Small impacts to land use would occur from routine operations.
Accidents		Depending on the type and extent of an accident during operations, impacts could be small, moderate, or large. Portions of the F-Area could be precluded from employee use until corrective cleanup and appropriate monitoring measures were implemented.
		Small, moderate, or large impacts to land use in the immediate vicinity of SRS could occur in the event that a highly unlikely accident results in radioactive material migrating off site.
Noise		
Construction	Not applicable	Small impacts would occur from noise levels generated during construction.
Equipment noise levels		Equipment and vehicle noise would reach levels of 85–90 dBA at distances of 15 m (50 ft) from the source.

Table 2.1. Continued

Impact area	Continued storage (no action)	Proposed action
		Noise levels at the SRS boundary could reach 38 dBA, which is below EPA guidance of 55 dBA for protection of the public.
Operations	No significant impacts would occur at DOE plutonium storage facilities above noise levels currently generated by traffic and worker activities.	Small impacts would occur from noise levels generated during operation.
Process equipment, diesel generators, air-conditioning noise		Noise levels could be as high as <29 dBA at the SRS boundary, which is well below the 55-dBA EPA guidance level.

Socioeconomics

Impact area	Continued storage (no action)	Proposed action
Construction	Not applicable	Impacts on the REA and ROI would be small.
Employment		1,010 direct jobs, 810 indirect jobs for peak construction year.
Income		$91.9 million in peak construction year.
In-migrating population		350
Operations	No impacts would occur from continued storage of surplus plutonium at DOE facilities.	Small impacts on the REA and ROI would occur during operations.
Employment		490 direct jobs, 780 indirect jobs.
Income		$64 million per year
In-migrating population		180

Cost-Benefit Impacts

Impact area	Continued storage (no action)	Proposed action
Construction	Continued storage of surplus plutonium at DOE facilities would not result in additional impacts to the REA and ROI.	No significant adverse impacts related to costs would occur from construction of the proposed facilities. Some beneficial impacts would occur. In general, the impacts would be considered small.
REA & ROI impacts		
Employment		1,020 average annual employment
Total income		$370 million
Total regional product		$760 million

Table 2.1. Continued

Impact area	Continued storage (no action)	Proposed action
Operations	Impacts related to costs and benefits from continued storage would be small.	Impacts related to costs and benefits from operation of the proposed facilities would be small.
REA & ROI impacts		
Employment		1,270 jobs
Total income		$640 million
Total regional product		$1,180 million
Net benefit		$1,940 million
National Impacts		
Costs		$4,064 million
Benefits		Economic benefits for materials supplied, services, new scientific knowledge, safe use of plutonium stockpile, generation of electricity from MOX fuel.
Transportation		
Radiological	No intersite transportation expected.	Radiological impacts would be small.
Routine dose to the public (person-Sv)		3.1-5.6
LCFs		0.2-0.3
Dose to the transportation crew (person-Sv)		2.1-5.3
LCFs		0.1-0.3
Accident dose risk to the public (person-Sv)		0.23
LCFs		0.01
Nonradiological	No intersite transportation expected.	Nonradiological impacts would be small.
Vehicle emissions (latent fatalities)		1-2
Accidents (fatalities)		0.078-0.20

[a]Some of the impacts for the no-action alternative are from the entire DOE site, not just activities associated with continued storage. Therefore, the impacts of the no-action alternative are overestimated.

expansion of storage facilities at the Pantex site in Texas. Construction impacts would be small or negligible at Pantex if storage facility expansion was necessary and would occur on previously disturbed land adjacent to the existing storage facilities (DOE 1999a). For all present DOE storage sites, radiological and nonradiological risks would be small. Continued storage would be expected to have no impacts on air quality, water quality, waste management systems, cultural resources, or soils, and the economic cost would be lower than that for the proposed action. However, continued storage would meet none of the DOE's goals for the plutonium disposition program.

Construction of the proposed MOX facility, PDCF, and WSB (hereafter referred to as the proposed facilities) would disturb up to 50.0 ha (123.4 acres) of land. Impacts to endangered or threatened species, wetlands, or aquatic or terrestrial habitats (including woodlands) at the SRS and the F-Area vicinity would be small. Impacts to two prehistoric archaeological sites eligible for listing on the *National Register of Historic Places* (NRHP) have been mitigated through data recovery, and the removal of the fill during construction will be monitored (see Section 5.2.9).

The primary benefit of operation of the proposed MOX facility would be the resulting reduction in the supply of weapons-grade plutonium available for unauthorized use once the plutonium component of MOX fuel has been irradiated in commercial nuclear reactors. Converting surplus plutonium in this manner is viewed as being a safer use/disposition strategy than the continued storage of surplus plutonium at DOE sites, as would occur under the no-action alternative, since it would reduce the number of locations where the various forms of plutonium are stored (DOE 1997). Further, converting weapons-grade plutonium into MOX fuel in the United States — as opposed to immobilizing a portion of it as DOE had previously planned to do — lays the foundation for parallel disposition of weapons-grade plutonium in Russia, which distrusts immobilization for its failure to degrade the plutonium's isotopic composition (DOE 2002a). Converting surplus plutonium into MOX fuel is thus viewed as a better way of ensuring that weapons-usable material will not be obtained by rogue states and terrorist groups. Implementing the proposed action is expected to promote the above nonproliferation objectives. Additionally, building and operating the proposed MOX facility is expected to result in a gain of scientific knowledge relative to the conversion of weapons-grade plutonium into reactor fuel.

In addition to the above primary benefits, there are secondary economic benefits of the proposed action. Impacts of construction on the regional economic area (REA) and region of influence (ROI) would be beneficial with respect to jobs and income. Direct construction jobs for the proposed action would total about 1,010 in the peak construction year. Although in-migration of workers during construction would be greater for the proposed action, no adverse impacts are anticipated to public services, schools, housing availability, or the local transportation network. Construction of the proposed facilities would be expected to generate 91.9 million in total income within the REA during the peak construction year.

During operations, the MOX facility, PDCF, and WSB would be expected to generate 490 direct and 780 indirect jobs, producing a total annual income of $64 million in the REA. Approximately 180 people would be expected to relocate to the SRS area during operation of the proposed facilities. No adverse socioeconomic impacts are expected as the result of proposed facility operations. Adequate public services, schools, and housing exist to satisfy needs of the in-migrating population.

The economic cost benefit analysis for the proposed action showed an overall net benefit to the ROI and REA of $1,940 million. National economic impacts (costs) for the proposed MOX facility, PDCF, and WSB are estimated to be $4,064 million. The economic benefits would include adding employment income in various national economic sectors and adding income to businesses from the purchase of related goods and services.

The following discussion compares the primary and secondary benefits set forth above to the environmental and economic costs of the proposed action.

Construction and routine operation of the proposed MOX facility would not be expected to cause any disproportionately high and adverse impacts to low-income or minority populations in the SRS vicinity. Of the accidents evaluated, a hypothetical tritium release accident at the proposed PDCF had the highest estimated short-term impacts, approximately 3 latent cancer fatalities (LCFs) among members of the off-site public. The same accident also had the highest 1-year exposure impact, up to 100 LCFs among members of the off-site public if ingestion of contaminated crops was considered. However, it is highly unlikely that such an accident would occur, and the risk to any population, including low-income and minority communities, is considered to be low. However, the communities most likely to be affected by a significant accident would be minority or low income, given the demographics and prevailing wind direction. The extent to which low-income or minority population groups would be affected would depend on the amount of material released and the direction and speed of the wind.

Continued storage of plutonium by the DOE at its present locations would not be expected to produce additional LCFs. (Annual LCFs of approximately 0.002 in the surrounding population of the storage sites [DOE 1999a] were estimated.) The annual collective dose to members of the public (i.e., those living and working within 80 km [50 mi] of the SRS) produced by routine operation of the proposed MOX facility, the PDCF, and the WSB would be expected to result in an LCF rate of approximately 0.0009/yr or less. Therefore, continued storage results in higher annual impacts.

No adverse impacts from chemical exposure of workers at the proposed MOX facility are anticipated. Less than one fatality, and approximately 120 worker injuries per year are anticipated during construction of the proposed facilities. Facility operations would result in about 40 injuries per year and less than one fatality per year.

Routine MOX facility operations are expected to produce small air quality impacts and would not result in concentrations above air quality standard levels for criteria pollutants at the SRS. Facility construction would contribute temporarily less than 0.1% of the $PM_{2.5}$ standard level, and facility operation would contribute about 0.01% or less of this level.

Water consumption during operation of the proposed facilities would be an increase of about 5% of the water demand for the A-Area loop in 2000 and about 2% of the excess A-Area loop capacity. Impacts to surface water are expected to be small during facility operations because the concentrations of nonhazardous wastes in the discharge produced by the proposed facilities would be within the guidelines of the existing NPDES permit.

Waste management systems at the SRS would not be adversely affected by wastes generated by the proposed MOX facility, PDCF, and the WSB. Adequate storage capacity and handling procedures are in place at the SRS to process hazardous wastes generated during both construction and facility operations. Nonhazardous liquid and solid wastes would not adversely affect operation of the Central Sanitary Waste Treatment Facility at the SRS.

Transportation of uranium and plutonium feedstock materials, transuranic waste, and fresh MOX fuel would result in approximately 3,300,000 to 8,200,000 km (2,050,000 to 5,100,000 mi) traveled by 1,497 to 3,512 truck shipments over the operations period of the proposed MOX facility. Up to 1 latent cancer fatality (LCF) might be expected because of the radioactive nature of the cargo. (Estimated LCFs for members of the public and the transportation crews were 0.2 to 0.4 and 0.1 to 0.3, respectively.) One to two latent fatalities from vehicle emissions were estimated, and no fatalities (0.078 to 0.20 fatality) from the physical trauma of potential vehicle accidents were estimated.

The use of sand filters was identified during the EIS scoping process as a potential substitute for final HEPA filters. The sand filter technology is described in Section 2.2.5. A comparison between sand filter and HEPA filter impacts is presented in Section 4.3.8. The NRC concludes that the technology option to install a sand filter poses no clear reduction in overall environmental impacts over the installation and use of HEPA filters.

A sand filter typically is designed to use locally available sand and gravel. The outer wall of the sand filter consists of reinforced concrete placed below or partially below grade. It is designed to withstand a design-basis earthquake and/or flood without cracking or leaking. A sand filter designed for the proposed MOX facility would be rectangular and would require a surface area of about 313 m^2 (33,650 ft^2). The filter would be about 3 m (10 ft) deep and would consist of gravel layers overlaid with sand layers arranged in order of decreasing particle size (Orr 2001). A coarse sand layer would be placed at the top of the filter to maintain the integrity of the lower sand layers during filter operations. No routine maintenance is required during operation of sand filters.

Use of the HEPA filters would result in a slightly higher radiological dose to facility workers during the course of normal operations, as discussed in Section 4.3.1.1.2, and the use of a sand filter might result in some accident impacts lower than those estimated in Section 4.3.5.2. As discussed in Section 4.3.2.2, the air filtration method would not have an impact on air quality. Both filter types have approximately the same efficiencies for particulates, and neither filter type is designed to trap gases. In addition, the disposal costs were estimated to be similar for each filter type (Section 4.3.4).

2.5 Recommendation Regarding the Proposed Action

After weighing the costs and benefits of the proposed action and comparing alternatives (see FEIS Sections 2.4 and 4.6), and after considering the comments received on the DEIS (see FEIS Appendix J), the NRC staff, in accordance with 10 CFR 51.91(d), sets forth below its NEPA recommendation regarding the proposed action. The NRC staff recommends that, unless safety issues mandate otherwise, the action called for is the issuance of the proposed license to DCS, with conditions to protect environmental values. In this regard, the NRC staff concludes that (1) the applicable environmental requirements set forth in FEIS Chapter 6 and (2) the proposed mitigation measures discussed in FEIS Chapter 5 would eliminate or substantially lessen any potential adverse environmental impacts associated with the proposed action.

The NRC staff has concluded that the overall benefits of the proposed MOX facility outweigh its disadvantages and costs, based upon consideration of the following:

- The national policy decision to reduce supplies of surplus weapons-grade plutonium, as reflected in agreements between the United States and Russia;

- The small radiological impacts on, and risk to, human health, that would be caused by constructing, operating, and decommissioning the proposed MOX facility;

- The small environmental impact the proposed action would have; and

- The economic benefit to the local community.

As discussed in FEIS Chapter 4, postulated severe accidents evaluated in connection with the proposed action would be expected to produce moderate to large impacts. While the consequences of these bounding accidents would be expected to produce moderate to large impacts, the likelihood of such accidents occurring is expected to be very low (highly unlikely). Accordingly, the NRC concludes in its NEPA analysis that the benefits of the proposed action outweigh its connected risks and costs.

2.6 References for Chapter 2

DCS (Duke Cogema Stone & Webster) 2000. *Mixed Oxide Fuel Fabrication Facility Environmental Report.* Docket No. 070-03098. Charlotte, NC.

DCS 2001. *Mixed Oxide Fuel Fabrication Facility Construction Authorization Request.* Docket No. 070-03098. Charlotte, NC.

DCS 2002. *Mixed Oxide Fuel Fabrication Facility Environmental Report, Revision 1 & 2.* Docket No. 070-03098. Charlotte, NC. July.

DCS 2003a. *Mixed Oxide Fuel Fabrication Facility Environmental Report, Revision 3.* Docket Number 070-03098. Charlotte, NC. June.

DCS 2003b. *Mixed Oxide Fuel Fabrication Facility Environmental Report, Revision 4.* Docket Number 070-03098. Charlotte, NC. Aug.

DCS 2004. *Mixed Oxide Fuel Fabrication Facility Environmental Report, Revision 5.* Docket Number 070-03098. Charlotte, NC. June 10.

DOE (U.S. Department of Energy) 1997. Record of Decision for the Storage and Disposition Programmatic Environmental Impact Statement. Office of Fissile Materials Disposition, Washington, DC. Nov.

DOE 1999a. *Surplus Plutonium Disposition Final Environmental Impact Statement.* DOE/EIS-0283. Office of Fissile Materials Disposition, Washington, DC. Nov.

DOE 1999b. "Finding of No Significant Impact in the Environmental Assessment for the Parallex Project Fuel Manufacture and Shipment." *Federal Register* 64(173):48810-48813, Sept. 8.

DOE 2000. "Record of Decision for the Surplus Plutonium Disposition Final Environmental Impact Statement." *Federal Register* 65(7):1608-1620, Jan. 11.

DOE 2002a. "Surplus Plutonium Disposition Program." (Amended Record of Decision) *Federal Register* 67(76):19432-19435, April 19.

DOE 2002b. *Supplement Analysis for Storage of Surplus Plutonium Materials in the K-Area Material Storage Facility at the Savannah River Site.* DOE/EIS-0229-SA-2. Assistant Secretary for Environmental Management, Washington, DC. Feb.

DOE 2003. *Changes Needed to the Surplus Plutonium Disposition Program, Supplement Analysis and Amended Record of Decision.* DOE/EIS-0283-SA1. Office of Fissile Materials Disposition, Washington, DC. April.

Framatone ANP 2001. *Fuel Qualification Plan.* Document No. 77-5005775-02. Prepared by Framatone ANP for U.S. Department of Energy, Office of Material Disposition, on behalf of Duke Cogema Stone & Webster. April.

Kolman, D.G., et al. 2000. "Thermally Induced Gallium Removal from Plutonium Dioxide for MOX Fuel Production." *Journal of Nuclear Materials* 282:245-254.

LANL (Los Alamos National Laboratory) 1998. *Pit Disassembly and Conversion Facility Environmental Impact Statement Data Report - Savannah River Site.* LA-UR-97-2910. Los Alamos, NM. June.

Macfarlane, A., et al. 2001. "Plutonium Disposal, the Third Way." *Bulletin of the Atomic Scientists* May/June, pp. 53-57.

NAS (National Academy of Sciences) 2000. *The Spent-Fuel Standard for Disposition of Excess Weapon Plutonium; Application to Current DOE Options.* National Academy Press, Washington, DC.

Orr, M.P. 2001. "Comparison of HEPA and Deep-Bed Sand Filters for Final Air Filtration at MOX Fuel Fabrication Facility." Letter report from Orr (Advanced Technologies and Laboratories, Inc., Rockville, MD) to W.C. Gleaves (U.S. Nuclear Regulatory Commission, Washington, DC). Nov. 15.

3 AFFECTED ENVIRONMENT

3.1 General Site Description

The Mixed Oxide (MOX) Fuel Fabrication Facility (the proposed MOX facility) and its support facilities, the Pit Disassembly and Conversion Facility (PDCF) and the Waste Solidification Building (WSB), are proposed for construction at the U.S. Department of Energy's (DOE's) Savannah River Site (SRS). The SRS is located in the southwestern portion of the state of South Carolina, as shown in Figure 3.1. The SRS is adjacent to the Savannah River, along the state border with Georgia, approximately 20 km (12 mi) southeast of Aiken, South Carolina, and 24 km (15 mi) east of Augusta, Georgia (Arnett and Mamatey 2001b). The U.S. Government owns the SRS, which was set aside in 1950 for the production of nuclear materials for national defense. Since the end of the Cold War in 1991, national priorities have shifted, and the site's priorities are now focused on waste management, environmental restoration, technology development and transfer, and economic development. The SRS covers approximately 803 km^2 (310 mi^2) in an approximately circular tract of land within Aiken, Barnwell, and Allendale Counties in South Carolina. Public access to the SRS is limited according to DOE security regulations.

The proposed facility sites are located adjacent to the north-northwest edge of F-Area near the center of the SRS (see Figure 1.2). F-Area contains facilities for chemical separations, including F Canyon, which is the main processing facility, and waste storage, which includes 20 of the 49 active liquid high-level (radioactive) waste (HLW) tanks on the SRS.

3.2 Geology, Seismology, and Soils

This section summarizes the geology, seismology, and soil conditions of the SRS and discusses site-specific conditions at F-Area. Geologic resources include mineral ores, fossil fuels, and aggregate (sand and gravel) materials that can have significant economic value. The value of soil resources depends upon the soil's ability to grow plants. Certain soils are classified by the U.S. Department of Agriculture, Natural Resources Conservation Service, as prime farmland or other important farmlands. The Farmland Protection Policy Act (*United States Code*, Title 7, Section 4201 *et seq.* [7 U.S.C. 4201] *et seq.*) and its implementing regulations (*Code of Federal Regulations*, Title 7, Part 658 [7 CFR Part 658]) require federal agencies as part of the National Environmental Policy Act (NEPA) process to consider the extent to which federal projects and programs contribute to the unnecessary conversion of important farmlands to nonagricultural uses. The site's geology and soil conditions are important in evaluating how water and potential contaminants move through the subsurface, in evaluating erosion impacts, and in predicting subsidence or landslides. Seismology is important in determining potential impacts from earthquakes.

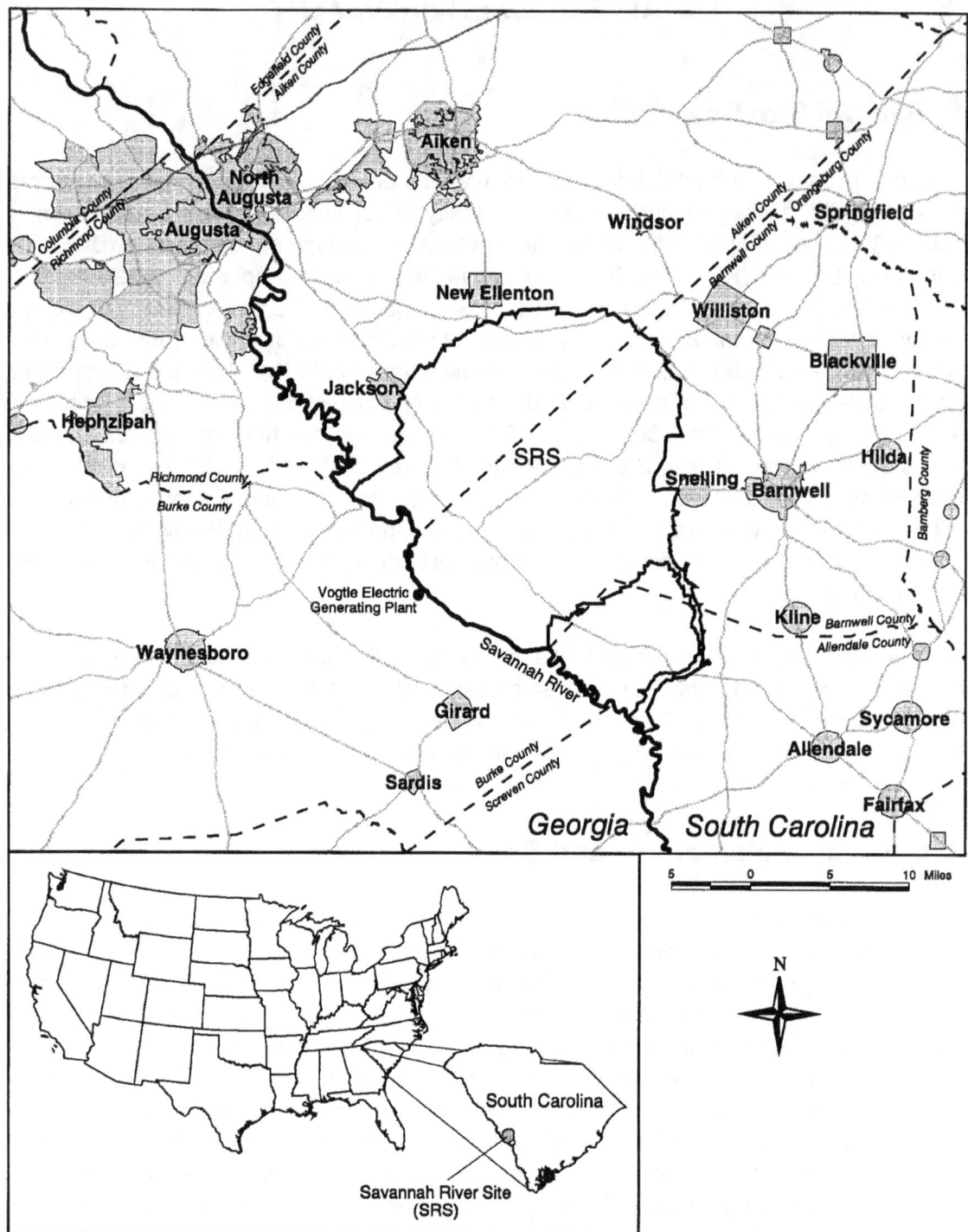

Figure 3.1. Regional location of the SRS.

3.2.1 Geology

The SRS is located in the Aiken Plateau portion of the Upper Atlantic Coastal Plain approximately 32 km (20 mi) east of the Fall Line. The Fall Line is a major physiographic and structural feature that separates the Piedmont and Coastal Plain physiographic provinces in southeastern South Carolina (DOE 1996). Soils within the Piedmont are predominantly derived from the weathering of bedrock. In contrast, soils within the Coastal Plain are predominantly sediments deposited by water. The Coastal Plain sediments are located above bedrock that consists of Paleozoic-age crystalline rock (such as granite) and Triassic-age sedimentary rock (such as siltstone) of the Dunbarton Basin. These sediments thicken from near zero at the Fall Line to about 1,220 m (4,000 ft) at the South Carolina coast (DCS 2003c). In general, the sediments have a regional dip (slant of the top surface) to the southeast. The Aiken Plateau is highly cut by narrow, steep-sided valleys separated by broad, flat areas.

Above the bedrock, the first layer of sediments at the SRS consists of about 210 m (700 ft) of Upper Cretaceous-age quartz sand, pebbly sand, and kaolinitic clay. The next ascending layer (known as the Tinker/Santee Formation) consists of 18 m (60 ft) of Paleocene-age clayey and silty quartz sand, and silt (DCS 2002). Within this layer, there are occasional beds of clean sand, gravel, clay, or carbonate. Deposits of pebbly, clayey sand, conglomerate, and Miocene- and Oligocene-age clay occur at higher elevations. This layer is noteworthy because it contains small, discontinuous, thin calcareous sand zones (i.e., sand containing calcium carbonate) that are potentially subject to dissolution by water. These "soft-zone" areas have the potential to subside, causing settling of the ground surface (WSRC 2000a; DCS 2003c). These areas were encountered in exploratory borings in F-, S-, H-, and Z-Areas of the SRS at depths between 33 and 45 m (100 and 150 ft) (DOE 1995).

The upper sediment layer in F-Area consists of primarily shallow marine quartz sand containing sporadic clay layers (known as the Barnwell Group) (DOE 1999). This layer is about 21 m (70 ft) thick near the western boundary of the SRS and about 52 m (170 ft) thick near the eastern boundary.

There are 11 deep boreholes at the SRS. The closest deep borehole is located just north of an unnamed tributary of Upper Three Runs Creek. The remaining 10 deep boreholes are not located in the vicinity of F-Area.

In 2000, 13 exploratory borings and 63 cone penetration test (CPT) holes were used to identify subsurface conditions at the proposed MOX facility site (DCS 2002). The CPT holes ranged from about 19.5 m (64 ft) to 42.7 m (140 ft) below the existing grade. Some soft zones related to past dissolution and deposition activity were identified at depth. The CPT holes were used to define the limits of the soft zones. The planned locations of heavily loaded structures, such as the MOX Fuel Fabrication Building and the Emergency Diesel Generator Building, were changed to minimize the potential impact of these underlying soft zones.

Except for some small gravel deposits, no economically viable geologic resources occur in the vicinity of F-Area (DOE 1995).

3.2.2 Seismology

On the basis of previous studies at the SRS and elsewhere, there are no known faults capable of producing an earthquake (referred to as capable faults) within the 320-km (200-mi) radius of the site that influence the seismicity of the region, except for poorly constructed faults associated with the Charleston seismic zone (DCS 2003c).

Capable Fault

A fault is described as capable if it has had movement at or near the ground surface at least once within the past 35,000 years, or recurrent movement within the past 500,000 years.

Several faults have been identified from subsurface mapping and seismic surveys beneath the SRS. The largest of these is the Pen Branch Fault. It passes through the SRS in a northeast-southwest direction and is located about 5.6 km (3.5 mi) southeast of F-Area (WSRC 2000a). Because there is no evidence of movement along this fault within the last 38 million years, the Pen Branch Fault is considered not capable.

Two large earthquakes have occurred within 300 km (186 mi) of the SRS. The larger of these was the Charleston earthquake of 1886. The Charleston earthquake is the most damaging earthquake known to have occurred in the southeastern United States and one of the largest historic shocks in eastern North America. This earthquake had an estimated Modified Mercalli Intensity of X (USGS 2001); it damaged or destroyed many buildings in the old city of Charleston, killed 60 people, and produced structural damage

Modified Mercalli Intensity Scale

The Modified Mercalli Intensity (MMI) Scale is a measure of the shaking strength of an earthquake at different locations in the region where an earthquake is felt. Earthquake intensities are characterized in terms of how the shaking affects people and buildings. The MMI Scale was originally developed in Italy nearly a century ago and includes 12 degrees of shaking. It was modified for use in the United States in 1931.

up to several hundred kilometers from its epicenter. At the SRS, this earthquake had an estimated Richter Scale magnitude ranging from 6.5 to 7.5. The SRS area experienced an estimated peak ground acceleration[1] of 0.10 g (1/10 the acceleration of gravity — 9.81 m/s/s [32.2 ft/s/s]) during this event (DCS 2002).

Three earthquakes have occurred at the SRS during recent years. They occurred on June 8, 1985, August 5, 1988, and May 17, 1997. These earthquakes were small, shallow events and were probably the result of strain release near intrusive bodies or the edges of metamorphic belts, typical of South Carolina Piedmont type seismic

Richter Scale

The magnitude of an earthquake is a measure of the energy released during the event. It is often measured on the Richter Scale, which runs from 0.0 upwards, with the largest earthquakes recorded having a magnitude of 8.6. The Richter Scale is logarithmic; a quake of magnitude 5 is 10 times more destructive than a quake of magnitude 4. Earthquakes greater than magnitude 6.0 can be regarded as significant, with the likelihood of damage and loss of life (Press and Siever 1982).

---

[1] Peak ground acceleration is the maximum acceleration amplitude (change in velocity with respect to time) measured by a seismic recording of an earthquake (called a strong motion accelerogram).

activity (WSRC 2000a). None of these earthquakes were associated with major faults (e.g., the Pen Branch Fault) in the area. Rather, these earthquakes are inferred to have seismic sources in the lower Paleozoic platform rock at a depth of about 12 km (7.5 mi) (DCS 2001a). These earthquakes had Richter Scale magnitudes of 3.2 or less and had epicenters that were within the SRS boundaries. Earthquakes of this magnitude are not felt, but do register on seismic instruments (Kirkham and Rogers 1981). Seismic alarms at the SRS reactor buildings were not triggered by any of these events (WSRC 2000a).

An earthquake with an average peak ground of 0.20 g is estimated to have an annual probability of exceedance of 1 in 10,000 (1×10^{-4}) at the SRS (DCS 2002, 2003b).

3.2.3 Soils

As discussed in Section 3.2.1, the surface soils at the SRS consist of Coastal Plain sediments. The surface soils are primarily sands and sandy loams with sporadic clay layers (DOE 1999). Currently, a stockpile of soils removed from the Actinide Packaging and Storage Facility (APSF) site on the SRS is mounded up to 15 m (50 ft) thick on the central portion of the proposed facility site in the F-Area. These soils are similar in texture to the natural soils at the site and would be removed from the site during construction.

The majority of soils in F-Area are classified by the U.S. Department of Agriculture, Natural Resources Conservation Service, as the Fuquay-Blanton-Dothan Association. These soils are nearly level to sloping and are well drained. Soils along stream floodplains are classified as the Troup-Pickney-Lucy Association. Both of these soil associations are subject to erosion. Slope stability, however, has not been a significant regional issue.

The surface soils allow precipitation to drain rapidly. Because of their sandy texture and drainage characteristics, some soil units at the SRS meet the requirements as prime farmland. However, the U.S. Department of Agriculture, Natural Resources Conservation Service, does not identify these areas as prime farmlands because they are not available for agricultural use.

Soil sampling was performed in the area of the proposed MOX facility and support buildings as part of a preconstruction baseline environmental monitoring survey conducted between September 2000 and March 2002 (SRS 2002). Fifty locations were identified for sampling by using a statistically based sampling grid. Samples were obtained from depths of between 0 and 30.5 cm (12 in.). Samples were analyzed for metals and radionuclides. None of the metal concentrations exceeded industrial use standards, and all of the radionuclides were well below SRS-developed scenario-specific radionuclide limits.

3.3 Hydrology

This section discusses the hydrologic environment of the SRS and the proposed site for the facilities. Hydrology deals with the properties, distribution, and circulation of water, particularly surface water and groundwater. The surface waters emphasized in this section are the

Savannah River and on-site streams, including treated effluent and runoff discharges to them. Groundwater resources are waters that occur within aquifers (e.g., water-bearing strata that can store and transmit water in significant quantities). These resources are discussed in relation to their use and potential contamination.

3.3.1 Surface Water

The principal surface water feature at the SRS is the Savannah River (see Figure 3.2). It borders the southwest boundary of the site for 32 km (20 mi) (DOE 1996). Six major streams flow through the SRS and discharge to the Savannah River: Upper Three Runs Creek, Beaver Dam Creek, Fourmile Branch, Pen Branch, Steel Creek, and Lower Three Runs Creek. Upper Three Runs Creek has two named tributaries, Tims Branch and Tinker Creek. Pen Branch has one tributary, Indian Grave Branch. Steel Creek also has one tributary, Meyers Branch. None of these bodies of water are federally designated as Wild and Scenic Rivers (DCS 2002). In the vicinity of the F-Area, Upper Three Runs Creek has two unnamed tributaries (see Figure 3.3) that flow to the northwest.

Two man-made lakes are located at the SRS: L Lake, which discharges to Steel Creek, and Par Pond, which discharges to Lower Three Runs Creek (DCS 2002). There are also about 50 other small man-made ponds and about 300 natural Carolina bays (closed depressions capable of holding water) at the SRS. The Carolina bays do not receive any direct effluent discharge; however, they do receive storm-water runoff.

The SRS withdraws surface water from the Savannah River mainly for industrial cooling. In 2000, the SRS withdrew about 49.7 billion L (13.1 billion gal) of water from the river. Most of this water is returned to the river through various discharges (DOE 1999).

The average flow in the Savannah River is 269 m^3/s (9,493 ft^3/s). The 7-day low flow, 10-year recurrence (referred to as "7Q10") flow is 123 m^3/s (4,332 ft^3/s) (WSRC 2000a). This flow is the lowest flow recorded over any 7 consecutive days within any 10-year period. Three large upstream reservoirs (Hartwell, Richard B. Russell, and Strom Thurmond/Clarks Hill) regulate flow in the Savannah River. This regulation is done to lessen the impacts of drought and flooding downstream. Several communities in the area use the Savannah River as a source for domestic water. The closest downstream water intake to the SRS is that of the Beaufort-Jasper Water Authority at Hardeeville, South Carolina, about 130 river miles downstream of the SRS (WSRC 2000a), which withdraws about 340 L/s (5,390 gpm) of water to service a population of 51,000 people.

Treated effluent is discharged to the Savannah River from upstream communities and from treatment facilities at the SRS. The average annual volume of flow discharged by the sewage treatment facilities at the SRS is about 700 million L (185 million gal). These effluents are released under National Pollutant Discharge Elimination System (NPDES) permits. The SRS has five NPDES permits, two (SC0000175 and SC0044903) for industrial wastewater discharges, two (SCR000000 and SCR1000000) for general storm-water discharges, and one (ND0072125) for land application (DOE 1999). Permit SC0000175 regulates 76 outfalls

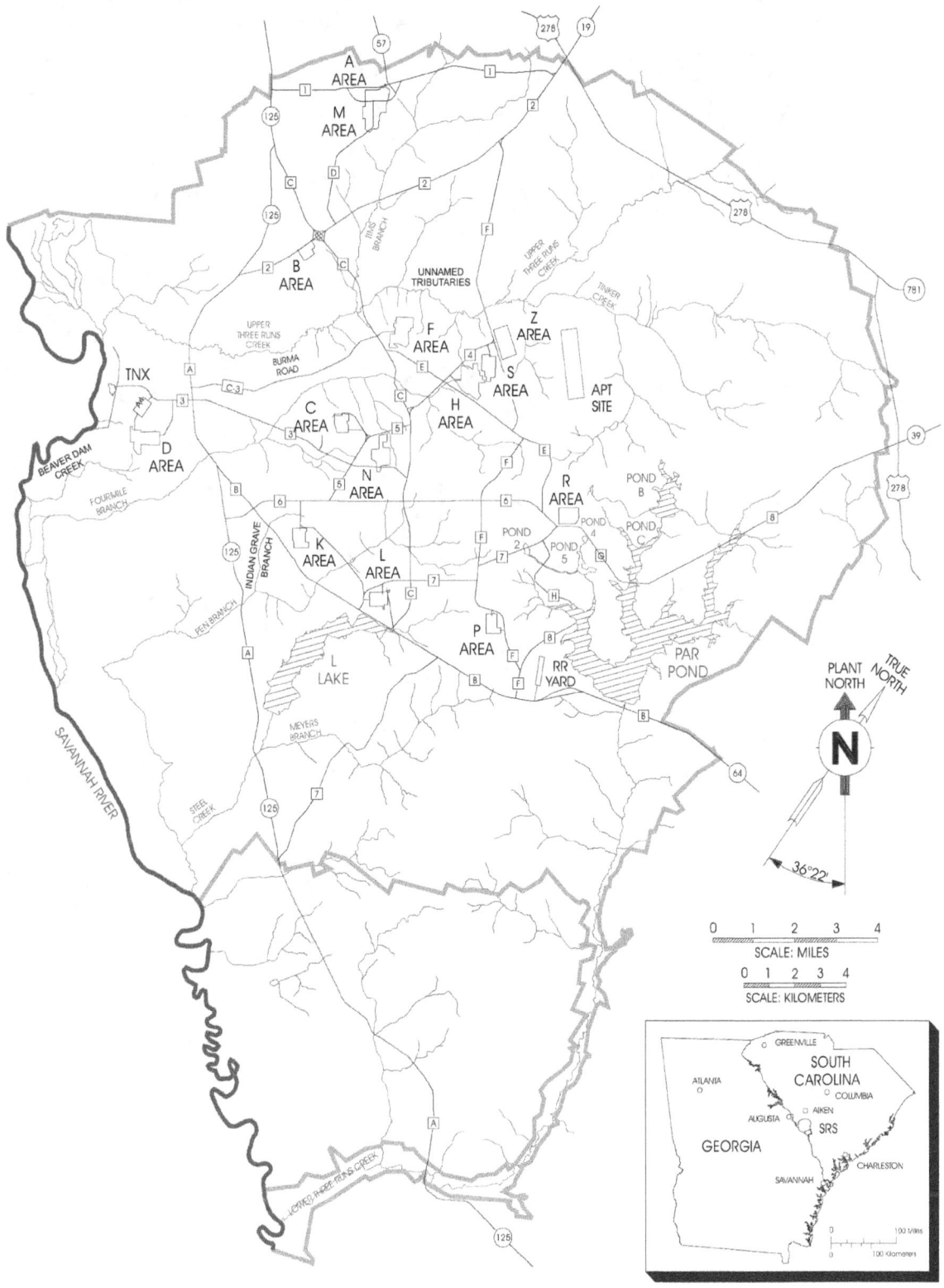

Figure 3.2. Locations of principal surface water features at the SRS (*Source*: DCS 2002).

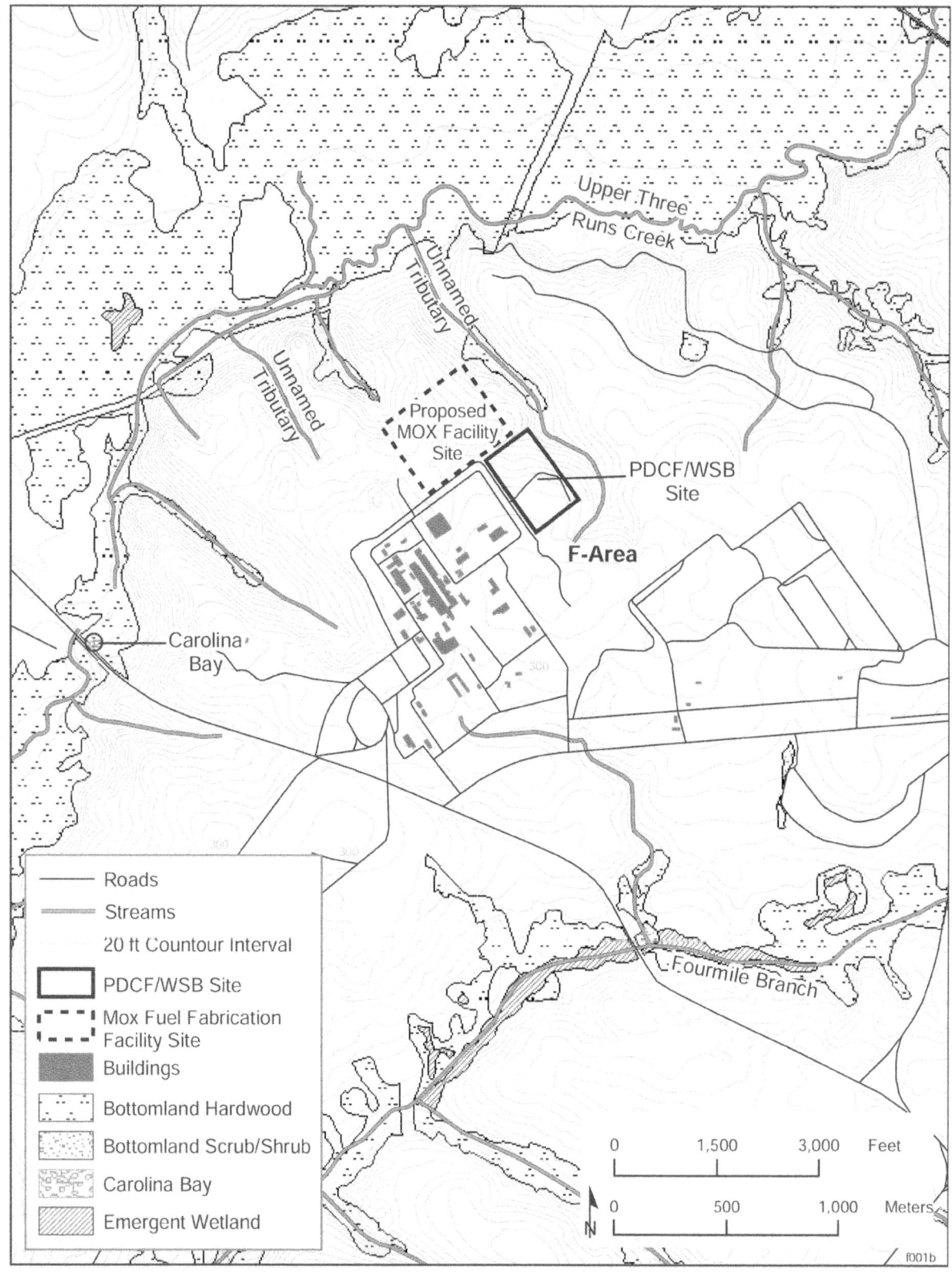

Figure 3.3. **Locations of surface water and wetlands in the F-Area**
(*Source*: Modified from DCS 2002).

(points of discharge); permit SC0044903 regulates another 7 outfalls. The 2000 compliance for these outfalls was 99.7%. The 48 storm-water-only outfalls regulated by the site's storm-water permits are monitored as required. A sediment reduction and erosion plan is required for storm-water runoff from any construction area that exceeds 2 ha (5 acres).

The Savannah River is classified as a freshwater source that is suitable for primary and secondary contact recreation, drinking after appropriate treatment, balanced native aquatic species development, and industrial and agricultural purposes. Primary contact means direct contact with the water, such as while swimming. Secondary contact means having some direct contact with the water but where swallowing is unlikely to occur, such as while fishing. Data from the river's monitoring locations generally indicate that South Carolina's freshwater standards are being met.

Runoff from the land area around F-Area drains to Upper Three Runs Creek and Fourmile Branch (DOE 1999). Runoff from the proposed facilities area drains into unnamed tributaries of Upper Three Runs Creek and flows to the northwest. Runoff from southern portions of the F-Area flow to the southeast into Fourmile Branch. The location for the proposed MOX facility is approximately 670 m (2,200 ft) southeast of Upper Three Runs Creek (WSRC 2000a). An unnamed tributary to Upper Three Runs Creek is located within about 150 m (500 ft) of the proposed MOX facility site (see Figure 3.3). The proposed MOX facility is located about 2,100 m (6,900 ft) north of Fourmile Branch.

Upper Three Runs Creek is a large, cool blackwater stream (i.e., a freshwater stream that has a dark color because of organic debris and tannin-containing compounds) that flows into the Savannah River along the western boundary of the SRS (see Figure 3.2). It drains an area of about 544 km^2 (210 mi^2) and had a mean discharge of 6.9 m^3/s (245 ft^3/s) near its mouth during water year 1995 (WSRC 2000a). A water year is measured from October 1 through September 30. The 7Q10 low-flow is about 2.8 m^3/s (100 ft^3/s). The stream is about 40 km (25 mi) long. It receives water from groundwater aquifer discharges and permitted discharges from several areas at the SRS, including F-Area, S-Area, the Central Sanitary Waste Treatment Facility, and treated industrial wastewater from the Chemical Waste Treatment Facility steam condensate. The stream, however, has never received heated discharges of cooling water from the former SRS production reactors. Flow from the sanitary wastewater discharge averages less than 0.001 m^3/s (0.035 ft^3/s).

Fourmile Branch is a blackwater stream that has been affected by past operational practices at the SRS (DOE 1999). Its headwaters are near the center of the SRS, and it flows southwesterly to the Savannah River. Until June 1985, it received large volumes of hot cooling water from the production reactor in C-Area. While the C-Area reactor was operational, the ambient temperature in Fourmile Branch was 60°C (140°F) (DOE 1999). It has a watershed area of about 54 km^2 (21 mi^2) and receives permitted effluent discharges from F-Area and H-Area. Average flow in the stream is approximately 1.8 m^3/s (64 ft^3/s). The 7Q10 low flow at the same location is about 0.23 m^3/s (8.2 ft^3/s) (WSRC 2000a). In its lower reaches, the stream widens and flows via braided channels through a delta. Downstream of the delta, it reforms into one main channel, with most of the flow discharging into the Savannah River at river mile 152.1; the remainder of the flow enters the Savannah River Swamp.

Under NPDES permit SC0000175, five outfalls discharge effluent to Fourmile Branch. Permitted discharges include 186 basin overflows, cooling water, floor drains, steam condensate, process wastewater, laundry effluent, storm water, sanitary treatment wastewater, ash basin runoff, and lab drains. Within the vicinity of F-Area, there are four permitted outfalls: F2, F3, F4, and F5. Discharge from the F2 outfall averages 0.0048 m^3/s (0.17 ft^3/s). F5 has a flow of 0.0013 m^3/s (0.046 ft^3/s). Outfall F3 is not currently used, but discharges storm water. Outfall F4 is an "administrative outfall" (i.e., an outfall with no pollutant load).

When the Savannah River floods, water from Fourmile Branch flows along the northern boundary of the floodplain and joins with other streams to exit the swamp via Steel Creek instead of flowing directly into the Savannah River. The location for the proposed facilities would not be within the 100-year floodplain of Upper Three Runs Creek (DCS 2002). Similarly, estimated water levels for the probable maximum flood (PMF) for Upper Three Runs Creek are about 15 m (50 ft) below the lowest elevation in F-Area (67 m [220 ft]).

3.3.2 Groundwater

Several aquifers occur at the SRS (see Figure 3.4). However, no federally designated sole-source aquifers occur there. The uppermost aquifer is known as the Upper Three Runs Aquifer. It occurs at an elevation of about 55 to 67 m (180 to 210 ft) above mean sea level (MSL) in F-Area (DCS 2002). The Upper Three Runs Aquifer lies on top of the leaky Gordon Confining Unit (Green Clay aquitard), which forms a confining layer for the Gordon Aquifer (Congaree

---

Sole Source Aquifer

- An aquifer that supplies at least 50% of the drinking water to the area above the aquifer.

- Areas that have no other water supply capable of physically, legally, or economically providing drinking water to local populations.

Aquifer). The Upper Three Runs Aquifer along with the Gordon Confining Unit and the Gordon Aquifer constitute the Floridan Aquifer System (WSRC 2000a). To the north, the Gordon Confining Unit is not present, and the Gordon and Upper Three Runs Aquifers merge to form the Steed Pond Aquifer. Beneath the Gordon Aquifer is the leaky Crouch Branch Confining Unit (Ellenton aquitard), which, in turn, confines the Crouch Branch Aquifer (Cretaceous Aquifer) (DOE 1999; WSRC 2000a).

Groundwater in aquifers predominantly flows horizontally to points of discharge, such as streams and swamps. In addition, some flow also occurs vertically to either underlying or overlying groundwater aquifers. Groundwater in the Upper Three Runs Aquifer, in general, flows horizontally and discharges to nearby streams. A small portion of the groundwater flows vertically downward to the Gordon Aquifer. Flow in the Gordon Aquifer is mostly horizontal to eventual stream discharge or discharge to the Savannah River, depending on location. Some of the water also flows downward to the underlying Crouch Branch Aquifer. Water in the Crouch Branch Aquifer primarily discharges to Upper Three Runs Creek and the Savannah River. Groundwater beneath the SRS flows slowly at rates that range from inches per year in the clay aquitards that confine the aquifers to several hundred feet per year in the sandy aquifers

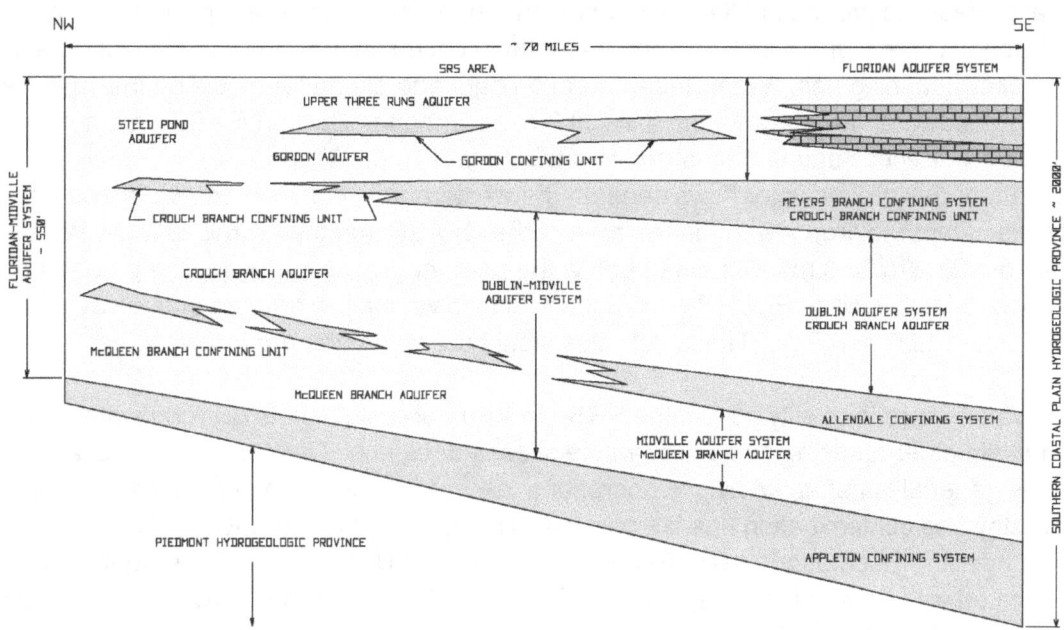

Figure 3.4. Aquifers at the SRS (*Source*: DCS 2001a).

(WSRC 2000c). Average annual recharge to the Upper Three Runs Aquifer is 35.6 cm (14 in.) (WSRC 1997a).

The Crouch Branch Aquifer is an abundant and important water resource for the SRS region. At the SRS, groundwater is the only source of domestic water. All groundwater at the SRS is classified by the U.S. Environmental Protection Agency (EPA) as a Class II water source (i.e., a current and potential source of drinking water). In 2000, the SRS withdrew 7.95 billion L (2.1 billion gal) of groundwater from the Crouch Branch Aquifer in support of site operations. Some nearby towns, such as Aiken, South Carolina, obtain groundwater from the Crouch Branch Aquifer, but most of the rural population draws water from the Gordon, Upper Three Runs, or Steed Pond Aquifers. About 8 billion L/yr (2.1 billion gal/yr) of groundwater is withdrawn from these upper aquifers within a 16-km (10-mi) radius of the site (DCS 2002).

F-Area is located on a groundwater divide between Fourmile Branch and Upper Three Runs Creek. Near-surface groundwater in the southern portion of the F-Area primarily moves laterally and discharges to Fourmile Creek and its tributaries to the south. In the northern portion of the F-Area, including the proposed location of the facilities, near-surface groundwater also primarily moves laterally, but discharges to Upper Three Runs Creek and its tributaries to the north (WSRC 2000c). F-Area is located in a region of groundwater recharge from precipitation.

Beneath the site for the proposed MOX facility, the Upper Three Runs Aquifer is divided into upper and lower zones by the Tan Clay confining unit of the Dry Branch Formation (DCS 2002).

In the area near the proposed MOX facility site, the topography drops sharply to the north toward Upper Three Runs Creek, and the water table occurs in the lower aquifer zone beneath the Tan Clay confining unit. Water table elevation data and computer modeling indicate that shallow groundwater flows away from the Old F-Area Seepage Basin (OFASB) in a north-northwesterly direction and is discharged to a tributary of Upper Three Runs. A small component of this groundwater flows beneath the westernmost corner of the proposed MOX facility site. Depth to groundwater in the area of the OFASB and the proposed MOX facility site ranges from 23.2 to 28.3 m (76 to 93 ft) below the present ground surface. Site preparation for the proposed MOX facility, PDCF, and WSB would involve shallow grading and excavation to a depth of about 12.2 m (40 ft). These activities would not encounter groundwater.

Groundwater varies in quality across the SRS. In some areas, it meets drinking water quality standards; in other areas, such as near waste sites, it does not. The deep Crouch Branch Aquifer is generally unaffected by site operations, except for a location near A-Area, where trichloroethylene contamination has been found. Tritium has been reported in the Gordon Aquifer under the Separation Areas (F- and H-Areas). The Upper Three Runs Aquifer is contaminated with solvents, metals, and low levels of radionuclides near several SRS areas and facilities, including the F-Area.

Groundwater is the only source of domestic water at the SRS. The existing capacity at the SRS is approximately 33.5 billion L/yr (8.9 billion gal/yr). Groundwater rights in South Carolina are associated with the absolute ownership rule. Owners of land overlying a groundwater resource are allowed to withdraw as much water as they desire; however, the state requires users who withdraw more than 138 million L/yr (36.5 million gal/yr) to report their withdrawals. Because the groundwater use at the SRS exceeds this value, DOE is required to report its usage to the state (DCS 2002).

Within F-Area, four groundwater wells are used for process water. Pumping capacities for these wells range from 1,500 to 3,800 L/min (400 to 1,000 gpm). They extract groundwater from the Crouch Branch Aquifer. Two of these wells were formerly used for domestic water supply. The current annual groundwater use at F-Area is 374 million L (98.8 million gal). The estimated capacity of the wells in F-Area is about 4.2 billion L/yr (1.1 billion gal/yr) (DCS 2002).

The F-Area wells are part of a SRS A-Area domestic water loop. The combined capacity of the F-Area and A-Area wells is about 11,360 L/min (3,000 gal/min) (DCS 2003a,b). Water consumption in 2000 averaged 2,850 L/min (754 gal/min). Therefore, an excess capacity of about 8,500 L/min (2,250 gal/min) exists for the A-Area loop. The A-Area loop supplies water to both A-Area and F-Area.

Groundwater quality in F-Area is not significantly different from that of groundwater throughout the rest of the SRS. It is abundant, usually soft, slightly acidic, and low in dissolved solids. F-Area groundwater can exceed drinking water standards for several contaminants. In 1999, 18% of 365 wells sampled at the General Separations and Waste Management Areas (Areas F, E, H, S, and Z) had metal concentrations that exceeded metal drinking water standards; 10% of 471 wells sampled had organic concentrations that exceeded organic drinking water standards; 53% of 483 wells sampled exceeded drinking water standards for tritium; 40% of 372 wells sampled exceeded drinking water standards for other radionuclides; and 31% of 307 wells sampled exceeded drinking water standards for other constituents. The sources of the detected groundwater contamination included burial grounds, waste management facilities, canyon buildings, seepage basins, and saltstone disposal facilities (WSRC 2000c).

Near the F-Area seepage basins and inactive process sewer line, there is widespread radionuclide contamination. Near the F-Area Tank Farm, tritium, mercury, nitrate-nitrite (as nitrogen), cadmium, gross alpha, and lead were detected in concentrations that exceeded drinking water standards in one or more wells. At the Sanitary Sludge Application Site, tritium, specific conductance, lead, and copper values exceeded their drinking water standards in one or more wells. In addition, a subsurface plume of tritium and strontium contamination has recently been found in F-Area. The source of groundwater contamination is from various heavy industrial and nuclear operations over the past 50 years in the F-Area. The contaminant plume appears to originate inside F-Area and extend beneath the MOX facility site, with movement in a fan-like direction of groundwater flow under the proposed MOX facility site.

Contaminated groundwater also exists beneath the OFASB. The OFASB is located about 180 m (600 ft) north of F-Area, immediately adjacent to the western boundary of the proposed MOX facility site. The OFASB has been remediated by filling the basin with clean soil, capping, and stabilizing the contaminated soil within the basin with grout (WSRC 1997a). The results of sampling in the compliance wells for the OFASB indicated that concentrations of several target constituents were above drinking water standards in several wells. These contaminants included iodine-129, nitrate, radium-226, radium-228, strontium-90, tritium, uranium (total), and lead. There is, however, some uncertainty about whether these exceedances are related entirely to OFASB, to upgradient F-Area facilities, or to both. A small component of the contaminant plume from OFASB flows beneath the westernmost corner of the proposed MOX site. Groundwater is monitored on a regular basis with 15 wells. Contaminant fate and transport models predict that the aquifer is expected to return to an uncontaminated state (i.e., a condition in which no maximum contaminant levels are exceeded) within 2 to 115 years, depending on the specific contaminant.

The results of recent groundwater sampling of nine wells distributed uniformly across the proposed MOX facility site indicate that shallow groundwater (i.e., groundwater in the Upper Three Runs Aquifer) is contaminated (SRS 2002). Gross alpha and beta activity, tritium, uranium, and trichloroethylene exceeded maximum contaminant levels for drinking water. Contamination is present beneath the entire MOX site, but is greatest beneath the western edge of the site. The contaminant plume appears to originate inside the F-Area fence and was and is related to F-Area nuclear operations and waste management practices at OFASB.

Groundwater in the Upper Three Runs Aquifer beneath the proposed MOX facility site is contaminated with various heavy industrial and nuclear contaminants. The proposed construction activities will take place at least 9.1 m (30 ft) above the zone of contaminated groundwater.

3.4 Meteorology, Emissions, Air Quality, and Noise

This section discusses the existing meteorology, current airborne pollutant emissions, air quality, and noise environment in the vicinity of the SRS. Section 3.4.1 describes the meteorology, or weather conditions, around the SRS. Meteorology includes the atmospheric conditions that determine where pollutants released into the atmosphere travel and how they are mixed with existing air and become diluted as they travel. Section 3.4.2 describes existing air emissions from the SRS and the surrounding area. Section 3.4.3 describes regional air quality and air quality standards. Air emissions from the proposed MOX facility, the PDCF, and the WSB would combine with existing emissions to affect local and regional air quality. Comparing the resulting combined air quality against the standard levels provides one measure of the facilities' impact on air quality. Section 3.4.4 describes the existing noise environment and applicable regulations. Noise generated by the facilities would combine with existing levels to produce the overall noise impact.

3.4.1 Meteorology[2]

To provide a thorough picture of weather conditions at a given location often requires the use of data from several locations. Different locations that record meteorological data may record different parameters. Data recorded near the site of the proposed action is generally considered most representative of the site. Meteorological data for F-Area (the site of the proposed facilities), H-Area, and Bush Field in Augusta, Georgia, were used to describe meteorological conditions of the affected environment.

Meteorology

Meteorology deals with weather conditions. Air pollution meteorology emphasizes weather conditions that determine how pollutants released into the air travel and mix with the air. The more important weather conditions involved in this process include wind speed and direction and atmospheric stability, a measure of how much mixing is occurring in the atmosphere.

The climate at the SRS is characterized by short, mild winters and long, humid summers (DCS 2002). Mountains to the north and west prevent or delay the approach of many cold air masses (Ruffner 1985).

The annual average wind speed is 2.8 m/s (6.2 mph) at Bush Field, which is located in Augusta, Georgia, about 24 km (15 mi) northwest of F-Area. Wind speed is highest in the spring, averaging 3.1 m/s (7.0 mph). March has the highest monthly average wind speed of 3.4 m/s (7.7 mph) and August the lightest, 2.3 m/s (5.1 mph). The prevailing monthly wind direction is from the west-northwest from November through February and variable for the rest of the year. On the basis of observations for 1995-1999, the highest 2-minute wind speed was 20 m/s

---

[2] Unless otherwise noted, the information presented in this section is based on meteorological data collected at Bush Field in Augusta, Georgia, about 24 km (15 mi) northwest of F-Area, and summarized by the National Climatic Data Center (NOAA 1999).

(45 mph) from the north-northwest in June 1998, and the maximum gust (5-second wind speed) was 25 m/s (55 mph) from the north-northwest in April 1997.

A wind rose based on data from the 5-year period 1992 through 1996 from the 62-m (200-ft) meteorological tower in H-Area at the SRS is presented in Figure 3.5. The wind rose indicates no strongly predominant prevailing wind direction, but the wind is from the northeast about 10% of the time and from the west-southwest over 9% of the time. Annual average wind speeds ranged from 3.6 to 4.2 m/s (8.0 to 9.4 mph) during the 5-year period (DCS 2002).

Wind Rose
A *wind rose* summarizes wind speed and direction graphically as a series of bars pointing in different directions. The direction of a bar shows the direction *from* which the wind blows. Each bar is divided into segments. Each segment represents wind speeds in a given range of speeds, for example, 6-8 m/s. The length of a given segment represents the percentage of the summarized hours that winds blew from the indicated direction with a speed in the given range.

The driest period occurs during the months of October and November, with rainfall increasing after then to a peak in March. A dry period extends from April through early June, followed by a wet period from late June through early September caused primarily by thunderstorms and showers (Ruffner 1985). Average annual precipitation at Bush Field is 114 cm (44.7 in.). Data from 1967 to 1996 at the SRS show an average annual precipitation of 126 cm (49.5 in.) (DCS 2002). Average monthly precipitation ranges from 6.30 cm (2.48 in.) in November to 11.8 cm (4.65 in.) in March. The greatest amount of precipitation recorded in a single month was 37.6 cm (14.8 in.) in October 1990, and the least amount was in October 1953, when only trace amounts of rainfall were recorded. The greatest amount of precipitation recorded in a 24-hour period was 21.8 cm (8.57 in.) in October 1990. Snowfall occurs only one to three times in the winter and usually remains on the ground for only a short period (Ruffner 1985). Annual snowfall averages 3.3 cm (1.3 in.). The greatest monthly snowfall occurred in February 1973, with 35.6 cm (14.0 in.), and the greatest 24-hour snowfall was 34.8 cm (13.7 in.) in the same month. Freezing rain may occur one to three times per winter (Ruffner 1985).

The average annual temperature at Bush Field is 17.5°C (63.5°F). At the SRS, the average annual temperature is 17.3°C (63.2°F) (DCS 2002). January is the coldest month, with an average temperature of 7.39°C (45.3°F), and July the warmest, averaging 26.7°C (80.1°F). Daily extreme temperatures have ranged from 42.2°C (108°F) in August 1983 to -18°C (-1°F) in January 1985. An average of 309 freeze-free days (days with a minimum temperature greater than 0°C [32°F]) occur per year. There are no freeze days from May through September. Temperatures above 32°C (90°F) occur about 73 days per year, with 56 of them occurring in June, July, and August.

Average annual relative humidity at Bush Field ranges from 83% in the early morning to 51% in the afternoon. In July and August, the early morning relative humidity averages 90%, with afternoons averaging 55-56%. At the SRS, comparable values for August are 97% and 50% (DCS 2002). Dew point temperatures at Bush Field range from 1.33°C (34.4°F) in January to 21.0°C (69.7°F) in July. Heavy fog with visibility less than 0.40 km (0.25 mi) occurs on an

SRS H-Area Meteorological Tower (200-ft level)
(Period: 1992-1996)

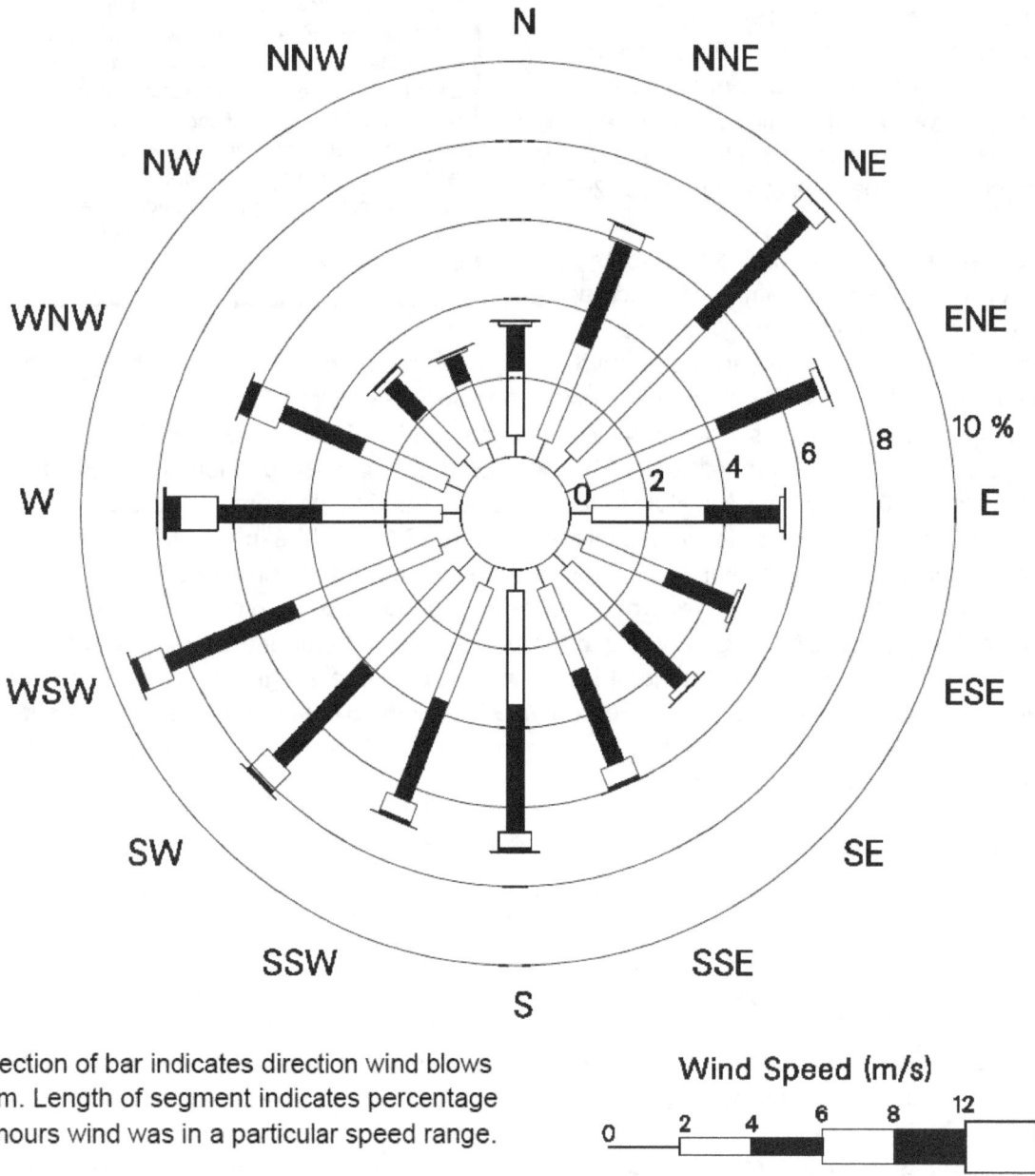

Direction of bar indicates direction wind blows from. Length of segment indicates percentage of hours wind was in a particular speed range.

Figure 3.5. Annual wind rose for the SRS (*Source*: Arnett and Mamatey 2000a, Table 31).

average of about 32 days per year. Heavy fog occurs throughout the year but is most likely in November and December.

Thunderstorms, tornadoes, and hurricanes provide occasional severe weather to South Carolina (Ruffner 1985). Thunderstorms occur on an average of 53 days per year at Bush Field. July averages 12.6 thunderstorm days, December 0.7. More than 70% of the thunderstorms occur in the four-month period from May through August. They are most common in the summer months, but the more violent storms generally occur along active cold fronts in spring (Ruffner 1985). Hail with thunderstorms is infrequent and occurs about once every 2 years on the average (DCS 2002).

Tornadoes are rare in South Carolina. Most that do occur are during the period March through June. April is the peak month for tornadoes, with a smaller peak in August and September (Ruffner 1985). For the 49-year period of 1950-1998, an average of 11 tornadoes per year occurred in South Carolina (Storm Prediction Center 2001). Between 1880 and 1995, a total of 17 significant tornadoes were reported in Aiken and Barnwell Counties, South Carolina, and Burke County, Georgia. Nine tornadoes have caused damage on the SRS, one with estimated wind speeds as high as 67 m/s (150 mph). None have caused damage to buildings on the SRS (DCS 2002).

Tropical storms or hurricanes affect South Carolina about once every 2 years. Most do little damage and affect only the costal areas, decreasing in intensity as they move inland. Those that do move far inland can cause considerable flooding (Ruffner 1985). Thirty-six hurricanes caused damage in South Carolina between 1700 and 1989, and the interval between them has ranged from 2 months to 27 years. About 80% have occurred in August and September. Between 1886 and the present, 17 storms (10 hurricanes and 7 tropical storms) have passed within 64 km (40 mi) of the proposed MOX facility site. All the hurricanes had been downgraded to tropical storms or tropical depressions before reaching SRS (Weather Site, Inc. 2003). The only hurricane-force winds measured at the SRS were associated with Hurricane Gracie on September 29, 1959, when wind speeds of 34 m/s (75 mph) were measured at F-Area (DCS 2002).

3.4.2 Emissions

The SRS is classified as a "major source" (of airborne pollutant emissions) under the Clean Air Act (CAA), with potential emissions of more than 227,000 kg/yr (250 tons/yr). The SRS has construction and operating permits from the South Carolina Department of Health and Environmental Control (SCDHEC), Bureau of Air Quality, for about 199 point sources. Thirty-eight of these sources are permitted for air toxics. During 2000, 137 sources operated at least part of the year, and 62 were on cold standby or under construction.

Significant sources of criteria air pollutants[3] or their precursors and toxic air emissions at the SRS include coal-fired powerhouse boilers (two in A-Area and three in H-Area) and No. 2 oil-fired package steam generators (two in K-Area and two portable units). Other facilities emitting nonradiological emissions include 128 pieces of equipment powered by diesel engines, the Defense Waste Processing Facility, groundwater air strippers, the Consolidated Incineration Facility, and controlled burning. During 2000, the SRS continued to be in compliance with permitted emission rates and special conditions (Arnett and Mamatey 2001b).

SRS point source emissions for 1999 are compared with point source and total emissions within the four surrounding counties — Aiken, Allendale, and Barnwell Counties in South Carolina and Burke County in Georgia — in Table 3.1. The SRS contributed less than 6% of the four-county point source emissions of nitrogen oxides (NO_x), sulfur dioxide (SO_2), volatile organic compounds (VOCs), and particulate matter less than 10 μm and less than 2.5 μm in diameter, (PM_{10} and $PM_{2.5}$, respectively) in 1999. The SRS contributed about 17% of the four-county area point source emission of carbon monoxide (CO). However, CO is generated primarily by mobile sources, and the SRS emitted only about 0.20% of the total point and nonpoint CO for the four-county area. Arnett and Mamatey (2001a) provide an inventory of about 200 toxic air pollutant emissions from the SRS for 1999. Table 3.2 lists the emissions that exceeded 0.9 MT (1 ton) per year.

3.4.3 Air Quality

The SRS is located in the Augusta-Aiken Interstate Air Quality Control Region (AQCR) #53, which comprises 6 counties in South Carolina and 13 in Georgia (see Figure 3.6) (EPA 1972). Both South Carolina and Georgia have adopted State Ambient Air Quality Standards (SAAQS) identical to the federal National Ambient Air Quality Standards (NAAQS) for the criteria pollutants (see adjacent text box). In addition, South Carolina has retained the annual standard for total suspended particulates (TSP) and adopted an additional standard for gaseous fluorides (SCDHEC 2000; GDNR 2000).

Air Quality Terms

Particulate matter (PM) is dust, smoke, other solid particles, and liquid droplets in the air. The size of the particulate is important and is measured in micrometers (μm). A micrometer is 1 millionth of a meter (0.000039 in.).

Total suspended particulate (TSP) is PM with a diameter less than 30 μm. PM_{10} is PM with a diameter less than 10 μm and $PM_{2.5}$ is PM with a diameter less than 2.5 μm. The U.S. Environmental Protection Agency (EPA) has set standards for PM_{10} and $PM_{2.5}$ designed to protect human health and welfare.

Criteria pollutants are pollutants for which the EPA has prepared documents detailing their health and welfare impacts and set standards specifying the air concentrations that avoid these impacts. The criteria pollutants are sulfur oxides, nitrogen dioxide, carbon monoxide, PM_{10}, $PM_{2.5}$, lead, and ozone.

Volatile organic compounds (VOCs) are organic vapors in the air that can react with other substances, principally nitrogen oxides, to form ozone. The reactions are energized by sunlight.

Background is a concentration value, usually based on measured pollutant data, that accounts for the impacts of emission sources not included explicitly in the air quality model.

---

[3] "Criteria" air pollutants are common air pollutants for which federal standards have been established.

Table 3.1. Estimated emissions from four counties around the SRS and SRS point sources in 1999[a]

| Pollutant[c,d] | Four-county area emissions (tons/yr)[b] | | SRS emissions | | |
| | | | | As percentage (%) of four-county area | |
	Point	Total	Total (tons/yr)	Point	Total
CO	712	62,300	124	17	0.20
NO_x	6,800	17,700	337	5.0	1.9
SO_2	14,600	15,400	346	2.4	2.3
PM_{10}	1,250	1,747	54.5	4.4	3.1
$PM_{2.5}$	696	1,120	37.9	5.4	3.4
VOCs	1,770	8,330	7.45	0.42	0.089

[a]Four SRS border counties: Aiken, Barnwell, and Allendale, South Carolina; and Burke, Georgia. "Point" values are for all point sources. "Totals" are for all sources, including point, area, and mobile.

[b]To convert tons to kilograms, multiply by 907.2.

[c]The reference does not include lead. Lead emissions have been lowered by reductions in the lead content of gasoline.

[d]Ozone is not emitted directly and is not listed in this table. It is formed in the air by chemical reactions involving VOCs and NO_x.

Source: EPA (2001).

South Carolina is currently designated as being in attainment (i.e., in compliance with standards) for all criteria pollutants (40 CFR 81.341). Georgia is designated as in attainment except for the 13-county area around Atlanta, which is designated as nonattainment for the 1-hour ozone standard (40 CFR 81.311). A list of the ambient standards and the high and low ambient concentrations at air quality monitoring stations within 80 km (50 mi) of the proposed MOX facility site is shown in Table 3.3. The regulations for Prevention of Significant Deterioration (PSD) of air quality (40 CFR 52.21) place limits on the total

National Ambient Air Quality Standards (NAAQS)

The EPA sets NAAQS for criteria pollutants (sulfur oxides, PM_{10}, $PM_{2.5}$, carbon monoxide, nitrogen dioxide, lead, and ozone). The primary NAAQS specify maximum ambient (outdoor air) concentrations of the criteria pollutants that would protect public health with an adequate margin of safety. Secondary NAAQS specify maximum concentrations that would protect public welfare. If both a primary and a secondary standard exist, the lower (more restrictive) standard is normally used for assessment purposes. Some of the NAAQS for an averaging time of 24 hours or less allow the standard values to be exceeded a limited number of times per year.

Table 3.2. Toxic air pollutant emissions at the SRS in 1999

Pollutant[a]	CAS number[b]	Emissions (tons/yr)[c]
Benzene	71-43-2	4.16
Chloroform	67-66-3	6.30
Formaldehyde	50-0-0	1.28
Formic acid	64-18-6	3.45
Hexane	110-54-3	1.14
Hydrochloric acid	7647-1-0	1.73
Hydrogen sulfide	7783-6-4	5.71
Methoxychlor	67-56-1	1.46
Nitric acid	7697-37-2	1.04
Sodium hydroxide	1310-73-2	1.32
Tetrachloroethylene	127-18-4	2.17
Toluene	108-88-3	1.87
Trichloroethylene	79-1-6	5.53
Xylenes	1330-20-7	4.96

[a]Only pollutants with emissions of more than 1 ton are listed.

[b]Chemical Abstract Services (CAS) number — a number assigned to a specific chemical by CAS. The number avoids the ambiguity associated with multiple names for the same chemical and also avoids problems associated with name differences between languages.

[c]To convert tons to kilograms, multiply by 907.2.

Source: Arnett and Mamatey (2001a, Table 45).

increase in ambient pollution levels above established baseline levels for SO_2, NO_2, and PM_{10}. Under those regulations, the allowable increases are smallest in Class I areas (national parks and wilderness areas). The rest of the country is subject to PSD II increments. States can choose a less stringent set of Class III increments, but no states have chosen to do so. The Cape Romain National Wildlife Refuge, the PSD Class I area closest to the SRS, is about 160 km (100 mi) to the east. The facilities at the SRS have not been required to obtain PSD permits (DCS 2002).

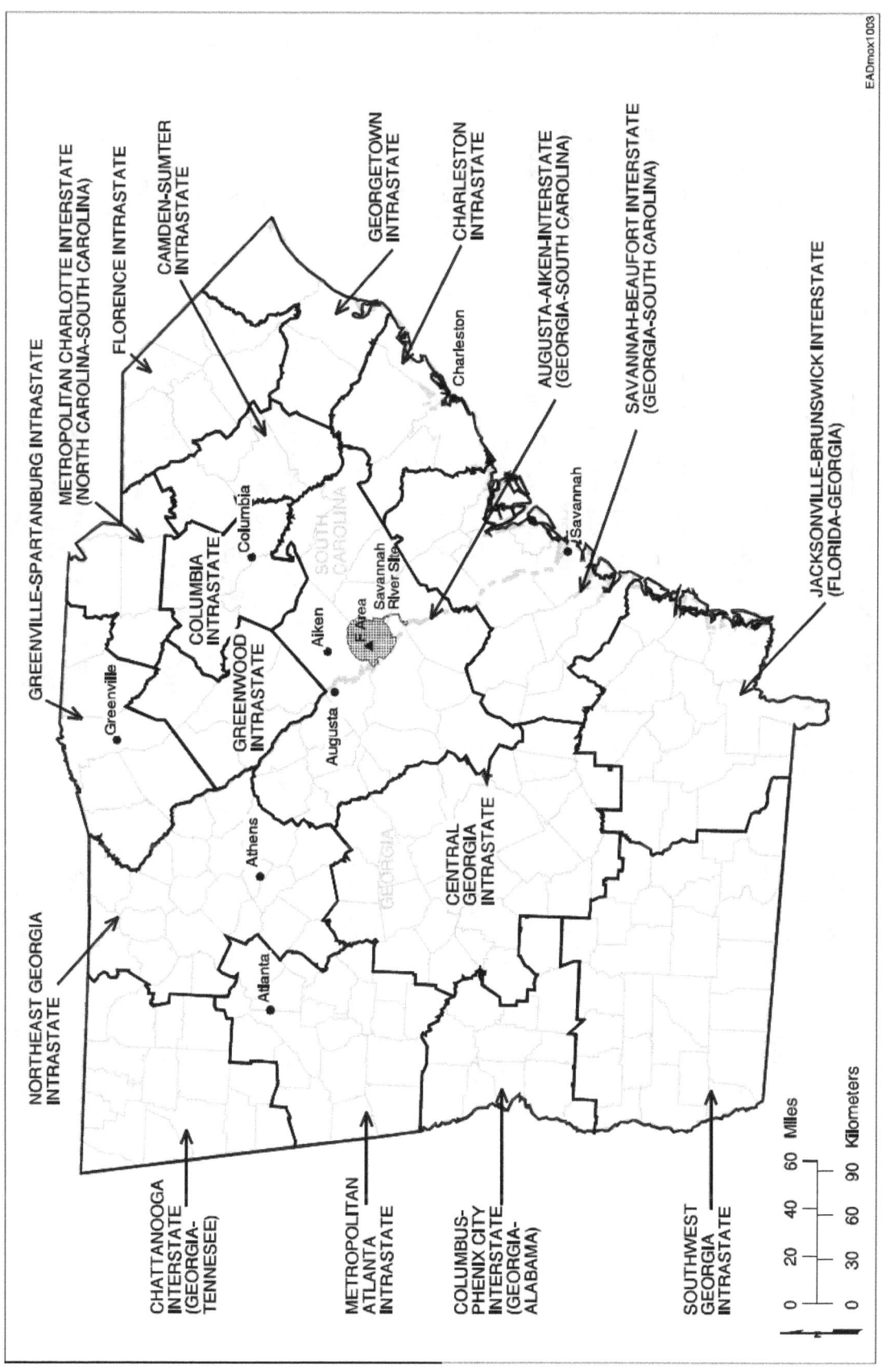

Figure 3.6. Air quality control regions, South Carolina and Georgia.

Table 3.3. Ambient air quality standards and range of pollutant levels in the vicinity of the SRS

Pollutant	Averaging time	Ambient standard ($\mu g/m^3$)[b,c]	Highest/lowest levels in vicinity of SRS[a]		
			Concentrations ($\mu g/m^3$)	Locations (city, county, state)	Years
SO_2	3 hours	1,300[d]	180	–, Barnwell, SC	1999
			58	–, Barnwell, SC	1997
	24 hours	365[e]	55	Augusta, Richmond, SC	1997-2000
			13	–, Barnwell, SC	1997-1998
	Annual	80[e]	5.2[f]	–, Aiken, SC/ Augusta, Richmond, GA	1999/ 1997-2000
			2.6[f]	–, Aiken, SC/ –, Barnwell, SC	1996-1998/ 1997-2001
NO_2	Annual	100[g]	9.4	–, Aiken, SC	1997-2000
			5.6	–, Barnwell, SC	1999, 2001
CO	1 hour	40,000[e]	–[h]		–
	8 hours	10,000[e]	–[h]		–
O_3	1 hour	235[g]	233[i]	Augusta, Richmond, SC/ –, Edgefield, SC	1998/1998
			165[i]	–, Edgefield, SC	2001
	8 hours[j]	157[g]	194[k]	Augusta, Richmond, SC	1998
			145[k]	–, Barnwell, SC	2001
PM_{10}	24 hours	150[g]	165[i]	–, Lexington, SC	1997
			36[i]	–, Aiken, SC	2001
	Annual	50[g]	29	–, Lexington, SC	1999
			17	–, Aiken, SC	2001

Table 3.3. Continued

Pollutant	Averaging time	Ambient standard (μg/m³)[b,c]	Highest/lowest levels in vicinity of SRS[a]		
			Concentrations (μg/m³)	Locations (city, county, state)	Years
PM$_{2.5}$	24 hours[j]	65[g]	42[l] 17[l]	Augusta, Richmond, GA –, Colleton, SC	1999 2000
	Annual[j]	15[g]	19.9 11.2	Augusta, Richmond, GA –, Colleton, SC	1999 2000
Pb	Calendar Quarter	1.5[g]	0.04 0.00	–, Lexington, SC Multiple	1999 1997–2001
TSP[m]	Annual	75[e]	41[n] 26[n]	–, Aiken, SC –, Lexington, SC	1998 2001

[a]Based on available data for 1997 through 2001 unless otherwise noted. The vicinity of the SRS was taken to be the area within 80 km (50 mi) of the proposed MOX facility and includes all or part of Aiken, Bamberg, Barnwell, Colleton, Edgefield, Hampton, Lexington, and Orangeburg Counties in South Carolina and Burke, Columbia, Jenkins, Richmond, and Screven Counties in Georgia. The listed concentrations are not always directly comparable to the ambient standards. Except for 13 counties around Atlanta, Georgia, that are nonattainment for 1-hour O$_3$, both South Carolina and Georgia have been designated as in attainment for all implemented standards. Footnote b summarizes criteria for determining standard attainment.

[b]Unless otherwise noted, South Carolina and Georgia SAAQS are the same as NAAQS. South Carolina has additional standards for gaseous fluorides that are not shown because they are not emitted by the proposed facility.

Footnotes continued on next page.

Table 3.3. Continued

[c]Methods of determining whether standards are attained depend on pollutant and averaging time. The 3-hour and 24-hour SO_2 standards and the 1-hour and 8-hour CO standards are not to be exceeded more than once per calendar year. The annual TSP, SO_2, and NO_2 standards are not to be exceeded in any calendar year. The lead standard is not to be exceeded in any calendar quarter. The 1-hour ozone (O_3) standard is attained when the expected number of days per calendar year with maximum hourly concentrations above the standard is less than or equal to one and applies only in areas designated nonattainment when the 8-hour O_3 standard was adopted in July 1977. The 8-hour O_3 standard is attained when the 3-year average of the annual fourth-highest daily maximum 8-hour concentrations is less than or equal to the standard. The 24-hour PM_{10} standard is attained when the expected number of days per calendar year with a 24-hour average concentration above the standard is less than or equal to one. In areas that meet certain criteria, attainment of the 24-hour PM_{10} standard is based on having the 3-year average of the 99th percentile 24-hour averages less than or equal to the standard. The annual PM_{10} standard is met when the 3-year average of the annual means is less than or equal to the standard. The 24-hour $PM_{2.5}$ standard is met when the 3-year average of the 98th percentile 24-hour averages is less than or equal to the standard. The annual $PM_{2.5}$ standard is met when the 3-year average of the annual means is less than or equal to the standard.

[d]Secondary (welfare-based) standard.

[e]Primary (health-based) standard.

[f]Years 2000 and 2001 data for Aiken County not available; years 1999 and 2000 data for Richmond County not available.

[g]Primary and secondary standard.

[h]No CO data in vicinity of SRS for 1997-2001.

[i]Second highest concentration.

[j]NAAQS only; implementation of the standard has been delayed, and states have not developed attainment plans.

[k]Fourth highest concentration.

[l]98th percentile concentration.

[m]South Carolina standard.

[n]Based on South Carolina data for 1998-2001.

Sources: 40 CFR 50; SCDHEC (2002a-d); EPA (2002, 2003a).

3.4.4 Noise

The *Noise Control Act of 1972* and subsequent amendments (*Quiet Communities Act of 1978,* 42 U.S.C. 2901-4918) delegate the authority to regulate noise to the states. However, South Carolina and Georgia do not have noise regulations. The Aiken County Zoning and Development Standards Ordinance limits noise levels by frequency band (see Table 3.4). The EPA guideline recommends an L_{dn}[4] of 55 dBA[5] to protect the public from the effects of noise in typically quiet outdoor and residential areas (EPA 1974). To protect the general population against hearing loss, the EPA guideline recommends an $L_{eq}(24)$[6] (L_{eq} averaged over 24 hours)

Table 3.4. Aiken County maximum allowable noise levels[a]

Frequency band (Hz)	Nighttime[b] sound pressure level at property boundary (dB)	
	Nonresidential	Residential
20-75	69	65
75-150	60	50
150-300	56	43
300-600	51	38
600-1,200	42	33
1,200-2,400	40	30
2,400-4,800	38	28
4,800-10,000	35	20

[a]This table gives nighttime sound pressure levels (SPLs). Allowable daytime levels are generally louder than nighttime levels.

[b]Nighttime: 9:00 p.m. to 7:00 a.m.

Source: DOE (1996).

---

[4] L_{dn} is a 24-hour average sound level that gives additional weight to noise that occurs during the night (10:00 p.m. to 7:00 a.m.).

[5] dBA is A-weighted decibels, a unit of weighted sound-pressure level measured by specific methods and using the A-weighting specified by the American National Standards Institute (ANSI). It duplicates the ear's sensitivity to sound.

[6] For sounds that vary with time, L_{eq} is the steady sound level that would contain the same total sound energy as the time-varying sound over a given time.

of 70 dBA or less over a 40-year period. The Federal Aviation Administration and the Federal Interagency Committee on Urban Noise have issued land use compatibility guidelines indicating that yearly day-night average sound levels (L_{dn}) of less than 65 dBA are compatible with residential land uses and that, if a community determines it is necessary, levels up to 75 dBA may be compatible with residential uses and transient lodgings (but not mobile homes) if such structures incorporate suitable noise reduction features (14 CFR 150, Appendix A).

Major noise sources in active areas at the SRS include industrial facilities and equipment such as cooling systems, transformers, engines, vents, paging systems; construction and materials-handling equipment; and vehicles. Outside of active operational areas, vehicles and trains generate noise. Most industrial facilities at the SRS are located far enough from the site boundary that the associated noise levels at the boundary would be barely distinguishable from background levels.

Noise impacts to the general public arise primarily from transportation of people and materials to and from the site by vehicles, helicopters, and trains (DCS 2002). A noise survey was conducted in the SRS area in 1989 and 1990 (NUS 1990). Seven off-site locations were selected along major routes used by SRS employees entering and leaving the site. Summer L_{dn} levels ranged from 62 to 72 dBA; winter L_{dn} levels ranged from 51 to 70 dBA. Summer 24-hr L_{eq} levels ranged from 60 to 67 dBA; winter values ranged from 54 to 65 dBA.

3.5 Ecology

This section describes the plant and animal resources at the SRS, with emphasis on those components that could be affected by the construction and operation of the proposed MOX facility and associated Pit Disassembly Conversion Facility/Waste Storage Building (PDCF/WSB) complex. Particular attention is given to species and special habitats protected

> **Ecological Resources**
>
> Ecological resources include plant and animal species and the habitats on which they depend (e.g., forests, fields, wetlands, streams, and ponds).

by the federal government under the Endangered Species Act, as well as species of special concern listed by the states of South Carolina (Aiken and Barnwell counties) and Georgia (Burke County). In addition to federal and state regulations, DOE protects plants, animals, and Carolina bays in DOE Research Set-Aside Areas. Unless otherwise cited, the information presented in this section has been abstracted from DCS (2002).

3.5.1 Terrestrial

This section describes the native plant communities and wildlife species at the SRS and in the F-Area where the proposed facilities would be constructed. Wildlife habitats, wildlife management areas, and ecological research sites are also described.

3.5.1.1 Vegetation

At the time land for the SRS was purchased by the government in 1950, about 40% of the site was old field, crop land, or developed by the former town of Ellenton. The remainder of the area was forested (WSRC 1994). As the DOE developed the SRS, the vegetation changed over time. Many of the old fields reverted back to forested areas. In addition, this increase in wooded area also resulted from timber and watershed protection management directed by the U.S. Forest Service (WSRC 1994; DOE 1999).

In 1972, the entire SRS was designated as the nation's first National Environmental Research Park (NERP). Thirty specified areas within the SRS are designated as DOE Research Set-Aside Areas that are reserved for ecological research. These areas total 5,672 ha (14,005 acres), or about 7% of the SRS (Davis and Janecek 1997), and are selected and managed by the Savannah River Ecology Laboratory (SREL) (WSRC 1994). They serve as control areas, providing a context for comparisons with other areas on the SRS that may be affected by human activities. The set-aside areas are located in each of the major vegetation communities characteristic of the SRS (DOE 2000b). The closest set-aside area to the proposed facilities is Set-Aside Area No. 13 (Organic Soils), located about 500 m (1,640 ft) northwest of the proposed facilities. Most of this 310.8-ha (767.3-acre) area is located on the north side of Upper Three Runs Creek. Set-Aside Area No. 15 (Whipple/Office of Health and Environmental Research [OHER] Study Site) is located about 1.8 km [1.1 mi] northeast of the proposed facilities, and three other set-asides (No. 1 [Field 3-412/Ellenton Bay], No. 6 [Beech-Hardwood Forest], and No. 14 [Mature Hardwood Forest]) are located more than 3.4 km (2.1 mi) southwest of the facility area. Upper Three Runs Creek borders or runs through these set-aside areas (Davis and Janecek 1997).

In June 1999, the DOE designated a 4,055-ha (10,012-acre) area of the SRS as a biological and wildlife refuge. This area, known as the Crackerneck Wildlife Management Area (WMA) and Ecological Preserve (Crackerneck WMA), is located in the western portion of the SRS. It is bordered by a narrow buffer zone along South Carolina State Route 125 and by Upper Three Runs Creek. The South Carolina Department of Natural Resources (SCDNR) manages this area (DOE 2000b).

Currently, nearly 90% of the land (72,900 ha [180,000 acres]) at the SRS is forested with upland pine, hardwood, mixed (pine and hardwood), and bottomland hardwood forests. The major upland and wetland forest types at the SRS (including major species and coverage) are listed in Table 3.5. Pine forests cover about 65% of the upland areas of the SRS (DOE 1999). These pine forests are managed by the U.S. Forest Service and have displaced much of the upland hardwood communities (DOE 1991a). Natural resource management is actively practiced on more than 80% of the SRS, including about 73,710 ha (182,000 acres) of commercial forests and 4,860 ha (12,000 acres) of nonforest lands (DOE 2000b; WSRC 1994).

Approximately 5% of land at the SRS is developed with industrial and transportation infrastructure and grassland, old fields, or shrub vegetation (WSRC 1994). This land is generally classified as "facility." The industrial and transportation development includes administrative and production facilities, electrical substations, roads, and railroads and occupies

Table 3.5. Major forest types at the SRS

Forest type	Canopy species	Midstory species	Coverage [hectares (acres)]
Upland Forests			
Dry longleaf pine-scrub oak	Longleaf pine (sparse)	Oaks, black cherry, common persimmon (continuous)	3,058 (7,551)
Longleaf pine	Longleaf pine, loblolly pine, water oak	Black cherry, common persimmon, sand hickory, sassafras, water oak	15,533 (38,353)
Mixed yellow pine	Loblolly, slash and/or longleaf pines	American holly, black cherry, common persimmon, sand hickory, sassafras, water oak, sweetgum, red maple, redbay, sweetbay magnolia	27,020 (66,716)
Southern mixed hardwood	Oaks (white, scarlet, laurel, post, southern red, turkey, bluejack, blackjack), hickories (mockernut, pignut), yellow poplar, blackgum, red maple, sweetgum, white ash, pines (loblolly, longleaf)	Sparkleberry, vaccinium, American holly, black cherry, mockernut hickory, white ash, sassafras, dogwood, Georgia hackberry	12,805 (31,618)
Wetland Forests			
Bottomland	Oaks (water, laurel, overcup, willow), southern magnolia, sweetgum, elms (American, winged), red maple, yellow poplar, river birch, tag alder, waxmyrtle, loblolly pine	American holly, redbay, sweetbay magnolia, ironwood, southern hackberry, red buckeye, honeysuckle	12,531 (30,941)
Southern swamp	Bald cypress, water tupelo, sweetgum	Ashes (water, red), sourgum, red maple, American elm	4,285 (10,581)
Total:			75,232 (185,760)

Sources: DOE (1991a, 2000b); Workman and McLeod (1990); WSRC (1994).

about 1,587 ha (3,919 acres). Vegetated areas associated with the developed areas are actively maintained (lawns and landscaped areas). These associated vegetated areas occur primarily on power line rights-of-way, roadsides, some borrow pits, some burial sites, and in forest openings and occupy about 1,345 ha (3,322 acres) (DOE 2000b). Unless managed, most scrub-shrub areas will develop into forest within 5 to 10 years (WSRC 1994). The vegetated areas also include permanent upland meadows, scrub-shrub areas, and SRS wildlife food plots. Controlled burns of 6,075 to 7,290 ha (15,000 to 18,000 acres) of pine-dominated uplands are conducted annually to reduce flammable materials and to enhance the development of fire-tolerant plant communities and improve wildlife habitat. Additionally, improved planting techniques and seedling survival have resulted in conversion of significant areas of loblolly and slash pine forests to young longleaf pine forests over the past 10 years (DOE 2000b).

Habitats in the 16.7-ha (41.3-acre) area proposed for the MOX facility include pine (or evergreen) forest (5.9 ha [14.6 acres]), mixed pine (combination of both pine and deciduous [hardwood] species, with pine trees predominant) (1.4 ha [3.4 acres]), mixed deciduous (0.3 ha [0.8 acre]), grassland (1.6 ha [3.9 acres]), "facility" (developed) (3.6 ha [9.0 acres]), old field (fields formerly used for agriculture but now undergoing natural succession) (1.1 ha [2.7 acres]), spoils (2.8 ha [6.8 acres]), and deciduous (hardwood trees, essentially the southern mixed hardwood forest type of Table 3.5) (0.04 ha [0.1 acre]) (see Figure 3.7). The grassland habitat occurs within the transmission line right-of-way that crosses the proposed MOX site. The spoils habitat originated from the excavation for the Actinide Packaging and Storage Facility (APSF) in the F-Area. Although soil was excavated, the APSF was not constructed. This area is covered primarily with various grass and forb species. The standard seed mixture used to establish a plant cover on such areas includes grass and forb species such as unhulled and hulled common Bermuda grass, browntop millet, and unscarified Appalachian lespedeza (Bowling 2001).

Habitats in the 9.1-ha (22.5-acre) area proposed for the PDCF and WSB include pine forest (0.8 ha [2.0 acres]), deciduous (2.5 ha [6.2 acres]), and facility (5.8 ha [14.3 acres]) (see Figure 3.7).

Forested and facilities areas primarily surround the immediate project area (see Figure 3.7). The forested areas are dominated by loblolly pine with some mixed hardwoods (e.g., sweetgum, turkey oak, and water oak). The sparse understory and shrub layers consist of sparkleberry, dogwood, jasmine, and wax myrtle. Also present are areas dominated by a closed canopy of longleaf pine with sweetgum and willow oak as minor components. Vegetation along the unnamed tributaries to Upper Three Runs Creek include loblolly pine, sweetgum, red oak, and sycamore in the canopy, with black cherry, dogwood, and young individuals of the canopy tree species in the understory (Wike and Nelson 2000). The grassland habitat associated with the transmission line also occurs in this area. The OFASB area located west of the proposed MOX facility site also contains a vegetated cover similar to that over the spoils area within the proposed MOX facility site.

3.5.1.2 Wildlife

Among the numerous wildlife species reported from the SRS are 44 species of amphibians, 59 species of reptiles, 258 species of birds, and 54 species of mammals. The SRS has among the highest biodiversity of herpetofauna (reptiles and amphibians) in the United States because of the area's warm, moist climate and its wide variety of habitats (DOE 2000b). Snakes that could occur in the project area include eastern hognose snake, eastern garter snake, eastern coachwhip, scarlet king snake, rat snake, corn snake, and pine snake. Lizards could include the green anole, southern fence lizard, several species of skinks, and the eastern glass lizard. Amphibians could include the southern toad and oak toad. The southern leopard frog, bullfrog, and other frogs and toads could occur in the small drainage basins near the site, while

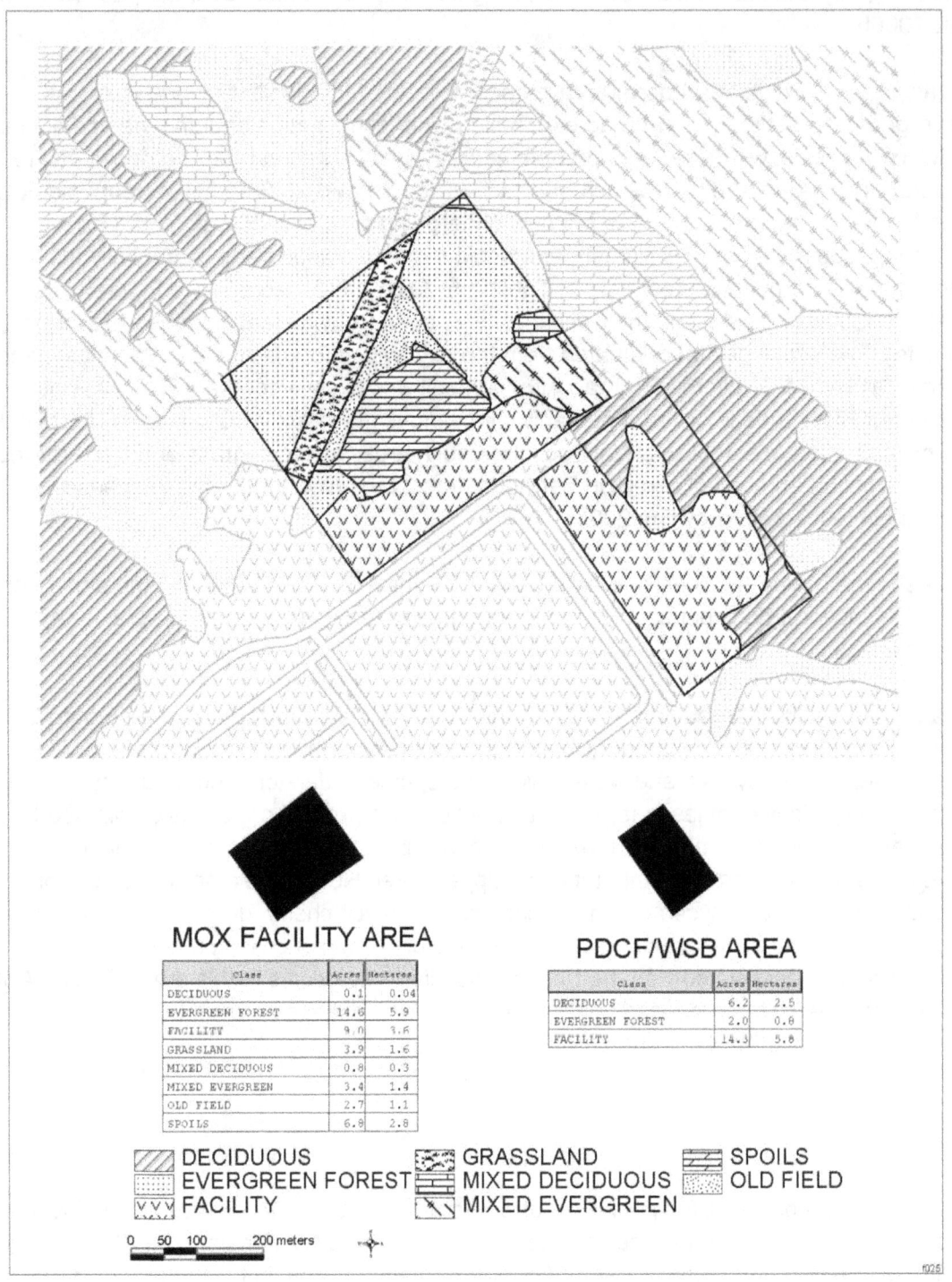

Figure 3.7. Current land cover in the area of the project site.

amphibians such as tree frogs and salamanders could occur within the unnamed tributary to Upper Three Runs Creek (Conant 1958; Mayer and Wike 1997).

Bird species at the SRS that are very common to abundant include black vulture, eastern kingbird, acadian flycatcher, common crow, northern mockingbird, blue-gray gnatcatcher, ruby-crowned kinglet, red-eyed vireo, northern parula, black-throated warbler, ovenbird, northern cardinal, savannah sparrow, white-throated sparrow, and song sparrow (WSRC 1994). As many as 17,000 ducks and coots are winter migrants at the SRS. Most of these congregate on Par Pond, L Lake, and other large ponds and Carolina bays (DOE 1991a). Wood ducks are the only waterfowl species that commonly breed on the SRS (WSRC 1994). Several mammal species can be found in old field/clearcuts, pine plantations, and scrub oak/longleaf pine habitats (these are the generalized habitat types that occur within the vicinity of the facilities). These species include southern short-tailed shrew, Virginia opossum, golden mouse, oldfield mouse, raccoon, eastern cottontail, and white-tailed deer. Other mammals that can occur within two of these habitat types include least shrew, striped skunk, raccoon, eastern harvest mouse, gray and fox squirrels, southeastern shrew, spotted skunk, feral hog, and gray fox. Several bat species also occur in one or more of these habitats (WSRC 1994).

Populations of white-tailed deer, feral hogs, and beaver are controlled through selective harvest strategies (DOE 2000b), which has included public hunts for white-tailed deer and feral hogs (Noah 1995; DOE 1996). The deer herd is estimated at about 3,000, with harvests averaging about 1,580 animals per hunting season. The feral hog population now exceeds 2,500 (DOE 2000b). The feral hogs originated from free-ranging domestic swine abandoned after resident farmers were relocated in 1952. They now occur over about 70% of the SRS (WSRC 1994). The hogs are trapped wherever they are found. Beavers are trapped where they compromise the safety or operations of roads, railroads, culverts, or research plots, or where they are causing significant resource damage. Increasing numbers of coyotes and armadillos may require the SRS to initiate control measures for these species in the future (DOE 2000b). Other commercial and recreational wildlife resources at the SRS are not exploited over most of the SRS because of restricted access and safety concerns. These species include bobcat, gray and red fox, mink, muskrat, Virginia opossum, river otter, eastern cottontail, raccoon, fox and gray squirrels, waterfowl, northern bobwhite, mourning dove, wild turkey, common snipe, and American woodcock (WSRC 1994). Hunting has been allowed for most of these species (except for bobcat, foxes, river otter, and fox squirrel) at the Crackerneck WMA (SCDNR 2000/2001). However, since late September 2001, hunting has been closed to the general public in this area. A controlled hunt was later allowed to help regulate the SRS deer herd.

The developed areas of the SRS include buildings, parking lots, infrastructure, and landscaped areas. Nevertheless, a number of wildlife species have been reported from these areas. A total of 43 species have been reported from the F-Area, including 4 species of amphibians, 12 species of reptiles, 18 species of birds, and 9 species of mammals. Several bird species are abundant: rock dove, common crow, northern mockingbird, American robin, and European starling. Common mammals include Virginia opossum, eastern cottontail, house mouse, feral cat, striped skunk, and raccoon. The densities of most wildlife species are higher in undeveloped areas then in developed areas. Exceptions include the house sparrow, house finch, rock dove, house mouse, Norway rat, and feral cat. Nevertheless, the use of developed areas of the SRS by wildlife is more common than previously reported, and these areas can be expected to contribute to the site's environmental diversity (Mayer and Wike 1997).

3.5.2 Aquatic

Six major streams and several associated tributaries flow through the SRS, and the Savannah River bounds the southwestern border of the SRS. More than 50 man-made ponds also occur at the SRS (DOE 1999). The two largest are L Lake (405 ha [1,000 acres]), which discharges into Steel Creek, and Par Pond (1,069 ha [2,640 acres]), which discharges into Lower Three Runs Creek (Section 3.3.1). These lakes do not have any direct interactions with the F-Area. Altogether, about 2,000 ha (4,940 acres) of open water occurs at the SRS (WSRC 1994).

At least 81 fish species have been identified at the SRS (DOE 2000b). Sport and commercial fishing on the SRS is allowed only within the Crackerneck WMA. Extensive fishing also occurs in the Savannah River. Commercial fish species include the American shad, hickory shad, and striped bass. Recreational species include largemouth bass, chain pickerel, crappie, bream, sunfish, and catfish (DOE 1996; WSRC 1994, 1997b). The man-made ponds support populations of bass and sunfish (DOE 1999).

Some SRS surface waters are classified as Category I resources. These waters are defined by the U.S. Department of the Interior as unique and irreplaceable on a national or eco-regional basis. These areas would include Carolina bays and cypress-tupelo swamps. Any surface waters supporting species of concern and areas containing high-quality wetlands or headwater streams (e.g., portions of Upper Three Runs Creek) would also be considered for Category I status (DOE 2000b).

The F-Area is drained by Upper Three Runs Creek and Fourmile Branch (see Figure 3.3). Upper Three Runs Creek is the most pristine stream at the SRS and would be considered a Category I resource. It contains more than 550 species of aquatic insects and supports about 60 fish species. The more abundant fish species include bowfin, American eel, redfin pickerel, dusky shiner, yellowfin shiner, coastal shiner, flat bullhead, tadpole madtom, mosquitofish, redbreast sunfish, warmouth, spotted sunfish, and blackbanded darter. More than 10 other fish species are common in Upper Three Runs Creek (Bennett and McFarlane 1983). Upper Three Runs Creek is an important spawning area for blueback herring and provides seasonal nursery habitat for American shad, striped bass, and other Savannah River species (DOE 1999). This stream also appears to be an important spawning site for the spotted sucker (WSRC 1994).

About 48 fish species have been collected from Fourmile Branch. Those in the stream's lower reaches include species common to the Savannah River. The only abundant fish species collected from Fourmile Branch are mosquitofish, redbreast sunfish, and spotted sunfish. Common species include longnose gar, bowfin, golden shiner, bluehead chub, creek chub, creek chubsucker, pirate perch, and brook silverside (Bennett and McFarlane 1983).

Water bodies in the vicinity of the proposed facilities include unnamed tributaries to Upper Three Runs Creek (see Figure 3.3) and small drainages and detention basins associated with permitted discharge outfalls. Macroinvertebrate (e.g., aquatic insects, snails, clams, and worms) and fish surveys indicate that Upper Three Runs Creek is unaffected by SRS NPDES-permitted discharges (Specht and Paller 2001).

3.5.3 Wetlands

More than 20% of the SRS consists of wetlands, including open waters. Most wetlands on the SRS are associated with floodplains, streams, and impoundments. Wetland types on the SRS include bottomland hardwoods, southern swamp (cypress-tupelo), freshwater marshes, and Carolina bays. Areal coverage of forested wetlands is given in Table 3.5. The freshwater marshes total 1,380 ha (3,407 acres), and the

---

Wetlands

Wetlands are areas that are inundated or saturated by surface water or groundwater at a frequency and duration sufficient to support, and that under normal circumstances do support, a prevalence of vegetation adapted for life in saturated soil conditions.

Carolina bays total about 785 ha (1,939 acres) (DOE 2000b). The conditions of many wetlands at the SRS are similar to conditions that existed before the government assumed control of the site, except for those wetlands along stream corridors and adjacent portions of the Savannah River swamp that were degraded by thermal releases from reactor operations. These areas have been recovering since cessation of cooling water releases (WSRC 1994).

Over 300 Carolina bays (closed depressions capable of holding water) occur on the SRS (DOE 2000b). Carolina bays are characterized by their elliptical or ovoid shape, with a northwest/southeast orientation of their long axis (WSRC 1994). The Carolina bays on the SRS have remained largely undisturbed since 1950 and thus are valuable examples of these regional wetlands (Schalles et al. 1989). The median size of the Carolina bays is about 0.8 ha (2.0 acres), and only 15 exceed 4 ha (10 acres). The Carolina bays have characteristics similar to other wetlands (e.g., shallow marshes, herbaceous bogs, shrub bogs, or swamp forests). They also have a xeric to hydric (dry to moist) gradient from their peripheral sand rim to the center depression (Schalles et al. 1989). More than 135 species of plants have been identified from these wetlands. Most are dominated by grasses and sedges (Schalles et al. 1989; WSRC 1994). Amphibians are the most prevalent vertebrates that utilize the Carolina bays, but many reptiles, birds, and mammals also have been observed at these wetlands (Schalles et al. 1989). Less than 20 of the Carolina bays contain permanent fish populations. Fish species include redfin pickerel, mud sunfish, lake chubsucker, and mosquito fish (DOE 1999). An accelerated program has been initiated at the SRS to restore impacted Carolina bays (DOE 2000b). No Carolina bays occur near the proposed facility sites.

No wetlands occur on the proposed facility sites. Wetland habitat does occur along the unnamed tributary to Upper Three Runs Creek located near the eastern border of the proposed facility site (see Figure 3.3). The dominant species of vegetation in this wetland are yellow poplar, laurel oak, red maple, red bay, and cherrybark oak. Maiden cane also occurs near the wetland boundary (Wike and Nelson 2000).

3.5.4 Protected Species

Table A.1 (Appendix A) lists the threatened, endangered, and other special status species that may be found in the vicinity of the SRS. Appendix A also discusses the federally and state-endangered red-cockaded woodpecker (*Picoides borealis*), which receives special attention at the SRS.

No federal- or state-listed wildlife species have been reported from the proposed project area, but several species may exist in the general

---

Protected Species

Endangered species. Any species in danger of extinction throughout all or a significant portion of its range.

Threatened species. Any species likely to become endangered within the foreseeable future throughout all or a significant portion of its range.

---

vicinity. The American alligator (*Alligator mississippiensis*) is federally threatened (by virtue of its similarity to the endangered American crocodile [*Crocodylus acutus*]). While it is fairly common at the SRS, it has only been recently observed near the F-Area, and its occurrence there is considered uncommon. The federally threatened (state-endangered) bald eagle (*Haliaeetus leucocephalus*) actively nests in the Pen Branch area and in an area south of Par Pond. These areas are 14 km (8.7 mi) and 12 km (7.5 mi) southwest and southeast of the proposed project area, respectively. The closest nesting area of the federally and state-endangered red-cockaded woodpecker to the proposed facility site is about 5 km (3.1 mi) away. The proposed area for the facilities does not occur within red-cockaded woodpecker management areas (see Appendix A). However, all areas containing pines, including those at the proposed sites, provide suitable forage areas for this species. The federally and state-endangered wood stork (*Mycteria americana*) has been observed near the Fourmile Branch delta, about 21 km (13 mi) from the proposed site. The federally endangered (state-endangered) shortnose sturgeon (*Acipenser brevirostrum*) occurs in the Savannah River as far upstream as the SRS.

Walk-through surveys did not reveal any federal- or state-listed wildlife species within the proposed facility area (USFS 2000). The Bachman's sparrow (*Aimophila aestivalis*) is adapted to open meadow and shrubby meadow habitats such as those that occur throughout F-Area. The eastern woodrat (*Neotoma floridana*) could inhabit the transitional areas between the hardwood forest and F-Area facilities, and the moist stream bottom area is suitable for the star-nosed mole (*Condylura cristata*). The upland pine and pine-oak ridge habitats are highly suitable for the southern hognose snake (*Heterodon simus*) and pine snake (*Pituophis melanoleucus*) (USFS 2000). The American sandburrowing mayfly (*Dolania americana*) is a relatively common aquatic insect in Upper Three Runs Creek (WSRC 1994). This species was formerly a candidate species for federal listing, but it is not currently listed by the U.S. Fish and Wildlife Service (USFWS) or State of South Carolina.

More than 1,300 species of plants occur at the SRS (WSRC 1994); however, only 53 species are considered to be sensitive, as determined by state, federal, and global ratings. The smooth coneflower (*Echinacea laevigata*) is the only federally listed (endangered) plant species at the SRS; it is also state endangered. Smooth coneflowers inhabit roadsides and open, sunny areas. The collection of plants from natural populations was a significant factor in the endangerment of the species (Arnold et al. 1998). Three populations of the smooth coneflower have been identified at the SRS. Activities near these known populations are highly restricted (DOE 2000b).

Nearly 300 populations of other sensitive plant species occur at the SRS (DOE 2000b). Included are three populations of the state-listed (species of concern) piedmont azalea (*Rhododendron flammeum*) that have been found along the steep slopes adjacent to the Upper Three Runs Creek floodplain in an area northwest of F-Area (DOE 1999).

Walk-through surveys of the proposed MOX facility site in October 1998 and March 2000 did not reveal any populations of the smooth coneflower (USFS 2000). Because this species is adapted to meadow and open forest habitats, the project area appears to be too disturbed or shady for the coneflower's establishment and successful survival. The survey did indicate that suitable habitat for several rare plant species exists in areas adjacent to the survey site. The hardwood slope provides habitat suitable for leech brush (*Nestronia umbellata*), piedmont azalea, and striped garlic (*Allium cuthbertii*). The moist bottom and lower slope sections are suitable for green-fringed orchid (*Platanthera lacera*) and least trillium (*Trillium pusillum* var. *pusillium*). The upland pine and pine-oak ridge areas are suitable for lance-leaf wild-indigo (*Baptisia lanceolata*) and bearded milk-vetch (*Astragalus villosus*) (USFS 2000).

3.6 Land Use

This section briefly describes land use patterns on and around the SRS. Land use is a classification of parcels of land relative to their suitability for or the actual presence of human activities (e.g., industry, agriculture, recreation, etc.) and natural uses. Natural resource attributes and other environmental characteristics could make a site more suitable for some land uses than for others. Changes in land use may have both beneficial and adverse effects on other resources (e.g., ecological, cultural, geological, and hydrological).

3.6.1 Savannah River Site Land Use

Existing land use at the SRS can be characterized into three main categories: (1) undeveloped/forest, (2) wetlands/water, and (3) developed. Approximately 73% of the SRS is undeveloped; 22% consists of wetlands, streams, and lakes; and 5% is developed (e.g., facilities, roads, and utility corridors). The forested areas are managed for timber production. The U.S. Forest Service, under an interagency agreement with DOE, harvests approximately 728 ha (1,800 acres) of timber from the SRS each year. Prime farmland soils exist at the SRS, but areas of prime farmland are not identified within the SRS because the land is not available for agricultural activities (DCS 2002). A portion of the SRS is open for fishing, as discussed below for the Crackerneck WMA. Since late September 2001, hunting has been closed to the general public in this area. A limited hunting period was later allowed to control the SRS deer herd.

As discussed in Section 3.5.1.1, the SRS has been designated a National Environmental Research Park by DOE. The scientific community can use the site to study the impacts of human activity on cypress swamp and hardwood forest ecosystems. Approximately 5,700 ha (14,085 acres) of land is set aside at the SRS for nondestructive environmental research (DOE 1999).

The F-Area is generally classified by the SRS land use plan as developed; some areas within F-Area are classified as industrial or heavy industrial.

Future land use at the SRS is determined by the DOE through site development, land use, and future planning processes (DCS 2002). SRS planners have developed a land use zone planning model for the site that is consistent with their past support of a multiple-use planning concept where compatible. Three principal planning zones have been established: Site Industrial, Site Industrial Support, and General Support. The *SRS Long Range Comprehensive Plan* includes the construction and operation of the proposed facilities as part of the plan for its Nuclear Materials Stewardship mission (DOE 2000b). New missions for the SRS in the 21st Century, as stated in the *Savannah River Site Strategic Plan*, include the construction and operation of new facilities for tritium extraction and the storage and disposal of surplus plutonium. In addition to these new facilities, the SRS plans to play an increased role in the advancement of nuclear materials protection, control, and accounting (DOE 2000a).

3.6.2 Off-Site Land Use

Predominant regional land uses in the vicinity of the SRS include urban and residential, industrial, agricultural, and recreational areas. Forest and agricultural land predominantly border the SRS, with only limited urban and residential development. The nearest residences are located to the west, north, and northeast, some within 60 m (200 ft) of the SRS boundary. Farming is diversified throughout the region and includes such crops as peaches, watermelon, cotton, soybeans, corn, and small grains. Incorporated and industrial areas are also present near the site, including textile mills, polystyrene foam and paper plants, chemical processing plants, and a commercial nuclear power plant. Open water and nonforested wetlands occur along the Savannah River Valley. Recreational areas within 80 km (50 mi) of the SRS include Sumter National Forest, Santee National Wildlife Refuge, and Clark's Hill/Strom Thurmond Reservoir. State, county, and local parks include Redcliffe Plantation, Rivers Bridge, Barnwell and Aiken County State Parks in South Carolina, and Mistletoe State Park in Georgia. The Crackerneck WMA, which includes a portion of the SRS along the Savannah River, is open to the public for fishing (DOE 1999).

3.7 Cultural and Paleontological Resources

Cultural resources include archaeological sites and historic structures and features that are protected under the National Historic Preservation Act of 1966, as amended.

Cultural and Paleontological Resources

Cultural resources include archaeological sites, historic structures and features, and traditional cultural properties.

Paleontological resources are the fossil remains of past life forms.

Cultural resources also include traditional cultural properties that are important to a community's practices and beliefs and that are necessary to maintain the community's cultural identity. Cultural resources that meet the eligibility criteria for listing on the *National Register of Historic Places* (NRHP) are considered "significant" resources and must be taken into consideration during the planning of federal projects. Federal agencies are also required to consider the effects of their actions on sites, areas, or other resources (e.g., plants) that are of religious significance to Native Americans as established under the American Indian Religious Freedom Act. Native American graves and burial grounds are protected by the Native American Graves Protection and Repatriation Act.

Paleontological resources are the fossil remains of past life forms. Paleontological resources with significant research potential are protected under the Antiquities Act.

3.7.1 Archaeological Resources

The Savannah River Archaeological Research Program (SRARP) of the South Carolina Institute of Archaeology and Anthropology, University of South Carolina, has been conducting archaeological investigations at the SRS since 1973 (SRARP 1989). The SRARP prepared an archaeological resource management plan for the SRS in 1989. The purpose of the plan is to provide the DOE with a means of addressing future archaeological resource management needs at the SRS and to establish a series of research directions to facilitate better management of these resources. The SRS currently manages its archaeological resources under the terms of a 1990 Programmatic Agreement among the DOE Savannah River Operations Office, the South Carolina State Historic Preservation Officer (SCSHPO), and the Advisory Council on Historic Preservation.

Over a period of more than 25 years, members of the SRARP have been very active in recording more than 850 archaeological sites at the SRS.[7] Although most of these sites have not been formally evaluated for eligibility for listing on the NRHP, 67 sites have been identified as potentially eligible (DOE 1999). In general terms, prehistoric sites within the SRS consist of village sites, base camps, limited-activity sites, quarries, and workshops. Nearly 800 prehistoric sites have been recorded at the SRS (DCS 2002). As detailed below, several prehistoric sites have been recorded within or near the proposed facilities. Two prehistoric sites within the footprints of the proposed facilities and their associated grading area have been determined to be eligible for listing on the NRHP.

Historic sites at the SRS include farmsteads, tenant dwellings, mills, plantations, slave quarters, rice farm dikes, dams, cattle pens, ferry locations, churches, schools, towns, cemeteries, commercial buildings, and roads. About 400 historic sites have been recorded to date at the SRS (DOE 1999). No historic sites have been recorded within the vicinity of the proposed facilities.

---

[7] Of the 850 plus sites that have been recorded at the SRS, some are prehistoric, some are historic, and some have both a prehistoric and historic component. For this reason, the sum of prehistoric sites plus historic sites is much greater than the approximate total of 850 sites.

Archaeological surveys have been conducted in the F-Area in the vicinity of the proposed facilities. Fifteen prehistoric sites have been identified. Nine of these sites were recorded during 1993 and 1994 (Cabak et al. 1996). Four sites were recorded during SRS surveys conducted between 1973 and 1977 (Hanson et al. 1978). One site was recorded in 1983 (as cited in Cabak et al. 1996), and the remaining site was recorded in a 1999 survey covering unsurveyed lands remaining for the proposed location of the surplus plutonium disposition facilities (King and Stephenson 2000).

Four sites are located within the area of direct project disturbance. Two of the four prehistoric sites (38AK546/547 and 38AK757) are eligible for listing on the NRHP. Site 38AK546/547, located within the area of the proposed MOX facility, is eligible because of its potential to provide significant information about the prehistory of the Aiken Plateau, in particular the use of ridge slope settings during the Early Mississippian period (King and Stephenson 2000). Site 38AK757 is located within the boundary of the proposed PDCF facility and is important for learning more about the use of upland settings by prehistoric inhabitants of the area during the Mississippian Period (King and Stephenson 2000). Two sites within the area of the proposed MOX facility, 38AK330 and 38AK548, were determined not eligible in consultation with the SCSHPO, and no further work is required for these two sites (Green 2000, as cited in DCS 2002).

Eleven prehistoric sites are located near the proposed facilities. Five of those sites (38AK106, 38AK155, 38AK563, 38AK564, and 38AK581) have been recommended eligible for listing on the NRHP. Site 38AK106 has been recommended eligible on the basis of its integrity, high density of artifacts, and research potential for providing information on the Early Archaic, Early Woodland, and Late Woodland time periods. Site 38AK155 is eligible because of its potential to yield important information on subsistence strategies and the use of

Date Ranges of Prehistoric Time Periods Used by Archaeologists at the SRS	
Mississippian	A.D. 1100 - 1450
Late Woodland	A.D. 500 - 1100
Middle Woodland	600 B.C. - A.D. 500
Early Woodland	1000 B.C. - 600 B.C.
Late Archaic	3000 B.C. - 1000 B.C.
Middle Archaic	6000 B.C. - 3000 B.C.
Early Archaic	8000 B.C. - 6000 B.C.

Source: SRARP (1989).

upland streamside settings between 3000 B.C. and A.D. 1450 (between the Late Archaic and Early Mississippian periods). Site 38AK563 is important because it contains cultural deposits ranging from the Early Archaic Period through the Late Woodland Period and has the potential to provide information on the changes in human use of the floodplain over a considerable time range. Site 38AK564 has been recommended eligible because it contains stratigraphically[8] separated evidence of site use from the Early Archaic and Late Archaic/Early Woodland time periods. Site 38AK581 contains evidence of numerous occupations by prehistoric people during the Woodland Period. The site has been recommended eligible on the basis that these

[8] Archaeologists look at the position of artifacts relative to layers of soil and other artifacts to help determine sequences of events. Objects found closer to the surface of an undisturbed site were deposited more recently than objects found below them (i.e., an archaeologist would expect to find Woodland Period artifacts in one or more layers of soil above Archaic Period artifacts in a stratigraphically preserved site).

various occupations appear in a well-defined stratigraphic sequence and potentially contain important information about changes that occurred during that time period (Cabak et al. 1996).

3.7.2 Historic Structures

No architectural inventories have been conducted to date at the SRS. The SRS has a number of nuclear production facilities, including facilities important to tritium and plutonium production, that may have historic value as related to events during the Cold War. Construction of the F-Area began in 1951 under the Atomic Energy Commission. The F-Area was historically used for plutonium recovery during DOE's plutonium production phase (DCS 2002). The areas of construction for the proposed facilities do not contain structures. No existing buildings within the F-Area have been identified for reuse, modification, or demolition related to MOX facility activities.

3.7.3 Traditional Cultural Properties

Traditional cultural properties are places and resources important to traditional American cultures, which include, but are not restricted to, Native American cultures. Village sites, ceremonial locations, burials, cemeteries, and natural areas containing important resources, such as traditional plants, are typical types of properties of concern to Native American cultures. Properties of traditional value to immigrant groups (e.g., from Europe and Africa), such as cemeteries, also can be considered as traditional cultural properties. Native American groups with traditional ties to the area include the Apalachee, Cherokee, Chicksaw, Creek, Shawnee, Westo, and Yuchi (DCS 2002). Many of these groups were relocated to the Oklahoma Territory in the 1800s. However, issues related to the American Indian Religious Freedom Act have surfaced within the central Savannah River valley. Native American representatives have expressed concern over traditional plant resources that could exist at the SRS (DOE 1991b; DCS 2002). None of the identified plant resources is currently known to exist in the F-Area. Consultations with appropriate Native American Tribes, Bands, and Nations are underway regarding the proposed MOX facility (Appendix B).

3.7.4 Paleontological Resources

While some fossil-bearing strata are known to exist at the SRS, none are known within the F-Area. Paleontological resources that have been recorded within the SRS area mostly date to 54 to 39 million years ago during the Eocene Age. Those resources include fossil plants, invertebrate fossils, giant oysters, other mollusks, and bryozoa. Most known paleontological resources in the area are considered common and of low research potential (DOE 1999). The discovery of paleontological resources within the area of the proposed facilities is not anticipated.

3.8 Infrastructure

This section briefly describes the existing infrastructure of the SRS as it pertains to the proposed action. Site infrastructure includes utilities, roads, and railroads needed to support construction and operation of the facilities. A detailed discussion of the SRS infrastructure is provided in the DOE Surplus Plutonium Disposition EIS (DOE 1999).

3.8.1 Electricity

The SRS uses a 115-kV power line system in a ring arrangement to supply electricity to the operations areas. Power is supplied by three transmission lines from the South Carolina Electric and Gas Company. The F-Area receives power from the 200-F power loop supplied by the 251-F electrical substation. The current F-Area power consumption rate is about 63,000 MWh/yr; the F-Area total capacity is about 700,000 MWh/yr (DCS 2002). The total SRS usage of electrical power is 370,000 MWh/yr out of a site capacity of 4,400,000 MWh/yr.

3.8.2 Water

Domestic water supplies at the SRS come from a system composed of several wells and water treatment plants. The system includes three wells and a water treatment plant in the A-Area and two wells and a backup water treatment plant in the B-Area. A 43-km (27-mi) piping loop provides domestic water from the A- and B-Areas to other SRS operations areas, including the F-Area (DCS 2002). Current domestic water usage in F-Area is 378 million L/yr (100 million gal/yr) compared with a capacity of 890 million L/yr (235 million gal/yr).

Within F-Area, four deep groundwater wells are used for process water. Pumping capacities for these wells range from 1,500 to 3,800 L/min (400 to 1,000 gpm), and they extract groundwater from the Crouch Branch Aquifer. Two of these wells were formerly used for domestic water supply. The current annual groundwater use at F-Area is 1.4 billion L (370 million gal) (DCS 2002). The estimated capacity of the wells in F-Area is about 4.2 billion L/yr (1.1 billion gal/yr).

3.8.3 Fuel

Coal and oil are used at the SRS to power steam plants located in A-, D-, H- and K-Areas. The produced steam is distributed across the site in an aboveground pipeline distribution system. Coal is delivered by rail and is stored at coal piles in A-, D-, and H-Areas. Number 2 grade fuel oil is delivered by truck and is used in the K-Area. Natural gas is not used at the SRS.

This page is being withheld pursuant to 10 CFR 2.390(a).

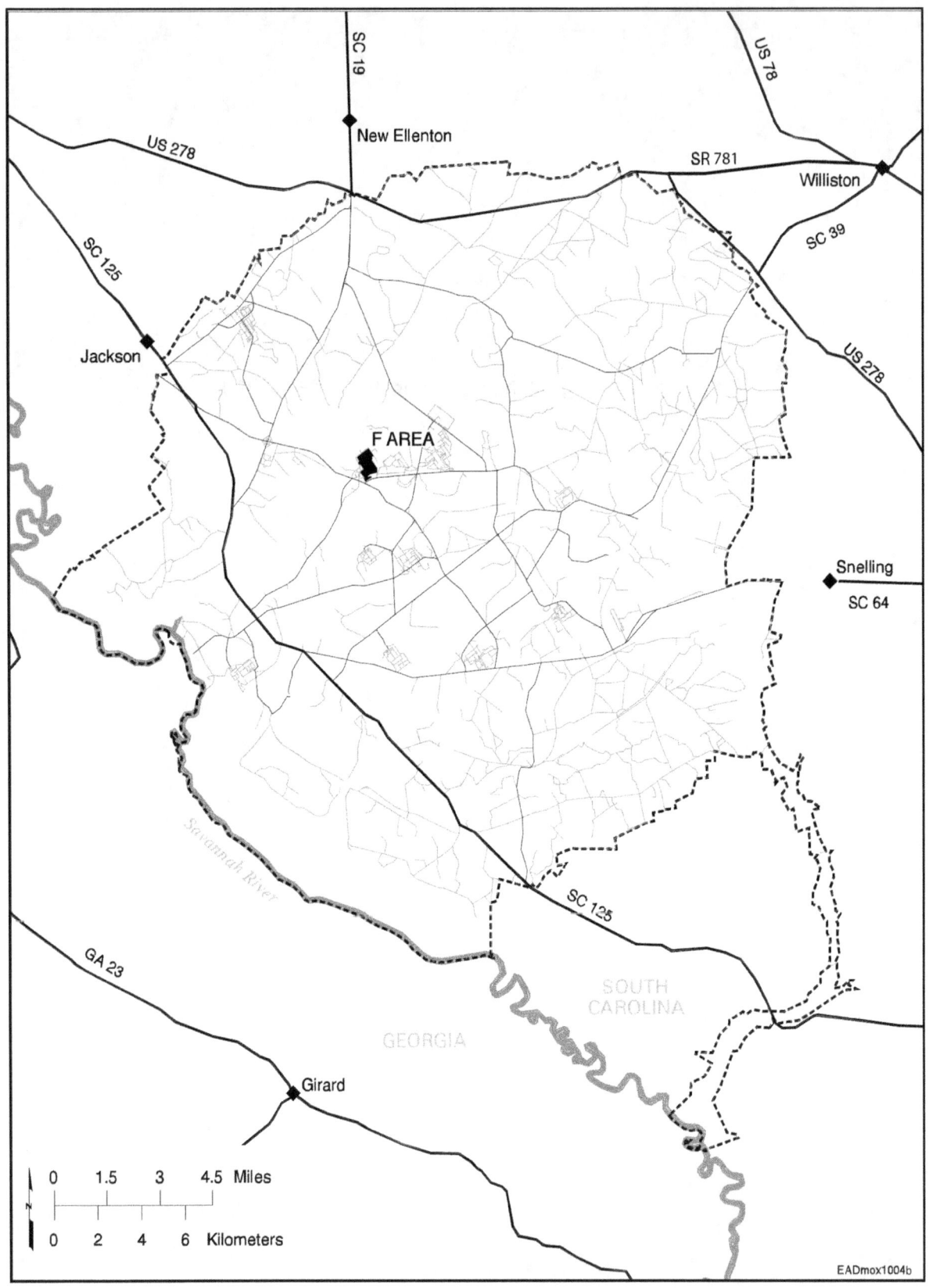

Figure 3.8. Roadways in the vicinity of the SRS.

Current waste generation rates and inventories at the SRS are presented in Table 3.6. Waste management practices at the SRS include minimization, characterization, treatment, storage, transportation, and disposal of waste generated from ongoing site activities. Waste minimization at the SRS is accomplished through source reduction, recycling, and employee participation in pollution prevention programs. Total solid waste volumes have decreased by 70% since 1991.

The types of waste currently managed at the SRS are high-level waste (HLW), transuranic (TRU) waste, mixed TRU waste, low-level waste (LLW), mixed LLW, hazardous waste, and

Table 3.6. Current waste generation rates and inventories at the SRS[a]

Waste type	Generation rate (m^3/yr)	Inventory[b] (m^3)
TRU[c]		
Contact handled	171	6,034
Remotely handled	0.6	1
LLW	8,195	1,616[d]
Mixed LLW		
RCRA	61	7,717
TSCA[e]	<1	3
Hazardous	74	1,416
Nonhazardous		
Liquid	416,100[f]	NA[g]
Solid	6,670	NA

[a]Sources for estimates presented in this table are DOE (1997) for TRU waste, LLW, and mixed LLW; DOE (1996) for hazardous and nonhazardous solid waste; and Sessions (1997) for nonhazardous liquid waste.

[b]Inventory projections were as of end of fiscal year 1996 for those presented in DOE (1997).

[c]Includes mixed TRU waste.

[d]LLW is disposed of on-site at the SRS. The estimated inventory shown is less than the generation rate (for FY1996) because it represents only LLW that had not been disposed of as of the end of FY 1996.

[e]TSCA = Toxic Substances Control Act.

[f]416,000 m^3/yr = 416,100,000 L/yr.

[g]NA = not applicable; nonhazardous wastes are not held in long-term storage.

nonhazardous waste. The first five types contain radioactive material. Of the seven waste types currently managed at the SRS, HLW would not be generated by the proposed MOX facility, the PDCF, or the WSB. The proposed MOX facility would generate a liquid high-alpha-activity waste that would be further processed, resulting in the generation of TRU waste and LLW (DCS 2002).

The TRU wastes generated at the SRS include contaminated equipment, protective clothing, and tools. Most of these wastes are stored on concrete pads that are not covered with soil. TRU waste generated before 1986 is stored on five concrete pads and one asphalt pad that have been covered with approximately 1.2 m (4 ft) of soil. TRU waste generated since 1986 is stored on 13 concrete pads that are not covered with soil. These storage pads are located in the Low-Level Radioactive Waste Disposal Facility, which is located in E-Area (DOE 1995). In 1996, it was decided to vent and purge all buried drums; this process was completed in 1999 (Arnett and Mamatey 2000b). A TRU waste characterization and certification facility to prepare TRU waste for treatment and to certify TRU waste for disposal at the Waste Isolation Pilot Plant (WIPP) is planned for 2007. This TRU waste facility would be built to manage other SRS TRU waste and is independent of the proposed action. In the interim, drums that are certified for shipment to WIPP will be stored on concrete pads in E-Area (DOE 1999).

Waste Types

Transuranic (TRU) waste: Refers to radioactive waste that contains more than 100 nanocuries per gram (nCi/g) of alpha-emitting isotopes with atomic numbers greater than 92 and half-lives greater than 20 years. Such waste results primarily from the fabrication of plutonium weapons and plutonium-bearing reactor fuel. Generally, little or no shielding is required.

Low-level waste (LLW): Refers to radioactive waste that is not classified as HLW, TRU, or spent nuclear fuel (SNF).

Hazardous waste: Refers to nonradioactive waste materials defined by the Resource Conservation and Recovery Act (RCRA) as hazardous wastes. These wastes are considered to pose potential hazard to human health when improperly treated, stored, disposed of, or otherwise managed because of their quantity, concentration, and physical and chemical characteristics. (Note: hazardous waste mixed with low-level [radioactive] waste or TRU waste is referred to as mixed low-level waste or mixed TRU waste, respectively.)

Liquid and solid LLW types are treated at the SRS. Aqueous LLW streams undergo filtration, reverse osmosis, and ion exchange at the F-and H-Area effluent treatment facility (ETF) to remove the radionuclide contaminants. The treated effluent is discharged to Upper Three Runs Creek.

Treatment residuals are eventually immobilized with grout for on-site disposal. Solid LLW is categorized into four groups: low-activity wastes (those that radiate less than 0.002 Sv/h [200 mrem/h] at 5.1 cm [2 in.] from the unshielded container); intermediate-activity wastes (those that radiate greater than 0.002 Sv/h [200 mrem/h] at 5.1 cm [2 in.]); intermediate-activity tritium waste (intermediate-activity waste with more than 3.7×10^{11} Bq [10 Ci] of tritium per container); and long-lived waste (waste contaminated with long-lived isotopes that exceed the waste acceptance criteria [WAC] for on-site disposal) (DCS 2002). Wastes in the first three categories are stored and disposed of in vaults, and wastes in the fourth category are placed in a waste storage building until treatment and disposal technologies are developed. Located in the E-Area, the vaults are below-grade concrete structures, and the storage building is a metal structure on a concrete pad. Disposal facilities at the SRS are projected to meet solid LLW disposal capacity needs for the next 20 years.

Mixed LLW is stored in various tanks and buildings located in the A-, E-, M-, N-, and S-Areas of the SRS. The current mixed waste program at the SRS primarily involves the safe storage of these wastes until treatment and disposal facilities become available. A site treatment plan (WSRC 2000b) for mixed wastes has been developed, as required by the Federal Facility Compliance Act, that specifies treatment technologies or technology development schedules for all SRS mixed waste. During 1999, plans for all mixed LLW were met in accordance with the site treatment plan (Arnett and Mamatey 2000b).

Hazardous waste is managed at the SRS either by accumulating the waste at the generating facility for a maximum of 90 days or storing it in Resource Construction and Recovery Act (RCRA)-permitted hazardous waste storage buildings or on interim storage pads located in the B- and N-Areas. Most of the waste is shipped off-site to commercial RCRA-permitted facilities. In 1999, 297 m^3 (388 yd^3) of hazardous waste was shipped off-site to commercial disposal facilities (Arnett and Mamatey 2000b).

The treatment of nonhazardous wastewater at the SRS has been centralized since 1994 with the completion and operation of the 2.8 million-L/day (0.75 million-gal/day) Central Sanitary Wastewater Treatment Facility. This facility treats sanitary wastewater by an extended aeration activated sludge process that separates the wastewater into clarified effluent and sludge.

The collection, hauling, and disposal of solid sanitary waste at the SRS is privatized, and the waste is sent to the Three Rivers Landfill southwest of the B-Area. Other nonhazardous waste consists of scrap metal, powerhouse ash, domestic sewage, scrap wood, construction debris, and used railroad ties. These wastes are disposed of by means appropriate to their nature.

3.10 Human Health Risk

Human health can be adversely affected by radioactive and hazardous chemical contaminants in the environment. This section discusses how humans can become exposed to these materials, the potential effects of this exposure, potential human receptors considered in this EIS, and the existing conditions at the SRS and the surrounding area. Methods used to estimate the potential for injuries or fatalities among workers are also discussed.

3.10.1 Hazard Exposure Pathways

3.10.1.1 Pathways for Human Exposure to Radiation and Radioactivity

Radioactivity released from the SRS reaches the environment and people in a variety of ways. The routes that radioactive materials follow to get from an SRS facility to the environment and then to people are called pathways. The primary human exposure pathways for these releases are discussed below:

- *Inhalation exposure pathway*: Individuals in the path of airborne emissions would receive a dose from breathing in the radioactive material. Some of this material also deposits on the ground and over time may become resuspended in the air, at which time it may also be inhaled.

- *Direct radiation from contaminated soil*: Material that is deposited on the ground from passing airborne emissions becomes an external exposure source of direct radiation.

- *Immersion in radioactive clouds*: Individuals in the path of radioactive airborne emissions would receive an external dose during immersion in the passing "cloud" of material.

- *Ingestion exposure pathway*: Radioactive materials can be transported through a variety of routes into the human diet. Airborne radioactive material may deposit directly on food crops or animal feed crops, resulting in potential exposure from human ingestion of the food crops or indirectly from ingestion of contaminated animal products. Material deposited on farmland may also be taken up through the roots by human and animal food crops. Material deposited on surface water or land may reach groundwater. Contaminated surface water or groundwater could be used for irrigating crops or direct consumption by humans. Contaminated surface water could also result in contamination of aquatic species, such as fish, which could subsequently be consumed by humans.

One important pathway of radioactive material released from the SRS in the form of particulate matter is the airborne pathway. After being discharged from a stack, the radioactive particulate matter will be carried by wind downwind of the facility, where it will either be inhaled by individuals or settle on the ground. Radioactivity in the soil will cause direct radiation exposures in individuals located near contaminated soil. Soil contamination may also be resuspended into the air by the wind and then inhaled farther downwind. Food produced on farmlands with contaminated soil will also contain this radioactivity. Precipitation runoff from downwind soil will carry radioactivity to local surface waters, such as lakes, rivers, and streams. Finally, radioactivity in surface water may accumulate in fish or other aquatic life that can be consumed by humans.

Radiation and Radioactivity

Radioactivity or *radioactive decay* is the process by which unstable atoms emit *radiation* to reach a more stable state.

Radiation is the movement of energetic particles or waves through matter and space. Radiation comes from radioactive material or from equipment such as x-ray machines. Radiation may be either ionizing radiation or non-ionizing radiation.

Ionizing radiation is radiation that has enough energy to cause atoms to lose electrons and become ions. For example, the radioactive decay of plutonium produces radiation that can ionize matter (e.g., tissue).

Radiation dose is the quantity of radiation energy that is deposited in a material. The radiation dose to humans is measured in units of sieverts (Sv). The unit of rem is also used. One sievert is equal to 100 rem.

Collective dose is the sum of the individual doses received in a given period of time by a specified population. The unit of collective dose is person-sieverts, or person-rem.

The DOE has determined the critical types of radioactivity and pathways for radioactive materials released from SRS operations. Tritium and cesium-137 are the primary contributors to doses to members of the public. The major pathways for tritium released into air were through breathing air and eating food, whereas the major pathway for tritium and cesium-137 released into site streams were through drinking river water and eating fish from the river (DOE 1999). Pathways or routes by which radioactive material moves through the environment to reach humans can be complex. For example, contaminants can settle on grass that is eaten by cows that produce milk that is consumed by humans. The meat of the cows can also be consumed by humans. Another example, more relevant to the SRS, would be game animals that consume contaminated vegetation and then are eaten by humans. A detailed discussion of the many pathways at the SRS is presented in the annual environmental report (Arnett and Mamatey 2001b).

3.10.1.2 Pathways for Human Exposure to Chemicals

Humans can also be exposed to nonradioactive chemicals released to the environment. The DOE has determined that the critical chemicals among those released from SRS operations to the environment are arsenic and benzene (Arnett and Mamatey 2000b). Exposures may occur primarily through inhaling pollutants released to air, drinking contaminated groundwater or surface water, ingesting contaminants in foodstuffs grown in contaminated soil or irrigated with contaminated groundwater, or ingesting contaminated soil.

3.10.1.3 Physical Hazards

Although not attributable to releases of contaminants to the environment, there is a risk of injuries and fatalities from physical hazards for construction and operation workers at any facility. The U.S. Bureau of Labor keeps statistics on the annual number of injuries and fatalities by industry type. Where possible, these statistics have been used to estimate the extent of physical hazard risk for the no-action and proposed action alternatives.

3.10.2 Receptors

Effects of radiation and chemical exposures for the no-action and proposed action alternatives during normal operations were estimated by first calculating the doses to relevant receptors. The analyses considered three groups of people: (1) members of the public, (2) SRS employees, and (3) facility workers. For purposes of this EIS, these three groups are defined as follows:

- *Members of the Public*: Individuals who live and work outside the SRS within 80 km (50 mi) of the proposed facilities:

  - Might be exposed to trace amounts of radioactive and chemical materials released to the environment through exhaust stacks.

  - Could receive radiation and chemical exposures primarily through inhalation of material in the air, external radiation from deposited radioactive material, and ingestion of contaminated food.

- *SRS Employees*: Individuals employed at the SRS who are not workers at the proposed MOX facility, the PDCF, or the WSB. SRS employees include those workers assigned radiological work at other nuclear facilities within the SRS boundary, as well as those who are not assigned radiological work, such as cafeteria workers or persons in administrative positions:

  - Might be exposed to direct radiation from radioactive materials (although at a great distance) and to trace amounts of plutonium or uranium released to the environment through site exhaust stacks.

  - Could receive radiation and chemical exposures primarily through inhalation of material in the air and external radiation from radioactive material deposited on the ground.

  - Work-related physical hazard risks are present.

  - Estimate of impacts to transient population groups (soda machine vendors, etc.) are bounded by impacts to this group.

- *Facility Workers*: Individuals who work at the proposed MOX facility, the PDCF, or the WSB and who receive a radiation dose in the course of employment in which the assigned duties of the individuals involve exposure to radiation or to radioactive material from licensed and unlicensed sources of radiation:

- Might be exposed to direct gamma radiation emitted from radioactive materials, such as depleted uranium compounds.

- Could receive small radiation doses from inhaling uranium, plutonium, or other radionuclides compared with the direct radiation doses resulting from enclosed processes; ventilation controls would be used to inhibit airborne emissions in facilities.

- Would be protected by a dosimetry program to control doses below the maximum regulatory limit of 0.05 Sv/yr (5 rem/yr) for workers (10 CFR 20.1201).

- For chemical exposures, facility workers are addressed under separate regulations (e.g., Occupational Safety and Health Act [OSHA]); their exposures are not quantitatively addressed in this FEIS. However, physical hazards (i.e., risks of injury and fatality) are addressed for both construction and operations workers.

Impacts to a maximally exposed individual (MEI) were also evaluated. The MEI is a hypothetical person who, because of proximity, activities, or living habits, could receive the highest possible dose of radiation or of a hazardous chemical from a given event or process. For members of the public, potential locations for an MEI would be at the site boundary, the closest possible public access points near the operations under consideration. For SRS employees not directly involved in facility operations, MEI locations are considered at distances of 100 m (330 ft) or more from a facility. An MEI for radiation exposure is not always considered for facility workers because these workers are monitored, and their exposure is expected to be kept as low as reasonably achievable (ALARA), with workers being rotated into and out of relatively higher exposure job functions. In such cases, an average worker dose was estimated.

3.10.3 Baseline Radiological Dose and Risk

The radiological baseline in the vicinity of the SRS includes background radiation, man-made (anthropogenic) sources, and radiation from ongoing SRS operations. Background radiation comes from natural sources, such as cosmic radiation and naturally occurring radioactive material, and from anthropogenic sources that cannot be controlled, such as global fallout from nuclear testing or nuclear accidents. Anthropogenic sources, including consumer products (e.g. television sets and smoke detectors) and medical procedures, account for additional exposure. Human exposure to radiation is measured in units of sieverts (Sv). Background radiation levels

What Is a Sievert?

A *sievert* is a unit of radiation dose. The effects of radiation exposure on humans depend on the kind of radiation received, the total amount absorbed by the body, and the tissues involved. A sievert (Sv) is calculated by a formula that takes these three factors into account. Another common unit of radiation dose is the rem (1 Sv = 100 rem). The U.S. average individual radiation dose is about 0.0036 Sv (0.36 rem) or 3.6 millisievert (mSv) [360 millirem (mrem)] from natural background and anthropogenic sources.

Latent Cancer Fatality (LCF)

What it is: The primary adverse health effect from the low-level radiation doses received from proposed MOX facility, PDCF, or WSB operations and potential accidents would be the possible induction of latent cancer fatalities (LCFs). LCFs are a measure of the expected number of additional cancer deaths in a population (or people dying of cancer) as a result of exposure to radiation. Death from cancer induced by exposure to radiation may occur at any time after the exposure takes place. However, latent cancers would be expected to occur in a population from one year to many years after the exposure takes place. To place the significance of these additional LCF risks from exposure to radiation into context, the average individual has approximately 1 chance in 4 of dying from cancer (LCF risk of 0.25).

How it is calculated: The U.S. Environmental Protection Agency has suggested (Eckerman et al. 1999) a conversion factor that for every 100 person-Sv (10,000 person-rem) of collective dose, approximately 6 individuals would ultimately develop a radiologically induced cancer. If this conversion factor is multiplied by the individual dose, the result is the individual increased lifetime probability of developing an LCF. For example, if an individual receives a dose of 0.00033 Sv (0.033 rem), that individual's LCF risk over a lifetime is estimated to be 2×10^{-5}. This risk corresponds to a 1 in 50,000 chance of developing a LCF during that individual's lifetime. If the conversion factor is multiplied by the collective (population) dose, the result is the number of excess LCFs. Because these results are statistical estimates, values for expected LCFs can be, and often are, less than 1.0 for cases involving low doses or small population groups. If a population group collectively receives a dose of 50 Sv (5,000 rem), which would be expressed as a collective dose of 50 person-Sv (5,000 person-rem), the number of potential LCFs experienced from within the exposure group is 3. If the number of LCFs estimated is less than 0.5, on average, no LCFs would be expected.

---

result in a national annual average individual exposure of approximately 3.0 mSv (300 mrem), with an additional 0.60 mSv (60 mrem) from other anthropogenic sources. A more detailed breakdown of these sources is presented in Table 3.7.

Radiation from SRS operations is estimated by analyzing monitoring data. The SRS has an extensive radiological monitoring network both on- and off-site to assess the effects of site operations on air, surface water, groundwater, soil, terrestrial and aquatic food products, and local game animals. These routine environmental surveillance activities include monitoring airborne and liquid effluent discharges from their points of origin at each operating facility on the SRS to determine compliance with applicable exposure standards. The results of the effluent monitoring and environmental surveillance and the potential radiation doses to members of the public in surrounding areas from those effluents are published annually by the Environmental Monitoring Section of Westinghouse Savannah River Company (e.g., Arnett and Mamatey 2001b).

Airborne emissions from the SRS operations for 2000 are summarized in Table 3.8. Liquid releases for 2000 are summarized in Table 3.9. The estimated off-site radiation doses from both airborne and liquid releases were below all applicable radiation exposure standards for humans and aquatic organisms (Arnett and Mamatey 2001b). The estimated exposures and the applicable standard for each exposure are summarized in Table 3.10. The estimated all-pathway dose to an MEI was 0.0018 mSv (0.18 mrem), which is 0.18% of the DOE's 1.0 mSv (100-mrem) all-pathway dose standard for annual exposure. For an NRC-licensed facility, such as the proposed MOX facility, a dose limit of 1.0 mSv/yr (100 mrem/yr) from operations for an individual member of the public is also applicable (10 CFR 20.1301).

**Table 3.7. Sources and contributions to the
U.S. average individual radiation dose[a]**

Source	Effective dose equivalent [mSv/yr (mrem/yr)]
Natural background radiation	
Cosmic radiation	0.27 (27)
Rocks and soil (external)	0.28 (28)
Internal to body	0.40 (40)
Radon (internal/inhalation)	2.0 (200)
Subtotal	≈2.95 (≈295)
Man-made background radiation	
Weapons test fallout	<0.01 (<1)
Consumer products	0.10 (10)
Medical	
Diagnostic X-rays	0.39 (39)
Nuclear Medicine	0.14 (14)
Subtotal	≈0.64 (≈64)
Total	≈3.60 (≈360)

[a]*Source*: Modified from Arnett and Mamatey (2001b) and
NCRP (1987).

Workers at the SRS with the potential to be exposed to external radiation or to inhale airborne radioactivity take part in a monitoring program in accordance with 10 CFR 835 ("Occupational Radiation Protection"). In 2000, 3,382 SRS workers had a measurable dose with a combined total effective dose equivalent (TEDE) of 1.632 person-sievert (person-Sv) (163.2 person-rem) for an average TEDE of 0.00048 Sv (0.048 rem) (DOE undated).

The primary health concerns attributed to radiation exposure are the development of cancer and hereditary (genetic) effects. Although radiation-induced genetic effects have been observed in laboratory animals (given very high doses of radiation), no evidence of genetic effects has been observed among the children born to atomic bomb survivors from Hiroshima and Nagasaki. Thus, latent cancer fatalities (LCFs) are the radiological health effect end point used in this EIS as a measure of human health impacts. A conservative assumption in this regard is that any amount of radiation may pose some risk for causing cancer, and that the risk is higher for higher radiation exposures. A linear, no-threshold dose response relationship is used to describe the relationship between radiation dose and the occurrence of cancer. This dose-response model suggests that any increase in dose, no matter how small, results in an incremental increase in risk. For the purposes of this EIS, the risk of a latent cancer fatality (LCF) is taken to be 0.06 LCF per person-Sv (0.0006 LCF per person-rem). (See the text box in this section for a discussion on LCFs.) This LCF risk factor is a gender- and age-averaged value that accounts for differences between male and female receptors from infancy through old age living in the United States (Eckerman et al. 1999). While female receptors were

**Table 3.8. Radioactive atmospheric releases
from SRS operations for 2000**

Radionuclide	Curies[a]	Radionuclide	Curies[a]
Gases and Vapors		*Particulates (cont.)*	
H-3 (oxide)	3.24×10^4	Eu-152	4.13×10^{-5}
H-3 (elem.)	1.24×10^4	Eu-154	1.64×10^{-5}
H-3 total	4.48×10^4	Eu-155	4.02×10^{-6}
C-14	1.33×10^{-1}	Hg-203	2.23×10^{-10}
Kr-85	5.28×10^4	Ra-226	1.74×10^{-5}
I-129	1.71×10^{-3}	Ra-228	2.74×10^{-5}
I-131	6.96×10^{-6}	Ac-228	1.80×10^{-6}
I-133	1.18×10^{-4}	Th-228	5.76×10^{-7}
		Th-230	1.74×10^{-5}
Particulates		Th-232	2.58×10^{-6}
Cr-51	1.21×10^{-4}	Th-234	1.04×10^{-4}
Co-57	3.26×10^{-7}	Ba-133	5.4×10^{-10}
Co-58	1.27×10^{-4}	U-233	1.50×10^{-8}
Co-60	8.60×10^{-4}	U-234	3.98×10^{-4}
Ni-59	4.17×10^{-13}	U-235	1.80×10^{-5}
Ni-63	5.09×10^{-6}	U-236	4.16×10^{-11}
Zn-65	2.23×10^{-5}	U-238	5.20×10^{-4}
Sr-89,90	3.89×10^{-3}	Np-237	2.26×10^{-10}
Zr-95	1.68×10^{-5}	Pu-238	3.59×10^{-4}
Zr-85	1.07×10^{-9}	Pu-239	2.05×10^{-3}
Nb-94	3.95×10^{-10}	Pu-240	1.99×10^{-7}
Nb-95	1.13×10^{-4}	Pu-241	4.09×10^{-6}
Tc-99	8.75×10^{-5}	Pu-242	7.03×10^{-9}
Ru-103	4.23×10^{-5}	Am-241	1.46×10^{-4}
Ru-106	1.04×10^{-5}	Am-243	6.02×10^{-6}
Sb-124	5.63×10^{-10}	Cm-242	4.47×10^{-7}
Sb-125	5.34×10^{-5}	Cm-244	7.68×10^{-5}
Sn-113	6.20×10^{-10}	Cm-245	1.04×10^{-13}
Sn-126	6.45×10^{-14}	Cm-246	3.98×10^{-6}
Cs-134	1.31×10^{-4}	Ar-39	3.30×10^{-5}
Cs-137	8.15×10^{-3}	Na-22	7.90×10^{-11}
Ce-141	4.16×10^{-5}	Mn-54	1.30×10^{-10}
Ce-144	1.44×10^{-4}	Se-79	4.47×10^{-9}
Pa-233	2.23×10^{-10}		
Pr-144	3.68×10^{-13}	Alpha	7.35×10^{-4}
Pr-144m	4.43×10^{-15}	Beta-Gamma	3.57×10^{-2}
Pm-147	1.30×10^{-5}		

[a]One curie (Ci) equals 3.7×10^{10} becquerels (Bq). One Bq equals one disintegration per second (dps).

Source: Modified from Arnett and Mamatey (2001b).

**Table 3.9. Radioactive liquid
releases from SRS operations
for 2000 (including direct
and seepage basin
migration releases)**

Radionuclide	Curies[a]
H-3	5.34×10^3
Sr-90	5.44×10^{-2}
Co-60	1.62×10^{-3}
I-129	7.82×10^{-2}
Cs-137	8.81×10^{-2}
U-234	2.87×10^{-5}
U-235	6.18×10^{-6}
U-238	1.97×10^{-4}
Pu-238	2.21×10^{-5}
Pu-239	1.68×10^{-5}
Am-241	1.19×10^{-5}
Cm-244	7.01×10^{-6}
Alpha	1.96×10^{-2}
Beta-Gamma	4.44×10^{-2}

[a]One Ci equals 3.7×10^{10} Bq.

Source: Modified from Arnett
and Mamatey (2001b).

estimated to have a slightly higher LCF rate than males, and infants a higher LCF rate than adults, the use of this risk factor for estimating collective LCF risks to the public in this EIS should provide a reasonable average based on current understanding of radiological effects in humans. On the other hand, the collective LCF risks to the facility workers and SRS employees evaluated in this EIS may be conservative (overestimated) because the more susceptible receptors, such as infants, considered in determining the LCF risk factor are not present in the SRS employee population.

3.10.4 Baseline Chemical Exposure and Risk

3.10.4.1 Chemical Risk Assessment Background

As stated in Section 3.10.2, human exposure to nonradioactive chemicals in air, water, or soil may occur through ingestion, inhalation, or contact with skin. Methods used to assess hazards associated with chemical exposures may simply involve a comparison of concentrations in air, water, or soil with health-risk based standards or guidelines available from state and federal agencies (see *SRS Baseline Risks* below). More detailed assessments estimate the extent of

Table 3.10. Estimated radiation exposures to the public from SRS emissions in 2000

Pathway/receptor	Dose	Standard
Air		
Maximally exposed individual [mSv (mrem)]	0.0004 (0.04)	0.10 (10)[a]
Collective population [person-Sv (person-rem)]	0.023 (2.3)	NA[b]
Liquid		
Maximally exposed individual [mSv (mrem)]	0.0014 (0.14)	0.04 (4)[c]
Collective population [person-Sv (person-rem)]	0.039 (3.9)	NA
Total		
Maximally exposed individual [mSv (mrem)]	0.0018 (0.18)[d]	1.0 (100)[e]
Collective population [person-Sv (person-rem)]	0.062 (6.2)[d]	NA

[a]Set by the EPA in "National Emission Standards for Hazardous Air Pollutants — Radionuclides," 40 CFR 61 Subpart H, December 15, 1989.

[b]NA = not applicable.

[c]Adopted from the EPA in DOE Order 5400.5 as set forth in "National Primary Drinking Water Standards," 40 CFR Part 141.11, July 9, 1976.

[d]Sum of the air and liquid pathways.

[e]All pathway dose standard from DOE Order 5400.5.

Source: Arnett and Mamatey (2001b).

human exposure due to a particular source and compare that exposure with benchmark levels for noncarcinogenic risks ("hazard index" approach) or benchmarks for carcinogenic risks.

In estimating either noncancer risks (that is, noncancer adverse health outcomes, such as liver damage or developmental impairment) due to chemical exposures or increased lifetime cancer risk, the first step is to estimate the chemical concentration in air, water, and/or soil, either present from natural sources or attributable to anthropogenic sources. The concentration estimate is combined with an estimate of the human intake level to produce a chemical-specific daily intake estimate. (The intake level is usually from the upper end of the expected range of possible intakes in order to make sure risk estimates take individuals who have unusually high intakes into account). Estimated intakes are compared with chemical-specific reference doses or cancer slope factors. The reference doses and cancer slope factors are developed by the EPA for many commonly used chemicals and are based on a broad range of toxicological data. See the text box for further information on risk estimation procedures.

3.10.4.2 SRS Chemical Baseline Risks

Public water supplies in the vicinity of the SRS are monitored and regulated to be in compliance with health-based federal standards, and remediation programs are underway at the SRS to

Concepts in Estimating Risks from Exposures to Chemicals in Air, Water, and Soil

Reference Dose: Intake level of a chemical that is very unlikely to have noncancer adverse effects; measured in units of milligrams per kilogram of body weight per day (mg/kg-d). Different reference doses often apply for oral and inhalation exposures.

Hazard Quotient: a comparison of the estimated intake level or dose of a chemical in air, water, or soil with its reference dose; expressed as a ratio.

   Example: If 5 parts per billion (0.005 mg/L) benzene is in groundwater used for drinking and 2 L is ingested daily by a 70-kg (150-lb) person over a period of 10 years, then
   Intake = (0.005 mg/L × 2 L/day)/70 kg = 0.00014 mg/kg-d.
   The reference dose for chronic ingestion of benzene is 0.0003 mg/kg-d.
   The benzene hazard quotient is 0.00014/0.0003 = 0.5. This hazard quotient is less than 1, indicating that the exposure is unlikely to cause adverse noncancer health effects.

Hazard Index: The sum of hazard quotients for all chemicals to which an individual is exposed. Used as a screening tool, a hazard index of less than 1 indicates that adverse health effects are unlikely. However, a hazard index of greater than 1 does not necessarily mean adverse health effects will occur, because different chemicals may react differently in the human body (that is, they may have different, nonadditive kinds of toxicity).

Slope Factor: an upper-bound estimate of a chemical's probability of causing cancer over a 70-year lifetime, based on the extent of intake during the exposure period and given in units of inverse intake [(mg/kg-d)$^{-1}$ or 1/(mg/kg-d)]. For a carcinogen, different slope factors often apply for oral and inhalation exposures.

Increased Lifetime Cancer Risk: an upper-bound estimate of the likelihood that an individual will develop cancer as a result of exposure to a cancer-causing chemical. It is the product of the intake level and the slope factor.

   Example: benzene is also a cancer-causing chemical with an oral slope factor of up to 0.055 (mg/kg-d)$^{-1}$.

   Assuming 5 parts per billion (0.005 mg/L) in water and calculating intake as above, but averaging over a lifetime of 70 years, the increased lifetime cancer risk for benzene ingestion would be:

   0.00014 mg/kg-d x 0.055 (mg/kg-d)$^{-1}$ x 10-yr exposure/70-yr lifetime = 0.0000011 (also can be stated as 1.1×10^{-6} or 1.1 in 1 million).

   This increased risk level would be considered to be small. It is at the lower end of the risk range of 0.000001 (10^{-6}, or 1 in 1 million) to 0.0001 (10^{-4}, or 1 in 10,000) which generally does not require mitigating actions.

control exposure to and eliminate areas of soil contamination. Therefore, the most important potential exposure pathway for workers and the general public would be through inhalation of contaminants released to air from ongoing SRS operations.

The SRS has approximately 200 regulated sources of air emissions. In 1991, the SCDHEC established Air Pollution Control Regulation 61-62.5, Standard No. 8, to regulate hazardous or toxic air pollutant emissions. To demonstrate compliance with this standard, the SRS completed an air emissions inventory and air dispersion modeling for all site sources in 1993, as summarized in Arnett and Mamatey (2001b). An update to the modeling was submitted in 1998 (Dukes 1998). The modeling effort provides estimates of maximum ambient concentrations at or beyond the SRS boundary due to SRS emission sources for about 200 toxic air pollutants (TAPs). The estimated maximum concentrations of the TAPs did not exceed values given in the 2001 version of the SCDHEC standard No. 8 (SCDHEC 2001).

Because regulatory standards are not developed exclusively on the basis of public health considerations, and because the basis for the SCDHEC standard concentrations is not described in available documentation (SCDHEC 2001), the potential for adverse human health impacts was assessed through comparison with health risk-based guideline levels. Specifically, the reported maximum ambient 24-hour average concentrations were modified by a factor of 0.2 to estimate annual average concentrations (based on EPA guidance [EPA 1992]). These estimated annual average concentrations were compared with health risk-based air concentrations developed by the EPA's Office of Air Quality Planning and Standards (OAQPS) (Smith et al. 1999) and with EPA-established reference concentrations for non-cancer effects (EPA 2003b). Although only two TAPs (TCDDs and tetrachloroethylene) exceeded the EPA guideline levels, 10 TAPs had estimated annual average concentrations between the EPA guideline cancer risk level values of 10^{-6} to 10^{-4} (see Table 3.11).

3.10.5 Baseline Physical Hazard Risks

Although worker physical hazard risks (i.e., risks of fatality or injury from on-the-job accidents) can be minimized when workers adhere to safety standards and use protective equipment as necessary, certain rates of accidents have been associated with all types of work. Risks can be calculated on the basis of historical industrywide statistics, as described below.

The expected annual numbers of worker fatalities and injuries for specific industry types are calculated on the basis of rate data from the Bureau of Labor Statistics, as reported by the National Safety Council (NSC 2001), and on the number of annual full-time equivalent (FTE) workers required for manufacturing activities. Employment at the SRS in 2000 was 13,227 people (DCS 2001b). It is assumed that, in general, the types of activities required for these employees would be similar to those for the manufacturing industrial sector, so those fatality and injury rates are used to estimate annual risks. A rate of 3.3 fatalities per 100,000 FTEs and 4.6 injuries per 100 FTEs is used. On the basis of these rates, the estimated annual number of fatalities for SRS workers is less than 1 (specifically, 0.44) per year. The estimated number of injuries is 610 per year (includes only injuries resulting in lost workdays, not including the day of injury). These physical hazard risks represent the baseline risks for existing SRS operations for comparison with impacts under the no-action and proposed action alternatives. However, actual injury and fatality risks over the past 10 years or more have been lower than those predicted on the basis of national statistics.

Table 3.11. Modeled site boundary ambient concentrations of select SRS toxic air pollutant (TAP) emissions in comparison with SCDHEC standards and EPA health risk-based guideline levels

Toxic air pollutant (TAP)	Number of SRS sources	SRS maximum modeled 24-hour average concentration ($\mu g/m^3$)[a]	SRS Estimated Annual Average Concentration ($\mu g/m^3$)[b]	SCDHEC standard ($\mu g/m^3$)	EPA guideline level ($\mu g/m^3$)[c]
TAPs with ambient level exceeding EPA guideline level					
TCDDs	1	0.00002	4×10^{-6}	0	3×10^{-8} to 3×10^{-6}
Tetrachloroethylene	36	99	20	3,350	0.17-17
TAPs with estimated annual ambient level between EPA Guideline 10^{-6} and 10^{-4} cancer risk level					
Arsenic	7	0.05	0.01	1.0	0.00023-0.023
Benzene	118	4.6	0.9	150	0.13-13 (30)
Beryllium	7	0.009	0.0020	0.01	0.00042-0.042 (0.02)
Bis(chloromethyl)ether	1	0.002	0.0004	0.03	2×10^{-5} to 2×10^{-3}
Carbon tetrachloride	16	4.2	0.84	150	0.067-6.7
Dimethyl benzidine	1	0.002	0.0004	NA	0.00038-0.038
Heptachlor	1	0.01	0.002	2.5	0.00077-0.077
Hydrazine	5	0.06	0.012	0.5	0.0002-0.02
Quinoline	1	0.004	0.0008	NA	0.00029-0.029
Trichloroethylene	38	23	5	6,750	0.5-50

[a] SCDHEC Standard No. 8 requires that the standards be compared with modeled maximum 24-hour average concentrations at or beyond the site boundary.

[b] EPA guideline values should be compared with annual average concentrations; these values were estimated as the maximum 24-hour ambient concentrations multiplied by 0.2.

[c] Where a range is given, the range corresponds to a 10^{-6} to 10^{-4} risk level (that is, the concentration that if inhaled for a lifetime would result in an increased individual risk of developing cancer of between 1 in 1 million and 1 in 10,000). Values in parentheses are verified reference concentrations established by the EPA (2003b), also recognized as important guidelines under SCDHEC Standard No. 8.

Sources: Dukes (1998); SCDHEC (2001); Smith et al. (1999, Table 2).

3.11 Socioeconomics

This section discusses existing socioeconomic conditions in the vicinity of the SRS as they relate to the proposed facilities. The socioeconomic data presented for the SRS describe a regional economic area (REA) comprising 15 counties around the site (see Appendix D) and a region-of-influence (ROI) surrounding the site comprising 4 counties — Columbia and Richmond Counties in Georgia and Aiken and Barnwell Counties in South Carolina. The REA is used to assess the potential regional economic impacts of site activities, specifically impacts on employment and unemployment and on personal income. The REA constitutes a broad market area defined by economic linkages between the various sectors in the regional economy.

The ROI was defined on the basis of the current residential locations of full-time SRS workers directly involved in the SRS activities and encompasses the area in which most of these workers spend their wages and salaries. The ROI is used to assess the impacts of site activities on population, housing, community services, and community fiscal conditions. More than 90% of SRS workers currently reside in these counties (DCS 2001b). In the following sections, data are presented for each of the counties in the ROI.

3.11.1 Population

The population of the ROI was at 475,095 in 2000 (U.S. Bureau of the Census 2002a) and was expected to reach 489,000 by 2001, as shown in Table 3.12. In 2000, 30% of the ROI total (142,552 people) resided in Aiken County (U.S. Bureau of the Census 2001), with 25,337 in the city of Aiken. Over the period 1990-2000, population in the ROI as a whole, in Aiken County, and in the city of Aiken grew slightly, with average growth rates of 1.4%, 1.7%, and 2.5%, respectively. Over the same period, population in South Carolina as a whole grew at a rate of 1.4%.

In 2000, 41% of the ROI population (195,182 persons) resided in the city of Augusta/ Richmond County, Georgia, with 19% (89,288) located in Columbia County, Georgia, and 5% (23,478) in Barnwell County, South Carolina (U.S. Bureau of the Census 2000). Growth in Augusta/Richmond County over the period 1990-2000 was slight at 0.3%, relatively high in Columbia County over the same period at 3.1%, and moderate in Barnwell County at 1.5%. Other incorporated places in the immediate vicinity of the SRS are Barnwell (population 5,035 in 2000), Blackville (2,973), Elko (212), Hilda (436), Jackson (1,625), New Ellenton (2,250), North Augusta (17,574), and Willston (3,307) (U.S. Bureau of the Census 2002a).

3.11.2 Employment and Unemployment

Employment in the REA totaled 207,660 people in 2000 and was expected to reach 214,000 in 2002. Employment grew at an annual average rate of 1.6% between 1990 and 2000 (U.S. Bureau of the Census 1992, 2002b). The economy of the REA is dominated by the trade

Table 3.12. ROI population statistics for selected years

Entity	1990[a]	2000[a]	Average annual growth rate (%), 1990-2000	2002 (projected)
Georgia				
Columbia County	66,031	89,288	3.1	95,000
Richmond County/City of Augusta	189,719	195,182	0.3	196,000
South Carolina				
Aiken County	120,991	142,552	1.7	147,000
City of Aiken	19,872	25,337	2.5	27,000
Barnwell County	20,293	23,478	1.5	24,000
ROI Total	415,394	475,095	1.4	489,000
Georgia	6,478,216	8,186,453	2.4	8,580,000
South Carolina	3,486,703	4,012,012	1.4	4,130,000

[a]*Source*: U.S. Bureau of the Census (2002a).

and service industries, with these activities currently contributing almost 63% of all employment in the REA (see Table 3.13). The manufacturing sector is also a significant employer in the REA, with 27% of total REA employment. Employment at the SRS in 2000 was 13,227 people (DCS 2001b).

Unemployment in the REA steadily declined during the late 1990s from a peak rate of 8.0% in 1993 to the 2002 rate of 5.7% (see Table 3.14) (U.S. Bureau of Labor Statistics 2002). Unemployment in Georgia was 4.7% in August 2002; in South Carolina the rate was 5.7% in that month.

3.11.3 Income

Personal income in the REA was $14.8 billion in 2000 and was expected to reach $15.6 billion in 2002. Personal income grew at an annual average rate of 1.8% over the period 1990-1999 (see Table 3.15). Personal income per capita in the REA also rose in the 1990s and was expected to reach $24,700 in 2002, compared with $23,146 at the beginning of the period.

3.11.4 Housing

Total housing in Columbia County grew at an annual rate of 3.5% over the period 1990-2000 (see Table 3.16), with total housing units expected to reach 35,400 in 2002, reflecting the relatively high growth in county population. About 9,580 new units were added to the existing housing stock in the county between 1990 and 2000. On the basis of annual population growth rates, there were expected to be 2,340 vacant housing units in the county in 2002, with 420 expected to be rental units available to construction workers at the proposed facilities.

Table 3.13. REA employment by industry, 2000

Sector	Employment	Percent of REA total
Agriculture[a]	6,250	3.0
Mining	877	0.4
Construction	11,399	5.5
Manufacturing	55,853	27.0
Transportation and Public Utilities	5,028	2.4
Trade	34,389	17.0
Finance, Insurance and Real Estate	7,783	3.7
Services	86,673	42.0
Other	193	0.1
Total	207,660	

[a]1997 data; U.S. Department of Agriculture (1999).

Source: U.S. Bureau of the Census (2002b), except as noted.

Total housing in the City of Augusta/Richmond County grew at an annual rate of 0.6% over the period 1990-2000 (see Table 3.16), with total housing units expected to reach 82,800 in 2002, reflecting the relatively slow growth in county population. Only 5,000 new units were added to the existing housing stock in the county between 1990 and 2000. On the basis of annual population growth rates, there were projected to be 8,440 vacant housing units in the county in 2002, with 3,550 of those expected to be rental units available to construction workers at the proposed facilities.

Total housing in Aiken County grew at an annual rate of 2.3% over the period 1990-2000 (see Table 3.16), with total housing units expected to reach 64,100 in 2002. Growth in the city of Aiken was 2.9% over this period, with 11,900 total housing units expected in 2002. Almost 12,700 new units were added to the existing housing stock in the county between 1990 and 2000, 2,830 of which were built in the city of Aiken. On the basis of annual population growth rates, there were expected to be 6,610 vacant housing units in the county in 2002, with 1,610 expected to be rental units available to construction workers at the proposed facilities.

Table 3.14. REA unemployment rates

Period	Rate (%)
REA	
1990-2000 average	6.7
2002[a]	5.7
Georgia	
1990-2000 average	5.0
2002[b]	4.7
South Carolina	
1990-2000 average	5.4
2002[b]	5.7

[a]Rate is for July 2002.

[b]Rate is for August 2002.

Source: U.S. Bureau of Labor Statistics (2002).

Table 3.15. REA personal income (2003 dollars)

Parameter	1990[a]	2000[a]	Average annual growth rate (%), 1990-2000	2002 (projected)
Total personal income ($ millions)	12,426	14,814	1.8	15,600
Personal income per capita ($)	23,146	24,681	0.6	25,300

[a]*Source*: U.S. Department of Commerce (2002).

Total housing in Barnwell County grew at an annual rate of 2.6% over the period 1990-2000 (see Table 3.16), with total housing units expected to reach 10,500 in 2002, reflecting the moderate growth in county population. About 2,300 new units were added to the existing housing stock in the county between 1990 and 2000. On the basis of annual population growth rates, there were projected to be 1,210 vacant housing units in the county in 2002, with 300 of those expected to be rental units available to construction workers at the proposed facilities.

Total housing in the ROI as a whole grew at an annual rate of 1.8% over the period 1990-2000 (see Table 3.16), with total housing units expected to reach 202,000 in 2002. About 31,600 new units were added to the existing housing stock in the ROI between 1990 and 2000. On the basis of annual population growth rates, there were projected to be 19,600 vacant housing units in the ROI in 2002, with 5,910 of those expected to be rental units available to construction workers at the proposed facilities.

3.11.5 Community Resources

Construction and operation of the proposed MOX facility, PDCF, and WSB would result in increased revenues and expenditures for local government jurisdictions, including counties, cities, and school districts. Revenues would come primarily from state and local sales taxes associated with employee spending during construction and operation and local property taxes.

Additional revenues would be used to support additional local community services currently provided by each jurisdiction.

Construction and operation of the proposed facilities would result in increased demand for community services in the counties, cities, and school districts likely to host relocating construction workers and operations employees. Additional demands would also be placed on local medical facilities and physician services.

Tables D.1 and D.2 in Appendix D present information on revenues and expenditures by the various local government jurisdictions in the ROI. Tables 3.17 and 3.18 present data on employment and levels of service (number of employees per 1,000 population) for public safety, general local government services, and physicians. Tables 3.19 and 3.20 provide staffing data for school districts and hospitals.

Table 3.16. City, county, and ROI housing characteristics[a]

Parameter	1990[b]	2000[c]	2002 (projected)
Georgia			
Columbia County			
Owner occupied	17,322	25,557	27,100
Rental	4,519	5,563	5,900
Total unoccupied units	1,904	2,201	2,340
Total units	23,745	33,321	35,400
Richmond County/City of Augusta			
Owner occupied	38,762	42,840	43,100
Rental	29,913	31,080	31,300
Total unoccupied units	8,613	8,392	8,440
Total units	77,288	82,312	82,800
South Carolina			
Aiken County			
Owner occupied	33,491	42,036	43,400
Rental	11,392	13,551	14,000
Total unoccupied units	4,383	6,400	6,610
Total units	49,266	61,987	64,100
City of Aiken			
Owner occupied	5,130	6,804	7,140
Rental	2,619	3,483	3,660
Total unoccupied units	794	1,086	1,140
Total units	8,543	11,373	11,900
Barnwell County			
Owner occupied	5,194	6,810	7,010
Rental	1,906	2,211	2,280
Total unoccupied units	754	1,170	1,210
Total units	7,854	10,191	10,500
ROI Total			
Owner occupied	99,673	123,902	128,000
Rental	49,250	54,016	55,200
Total unoccupied units	16,520	19,116	19,600
Total units	165,443	197,034	202,000

[a]Column entries may not add up due to independent rounding.

[b]*Source*: U.S. Bureau of the Census (1994).

[c]*Source*: U.S. Bureau of the Census (2002a).

Table 3.17. Local public service employment (2001)

Part A: Georgia

	Columbia County		Grovetown		Harlem	
	Number	**Level of service**[a]	**Number**	**Level of service**[a]	**Number**	**Level of service**[a]
Police protection	147	1.8	17	2.8	7	3.9
Fire protection[b]	3	0	4	0.7	1	0.6
General	435	5.3	33	5.4	14	7.7
Total	585	7.2	54	8.9	22	12.1

	Augusta-Richmond County		Blythe		Hephzibah	
	Number	**Level of service**[a]	**Number**	**Level of service**[a]	**Number**	**Level of service**[a]
Police protection	357	1.8	1	1.4	4	1.0
Fire protection[b]	283	1.4	0	0	7	1.8
General	1,673	8.6	1	1.4	4	1.0
Total	2,313	11.9	2	2.8	15	3.9

	State of Georgia level of service[a,c]
Police protection	2.4
Fire protection[b]	1.1
General	52.0
Total	55.4

Part B: South Carolina

	Aiken County		Aiken		Jackson	
	Number	**Level of service**[a]	**Number**	**Level of service**[a]	**Number**	**Level of service**[a]
Police protection	131	1.4	54	2.1	4	2.5
Fire protection[b]	78	0.8	-[d]	-[d]	-[d]	-[d]
General	60	0.6	239	9.4	7	4.3
Total	269	2.8	347	13.7	11	6.8

	New Ellenton		North Augusta		Wagener	
	Number	**Level of service**[a]	**Number**	**Level of service**[a]	**Number**	**Level of service**[a]
Police protection	4	1.8	48	2.7	3	3.5
Fire protection[b]	-[d]	-[d]	6	0.3	-[d]	-[d]
General	5	2.2	125	7.1	5	5.8
Total	9	4.0	179	10.2	8	9.3

Table 3.17. Continued

	Barnwell County		Barnwell		Blackville	
	Number	Level of service[a]	Number	Level of service[a]	Number	Level of service[a]
Police protection	26	2.1	13	2.6	8	2.7
Fire protection[b]	-[d]	-[d]	3	0.6	1	0.3
General	150	12.3	22	4.4	11	3.7
Total	176	14.5	38	7.6	20	6.7

	Williston		State of South Carolina level of service[a,c]
	Number	Level of service[a]	
Police protection	9	2.7	2.5
Fire protection[b]	1	0.3	0.8
General	12	3.6	54.9
Total	22	6.7	58.2

[a]Level of service represents the number of employees per 1,000 persons in each jurisdiction.

[b]Does not include volunteers.

[c]2000 data.

[d]Police and fire services are provided by a combined department.

Sources: Aiken County: Powell (2001); Barnwell County: Aguilar (2001); Columbia County: J. Johnson (2001); Edgefield County: Harling (2001); Richmond County: Colliander (2001); City of Aiken: Rideout (2001); City of Jackson: S. Johnson (2001); Town of New Ellenton: Bledsoe (2001); City of North Augusta (2000); Town of Wagener: Salley (2001); City of Barnwell: Vargo (2001); Town of Blackville: McDonald (2001); Town of Williston: Fowler (2001); Town of Grovetown: Kent (2001) and Capatillo (2001); Town of Harlem: Moore (2001); City of Augusta (1999); Town of Blythe (2000); Town of Hephzibah (2000); U.S. Bureau of the Census (2000).

3.11.6 Traffic

Vehicular access to the SRS is provided from South Carolina SCs 19, 64, 125, 781, and U.S. Highway 278, as shown in Figures 3.1 and 3.8. Highway 19 runs north from the site through New Ellenton towards Aiken; SC 64 runs in an easterly direction from the site towards Barnwell; SC 125 runs through the site itself in a southeasterly direction between North Augusta and Allendale, passing through Beech Island and Jackson. U.S. 278 also runs through the site, in a southeasterly direction between North Augusta and Barnwell. SC 781 connects U.S. 278 with Willston to the northeast of the site. The northern perimeter of the site is about 16 km (10 mi) from downtown Aiken. Table 3.21 shows average annual daily traffic (AADT) flows over these road segments, together with congestion level designations (levels of service). Levels of service designations were developed by the Transportation Research Board (1985) and range from A to F. Designations A through C represent good traffic operating conditions with some minor delays experienced by motorists; F represents jammed roadway conditions.

Table 3.18. Local physicians data (1997)

County	Number of physicians	Level of service[a]
Georgia		
Columbia County	324	4.0
Richmond County	1,189	6.1
South Carolina		
Aiken County	190	1.4
Barnwell County	14	0.6

[a]Level of service represents the number of physicians per 1,000 persons in each county.

Source: American Medical Association (1999).

Table 3.19. Local school district data (2001)

School district	Number of teachers	Student-to-teacher ratio[a]
South Carolina		
Aiken County	1,486	17.0
Barnwell County		
School District 19	80	14.4
School District 29	70	14.9
School District 45	183	15.3
State total	44,967	15.2
Georgia		
Columbia County	1,064	17.0
Richmond County	2,200	16.0
State total	89,561	16.0

[a]The number of students per teacher in each school district.

Sources: Ferriter (2001); Georgia Department of Education (2000).

Table 3.20. Local medical facility data (2001)

Hospital	Number of staffed beds	Occupancy rate (%)[a]
Aiken Regional Medical Centers	245	56
Barnwell County Hospital	33	37
Georgia Regional Hospital at Augusta	196	79
Medical College of Georgia Hospital	446	56
Select Specialty Hospital	17	NA[b]
St. Joseph Hospital	151	48
University Hospital	553	50
Walton Rehabilitation Institute	58	78
ROI Total	1,699	-

[a]Percent of staffed beds occupied.

[b]NA = not available.

Source: SMG Marketing Group Inc. (Copyright 2001, used with permission).

Table 3.21. Average annual daily traffic (AADT) in the vicinity of the SRS (2000)

Road segment[a]	Traffic volume (AADT)	Level of service[b]
SC 125 in the vicinity of Jackson	13,400	B
U.S. 278 between SC 302 and Barnwell county line	5,400	A
SC 19 in the vicinity of New Ellenton	13,900	B
SC 781 between U.S. 278 and U.S. 78	2,700	A
U.S. 278 to SC 37	2,500	A
SC 64 between SC 20 and Barnwell	6,900	A
SC 125 between SC 17 and Martin	2,100	A

[a]SC = state route (highway); U.S. = U.S. highway.

[b]Level of service designations as developed by the Transportation Research Board (1985). Levels range from A to F, with A representing the best traffic operating conditions and F representing jammed roadway conditions.

Source: McCoy (2001), except as noted.

3.12 Aesthetics

Natural and man-made features give a landscape character and aesthetic quality. The character of a landscape is determined by the elements of form, line, color, and texture; each may influence the character of a landscape to a varying degree. The stronger the influence of any one or all of these elements, and the more visual variety that can successfully coexist in the landscape, the more aesthetic quality present in the landscape

3.12.1 General Description of the Site

The viewshed within the vicinity of the SRS consists principally of agricultural and forested land, with some residential and industrial development. The landscape is characterized mainly by wetland or forest on low mountains and hills with intermittent open land. Vegetation consists of hardwood forests in the low-lying areas and wetland forests, with oak and pine forests on higher ground.

3.12.2 Description of the Location of the Proposed Facilities

Various concrete industrial buildings and other structures, administrative and support buildings, and parking areas are located within the F-Area at the SRS. The largest structures are approximately 30 m (100 ft) high, with some stacks and towers reaching 60 m (200 ft) high. All of the industrial and administrative areas are brightly lit at night and are visible when approached on SRS access roads. The industrial and other developed areas in the vicinity of F-Area, including utility corridors, are generally consistent with a Bureau of Land Management visual resource management (VRM) Class IV designation (activities that lead to major modification of the existing character of the landscape). The remainder of the site fits a VRM Class III (hosting activities which at most only moderately change the existing character of the landscape) or IV designation (DOI 1986a,b).

The closest publicly accessible viewing location is from State Highway 125, about 6 km (4 mi) to the southwest. Public view of F-Area is restricted by the heavily wooded terrain between Route 125 and the site.

3.13 References for Chapter 3

Aguilar, N. 2001. Personal communication from Aguilar (Barnwell County, SC) to L. Nieves (Argonne National Laboratory, Argonne, IL). May.

American Medical Association 1999. *Physician Characteristics and Distribution in the U.S.* Chicago, IL.

Arnett, M.W., and A.R. Mamatey (eds.) 2000a. *Savannah River Site Environmental Data for 1999.* WSRC-TR-99-00301. Westinghouse Savannah River Company, Aiken, SC.

Arnett, M.W., and A.R. Mamatey (eds.) 2000b. *Savannah River Site Environmental Report for 1999*. WSRC-TR-99-00299. Westinghouse Savannah River Company, Aiken, SC.

Arnett, M.W., and A.R. Mamatey (eds.) 2001a. *Savannah River Site Environmental Data for 2000*. WSRC-TR-2000-0329. Westinghouse Savannah River Company, Aiken, SC.

Arnett, M.W., and A.R. Mamatey (eds.) 2001b. *Savannah River Site Environmental Report for 2000*. WSRC-TR-2000-00328. Westinghouse Savannah River Company, Aiken, SC.

Arnold, J.E., et al. 1998. *Efforts to Save an Endangered Species* - Echinacea laevigata *(smooth coneflower)*. Horticulture Department, Clemson University, Clemson, SC. Available at http://virtual.clemson.edu/groups/hort/sctop/bsec/bsec-13.htm (last updated July 16).

Bennett, D.H., and R.W. McFarlane 1983. *The Fishes of the Savannah River Plant: National Environmental Research Park*. SRO-NERP-12. U.S. Department of Energy, Savannah River Ecology Laboratory, National Environmental Research Park Program, Aiken, SC. Aug.

Bledsoe, K. 2001. Personal communication from Bledsoe (Town of New Ellenton, SC) to J. Jackson (Argonne National Laboratory, Argonne, IL). July.

Bowling, T.J. 2001. E-mail from Bowling (Duke Cogema Stone & Webster, Charlotte, NC) to M. L. Birch (Duke Cogema Stone & Webster, Charlotte, NC). June 6.

Cabak, M.A., et al. 1996. *Distributional Archaeology in the Aiken Plateau: Intensive Survey of E Area, Savannah River Site, Aiken County, South Carolina*. Savannah River Archaeological Research Papers 8.

Capatillo, V. 2001. Personal communication from Capatillo (Town of Grovetown, GA) to J. Jackson (Argonne National Laboratory, Argonne, IL). July.

City of Augusta 1999. *City of Augusta Annual Financial Statement*. Augusta, GA. Dec.

City of North Augusta 2000. *City of North Augusta Annual Financial Statements*. North Augusta, SC. Dec.

Colliander, B. 2001. Personal communication from Colliander (Richmond County, SC) to L. Nieves (Argonne National Laboratory, Argonne, IL). May.

Conant, R. 1958. *A Field Guide to Reptiles and Amphibians of the United States and Canada East of the 100th Meridian*. Houghton Mifflin Company, Boston, MA.

Davis, C.E., and L.L. Janecek 1997. *DOE Research Set-Aside Areas of the Savannah River Site*. SRO-NERP-25. U.S. Department of Energy, Savannah River Ecology Laboratory, National Environmental Research Park Program, Aiken, SC. Aug.

DCS (Duke Cogema Stone & Webster) 2001a. *Mixed Oxide Fuel Fabrication Facility Construction Authorization Request*. Docket No. 070-03098. Charlotte, NC.

DCS 2001b. *Responses to Request for Additional Information for the Duke Cogema Stone & Webster (DCS) Mixed Oxide (MOX) Fuel Fabrication Facility (FFF) Environmental Report (ER)*. Report with letter and attachments on CD-ROM, submitted by P. S. Hastings (DCS, Charlotte, NC) to U.S. Nuclear Regulatory Commission, Washington, DC. July 12.

DCS 2002. *Mixed Oxide Fuel Fabrication Facility Environmental Report, Revision 1 & 2*. Docket Number 070-03098. Charlotte, NC. July.

DCS 2003a. *Mixed Oxide Fuel Fabrication Facility Environmental Report, Revision 3*. Docket Number 070-03098. Charlotte, NC. June.

DCS 2003b. *Mixed Oxide Fuel Fabrication Facility Environmental Report, Revision 4*. Docket Number 070-03098. Charlotte, NC. Aug.

DCS 2003c. *MOX Fuel Fabrication Facility Site Geotechnical Report.* DCS01-WRS-DS-NTE-G-00005-E, Charlotte, NC. June

DOE (U.S. Department of Energy) undated. *DOE Occupational Radiation Exposure 2000 Report.* Assistant Secretary for Environment, Safety and Health, Office of Safety and Health, Washington, DC.

DOE 1991a. *Draft Environmental Impact Statement for the Siting, Construction, and Operation of New Production Reactor Capacity.* DOE/EIS-0144D. Office of New Production Reactors, Washington, DC. April.

DOE 1991b. *American Indian Religious Freedom Act (AIRFA) Compliance at the Savannah River Site.* Prepared by NUS Corporation and RDN, Inc., for Savannah River Operations Office, Environmental Division, Aiken, SC. April.

DOE 1995. *Savannah River Site Waste Management Final Environmental Impact Statement.* DOE/EIS-0217. Savannah River Operations Office, Aiken, SC. July.

DOE 1996. *Storage and Disposition of Weapons-Usable Fissile Materials Final Programmatic Environmental Impact Statement.* DOE/EIS-0229. Office of Fissile Materials Disposition, Washington DC. Dec.

DOE 1997. *Integrated Data Base Report — 1996: U.S. Spent Nuclear Fuel and Radioactive Waste Inventories, Projections, and Characteristics.* DOE/RW-0006, Rev. 13. Office of Environmental Management, Washington, DC. Dec.

DOE 1999. *Surplus Plutonium Disposition Final Environmental Impact Statement.* DOE/EIS-0283. Office of Fissile Materials Disposition, Washington DC. Nov.

DOE 2000a. *Savannah River Site Strategic Plan.* Savannah River Site, Aiken, SC. March.

DOE 2000b. *Savannah River Site Long Range Comprehensive Plan.* Savannah River Site, Aiken, SC. Dec.

DOI (U.S. Department of the Interior) 1986a. *Visual Resource Inventory.* BLM Manual Handbook H-8410-1. Bureau of Land Management, Washington, DC. Jan.

DOI 1986b. *Visual Resource Contrast Rating.* BLM Manual Handbook H-8431-1. Bureau of Land Management, Washington, DC. Jan.

Dukes, M.D. 1998. "Savannah River Site (SRS) Air Dispersion Modeling Update." Letter from Dukes (Environmental Protection Department, Westinghouse Savannah River Co., Aiken, SC) to C.W. Richardson (Director of Bureau of Air Quality, South Carolina Department of Health and Environmental Control, Columbia, SC). Oct. 13.

Eckerman, K.F., et al. 1999. *Cancer Risk Coefficients for Environmental Exposure to Radionuclides, Federal Guidance Report No. 13.* EAP 402-R-99-001. Prepared by Oak Ridge National Laboratory, Oak Ridge, TN, for U.S. Environmental Protection Agency, Office of Radiation and Indoor Air, Washington, DC. Sept.

EPA (U.S. Environmental Protection Agency) 1972. *Federal Air Quality Control Regions.* Publication No. AP-102. Office of Air Programs, Rockville, MD. Jan.

EPA 1974. *Information on Levels of Environmental Noise Requisite to Protect Public Health and Welfare with an Adequate Margin of Safety.* EPA-550/9-74-004. Washington, DC.

EPA 1992. *Screening Procedures for Estimating the Air Quality Impact of Stationary Sources, Revised.* EPA-454/R-92-019. Office of Air and Radiation, Office of Air Quality Planning and Standards, Research Triangle Park, NC. Oct.

EPA 2001. *National Emission Inventory (NEI) 1999 Version 1 for Criteria Pollutants.* Office of Air Quality Planning and Standards, Research Triangle Park, NC. March 20. Available at ftp://ftp.epa.gov/EmisInventory/net_99.

EPA 2002. *AirData — Reports and Maps.* Office of Air Quality Planning and Standards, Research Triangle Park, NC. Aug. Available at http://www.epa.gov/air/data/reports.html.

EPA 2003a. *Air Data — Reports and Maps.* Office of Air Quality and Standards, Research Triangle Park, NC. June. Available at http://www.epa.gov/air/data/reports.html.

EPA 2003b. "Integrated Risk Information System Database." Available at http://www.epa.gov/iriswebp/iris/index.html. Accessed June 2003.

Ferriter, M. 2001. Personal communication from Ferriter (South Carolina Department of Education) to L. Nieves (Argonne National Laboratory, Argonne, IL). May.

Fowler, P. 2001. Personal communication from Fowler (Town of Williston, SC) to J. Jackson (Argonne National Laboratory, Argonne, IL). July.

GDNR (Georgia Department of Natural Resources) 2000. *Rules for Air Quality Control.* Chapter 391-3-1. Environmental Protection Division, Air Protection Branch, Atlanta, GA. Dec. 28.

Georgia Department of Education 2000. *1999-2000 Public Education Report Card.* Available at http://accountability.doe.k12.ga.us/report2000.

Hanson, G.T., et al. 1978. "The Preliminary Archaeological Inventory of the Savannah River Plant, Aiken and Barnwell Counties, South Carolina." Research Manuscript Series 134. South Carolina Institute of Archaeology and Anthropology, University of South Carolina, Columbia, SC.

Harling, M. 2001. Personal communication from Harling (Edgefield County, SC) to L. Nieves (Argonne National Laboratory, Argonne, IL). May.

Johnson, J. 2001. Personal communication from Johnson (Columbia County, GA) to L. Nieves, (Argonne National Laboratory, Argonne, IL). May.

Johnson, S. 2001. Personal communication from Johnson (Town of Jackson, SC) to L. Nieves (Argonne National Laboratory, Argonne, IL). May.

Kent, S. 2001. Personal communication from Kent (Town of Grovetown, GA) to J. Jackson (Argonne National Laboratory, Argonne, IL). July.

King, A., and K. Stephenson 2000. *Archaeological Survey and Testing of the Surplus Plutonium Disposition Facilities.* Technical Report Series No. 24. Savannah River Archaeological Research Program, South Carolina Institute of Archaeology and Anthropology, University of South Carolina, Columbia, SC. April.

Kirkham, R.M., and W.P. Rogers 1981. *Earthquake Potential in Colorado.* Colorado Geological Survey Bulletin 43. Department of Natural Resources, Denver, CO.

Mayer, J.J., and L.D. Wike 1997. *SRS Urban Wildlife: Environmental Information Document.* WSRC-TR-97-0093. Westinghouse Savannah River Company, Savannah River Site, Aiken, SC. May.

McCoy, N. 2001. Personal communication from McCoy (South Carolina Department of Transportation) to T. Allison (Argonne National Laboratory, Argonne, IL). May.

McDonald, J. 2001. Personal communication from McDonald (Town of Blackville, SC) to J. Jackson (Argonne National Laboratory, Argonne, IL). July.

Moore, D. 2001. Personal communication from Moore (Town of Harlem, GA) to J. Jackson (Argonne National Laboratory, Argonne, IL). July.

NCRP (National Council on Radiation Protection and Measurements) 1987. *Ionizing Radiation Exposure of the Population of the United States; Recommendations of the National Council on Radiation Protection and Measurements.* NCRP Report No. 93. Bethesda, MD.

NOAA (National Oceanic and Atmospheric Administration) 1999. *Local Climatological Data: Annual Summary with Comparative Data for Augusta, Georgia (AGS).* National Climatic Data Center, Asheville, NC.

Noah, J.C. (compiler) 1995. *Land-Use Baseline Report, Savannah River Site.* WSRC-TR-95-0276. Westinghouse Savannah River Company, Savannah River Site, Aiken, SC. June.

NSC (National Safety Council) 2001. *Injury Facts.* 2000 Edition. Itasca, IL.

NUS 1990. *Sound-Level Characterization of the Savannah River Site.* Report No. NUS-5251. NUS Corp. Aug.

Powell, D. 2001. Personal communication from Powell (Aiken County, SC) to L. Nieves (Argonne National Laboratory, Argonne, IL). May.

Press, F., and R. Siever 1982. *Earth.* W.H. Freeman and Company, San Francisco, CA.

Rideout, S. 2001. Personal communication from Rideout (City of Aiken, SC) to L. Nieves (Argonne National Laboratory, Argonne, IL). May.

Ruffner, J.A. 1985. *Climates of the States.* Third Edition. Gale Research Company, Book Tower, Detroit, MI.

Salley, T. 2001. Personal communication from Salley (Town of Wagener, SC) to J. Jackson (Argonne National Laboratory, Argonne, IL). July.

SCDHEC (South Carolina Department of Health and Environmental Control) 2000. *Air Quality Modeling Guidelines.* Bureau of Air Quality, Columbia, SC. Jan.

SCDHEC 2001. *Air Pollution Control Regulations and Standards, Regulation 61-26.5, Air Pollution Control Standards, Standard No. 8, Toxic Air Pollutants.* Oct. 26. Available at http://www.scdeh.net/eqc/baq/html/regulatory.html.

SCDHEC 2002a. *Ambient Air Quality Data Summary for 1998.* July. Available at http://www.scdhec.net/baq/html/modeling.htm.

SCDHEC 2002b. *Ambient Air Quality Data Summary for 1999.* July. Available at http://www.scdhec.net/baq/html/modeling.htm.

SCDHEC 2002c. *Ambient Air Quality Data Summary for 2000.* July. Available at http://www.scdhec.net/baq/html/modeling.htm.

SCDHEC 2002d. *Ambient Air Quality Data Summary for 2001.* July. Available at http://www.scdhec.net/baq/html/modeling.htm.

SCDNR (South Carolina Department of Natural Resources) 2000/2001. "Hunting Regulations in South Carolina, Wildlife Management Areas Game Zone 3 Aiken, Lexington & Richland Counties." Wildlife Management Offices, New Ellenton, SC. Available at http://water3.dnr.state.sc.us/dnr/etc/rulesregs/img/zone3.pdf.

Schalles, J.F., et al. 1989. *Carolina Bays of the Savannah River Plant.* Savannah River Plant, National Environmental Research Park Program, Aiken, SC. March.

Sessions, J. 1997. "Request for Waste Management Information." Personal communication from Sessions (Westinghouse Savannah River Company, Aiken, SC) to J. Dimarzio (SAIC). Oct. 1.

SMG Marketing Group Inc. 2001. Data downloaded from the web site www.hospitalselect.com and used with permission. Chicago, IL. Nov 20.

Smith, R.L., et al. 1999. *Ranking and Selection of Hazardous Air Pollutants for Listing Under Section 112(k) of the Clean Air Act Amendments of 1990, Technical Support Document.* U.S. Environmental Protection Agency, Office of Air Quality Planning and Standards. Available at http://www.epa.gov/ttn/atw/urban/main_txt.pdf (accessed on April 12, 2001).

Specht, W.L., and M.H. Paller 2001. *Instream Biological Assessment of NPDES Point Source Discharges at the Savannah River Site, 2000.* WSRC-TR-2001-00145. Westinghouse Savannah River Company, Savannah River Site, Aiken, SC. May.

SRARP (Savannah River Archaeological Research Program) 1989. *Archaeological Resource Management Plan of the Savannah River Archaeological Research Program.* South Carolina Institute of Archaeology and Anthropology, University of South Carolina, Columbia, SC. Dec.

SRS (Savannah River Site) 2002. *Plutonium Disposition Program (PDP) Preconstruction Environmental Monitoring Report.* ESH-EMS-2002-1141, Rev. 0. Westinghouse Savannah River Company, Aiken, SC. June.

Storm Prediction Center 2001. *Historical Tornado Data Archive, Graphs and Charts on U.S. Tornadoes: 1950-1998 Tornadoes by State.* April 26. Available at http://www.spc.noaa.gov/archive/tornadoes/index.html.

Town of Blythe 2000. *City of Blythe Financial Report.* Blythe, GA. Dec.

Town of Hephzibah 2000. *Town of Hephzibah Financial Report.* Hephzibah, GA. Dec.

Transportation Research Board 1985. *Highway Capacity Manual.* National Research Council, Washington DC.

U.S. Bureau of Labor Statistics 2002. *Local Area Unemployment Statistics.* Washington, DC. Available at ftp://ftp.bls.gov/pub/time.series/la/.

U.S. Bureau of the Census 1992. *County Business Patterns, 1990.* Washington, DC. Available at http://www.census.gov/ftp/pub/epcd/cbp/view/cbpview.html.

U.S. Bureau of the Census 1994. *City and County Data Book, 1994.* Washington, DC.

U.S. Bureau of the Census 2000. *Census of Governments 2000.* Washington, DC. Available at http://www.census.gov/govs/www/cog.html.

U.S. Bureau of the Census 2002a. *U.S. Census American Fact Finder.* Washington, DC. Available at http://factfinder.census.gov/.

U.S. Bureau of the Census 2002b. *County Business Patterns, 2000.* Washington, DC. Available at http://www.census.gov/ftp/pub/epcd/cbp/view/cbpview.html.

U.S. Department of Agriculture 1999. *Census of Agriculture — County Data, 1997.* National Agricultural Statistics Service, Washington, DC. Available at http://www.nass.usda.gov/census/census97/volume1/vol1pubs.htm.

U.S. Department of Commerce 2002. *Regional Accounts Data — Local Area Personal Income.* Bureau of Economic Analysis, Washington, DC. Available at http://www.bea.doc.gov/bea/regional/reis/.

USFS (U.S. Forest Service) 2000. *Threatened, Endangered, and Sensitive Species Survey and Evaluation of F-Area Plutonium Disposition Mission Area at the Savannah River Site.* U.S. Forest Service, Savannah River Institute, Savannah River Site, Natural Resource Management Program, Aiken, SC. March 29.

USGS (U.S. Geological Survey) 2001. *Large Earthquakes in the United States, Charleston, South Carolina.* Available at http://wwwneic.cr.usgs.gov/neis/eqlists/USA/1886_09_01.html, accessed April 15, 2001.

Vargo, K. 2001. Personal communication from Vargo (City of Barnwell, SC) to J. Jackson (Argonne National Laboratory, Argonne, IL). July.

Weather Site, Inc. 2003. *Hurricane* Site dot Com. Available at http://www.hurricanesite.com. May 7.

Wike, L.D., and E.A. Nelson 2000. Memo from Wike and Nelson to J.S. Roberts (Westinghouse Savannah River Company, Aiken, SC). March 30.

Workman, S.W., and K.W. McLeod 1990. *Vegetation of the Savannah River Site: Major Community Types.* SRO-NERP-19. Savannah River Site, National Environmental Research Park Program, Aiken, SC. April.

WSRC (Westinghouse Savannah River Company) 1994. "SRS Ecology Environmental Information Database." CD database prepared for the U.S. Department of Energy by Westinghouse Savannah River Company under Contract No. DE-AC09-89SR18035. May 5.

WSRC 1997a. *Groundwater Mixing Zone Application for the Old F-Area Seepage Basin (U).* WSRC-RP-97-39, Rev. 1. Savannah River Site, Aiken, SC. March.

WSRC. 1997b. *SRS Ecology Environmental Document.* WSRC-TR-97-0023. Savannah River Site, Aiken, SC.

WSRC 2000a. *Natural Phenomena Hazards (NPH) Design Criteria and Other Characterization Information for the Mixed Oxide (MOX) Fuel Fabrication Facility at Savannah River Site (U).* WSRC-TR-2000-00454. Savannah River Site, Aiken, SC. Nov.

WSRC 2000b. *Savannah River Site Approved Site Treatment Plan, 2000 Annual Update.* WSRC-TR-0608, Rev. 8. Savannah River Site, Aiken, SC. March.

WSRC, 2000c, *Savannah River Site Environmental Report for 1999.* WSRC-TR-99-00299. Savannah River Site, Aiken, SC.

4 ENVIRONMENTAL CONSEQUENCES

4.1 Introduction

This final environmental impact statement (FEIS) evaluates the potential impacts of the construction, operation, and decommissioning of the Mixed Oxide Fuel Fabrication Facility (the proposed MOX facility) proposed for construction at the Savannah River Site (SRS). Operation of the proposed MOX facility would also require the construction of two support facilities, the Pit Disassembly and Conversion Facility (PDCF) and the Waste Solidification Building (WSB).

Construction of the facilities would involve site preparation, including the clearing and grading of land, realignment of electrical utilities, and addition of access roads. After site preparation, the remaining construction activities would involve excavation for the foundation and erection of the buildings, connection of SRS utilities to the facilities, and final landscaping. Details of the construction and operational impacts are provided in Sections 4.3 and Appendix H. Operational impacts would include routine facility emissions, waste management, and potential accidents. The impacts of the transportation of the MOX feed materials, the fresh MOX fuel, and spent MOX fuel are discussed collectively with the transport of transuranic (TRU) waste generated by MOX fuel production in Section 4.4.1.[1]

Once the fresh MOX fuel was manufactured and transported, it would be irradiated in authorized nuclear reactors as part of the power generation process. Following irradiation, the spent fuel would be temporarily stored at the reactor sites until shipped to a final disposal repository. The potential indirect impacts for the use of MOX fuel in a nuclear reactor are discussed in Section 4.4.3.

An initial evaluation of projected decommissioning impacts is provided in Section 4.3.6. However, the exact nature and scope of these impacts are uncertain because only present-day technologies are considered, and decommissioning of the facilities would occur well into the future.

In addition to considering the proposed action, this FEIS, in Section 4.2, considers the no-action alternative should the U.S. Nuclear Regulatory Commission (NRC) either not authorize construction of the proposed MOX facility, or not license its operation. Under the no-action alternative, the surplus plutonium would continue to be stored at its current storage locations.

As stated in Section 1.4.2, this chapter presents significant or more important environmental impacts of the proposed action and no-action alternative. Impacts considered to be less significant are presented in Appendixes G and H. The technical areas discussed in this chapter include human health, air quality, surface water and groundwater, waste management, and decommissioning. Impacts from potential accidents at the proposed MOX facility, the PDCF,

---

[1] Definitions of descriptive terms used to categorize the magnitudes of impacts are provided in Section 2.4.

and the WSB are discussed in Section 4.3.5. Environmental justice is discussed in detail in Section 4.3.7. In addition, transportation impacts are discussed in detail for the proposed action in Section 4.4.1.

Human health impacts include potential exposure to radiological and chemical materials via pathways associated with air, water, soil, and the food chain. Air quality impacts relate to compliance with National Ambient Air Quality Standards (NAAQS) from emissions of chemical pollutants. Surface and groundwater impacts relate to capacity effects from using these waters and to potential changes in quality of these waters. Waste management impacts relate to the types and quantities of both radiological, hazardous, and nonhazardous wastes generated and how those wastes would be handled. Generally technical terms used in this chapter are defined and discussed in Chapter 3. In those cases, the reader is referred back to specific areas of Chapter 3.

4.2 Impacts of the No-Action Alternative

4.2.1 Introduction

As described in Section 2.1, the no-action alternative would be a decision by the NRC not to approve the proposed MOX facility. If such a decision is made, the 34 MT (37.5 tons) of weapons-useable fissile nuclear materials would remain in storage at DOE sites. The impacts of the continued storage of surplus plutonium would be essentially the same as those discussed under the no-action alternative of the *Surplus Plutonium Disposition Final Environmental Impact Statement* (SPD EIS) (DOE 1999a, Section 4.2) and are summarized in the following sections. Some of the impacts for the no-action alternative presented in this EIS represent impacts for the entire DOE site at which the surplus plutonium is currently being stored.

It is possible that limited new construction would be required at one or more sites to upgrade surplus plutonium storage conditions. For example, previous analyses assumed that surplus pits[2] at the Pantex site in Texas would be moved from Zone 4 to Zone 12, but DOE decided to leave the surplus pits in Zone 4 for long-term storage (DOE 2002a). If new construction is required to accommodate continued storage, the impacts of that construction would be addressed under a separate environmental review required by the DOE regulations for implementation of the National Environmental Policy Act (NEPA) (*Code of Federal Regulations* Title 10, Part 1021 [10 CFR 1021]).

The SPD EIS discusses plans to build an Actinide Packaging and Storage Facility (APSF) at the SRS and to move SRS surplus plutonium to that facility for continued storage (DOE 1999a). After publication of the SPD EIS, the APSF project was canceled. Surplus plutonium at the SRS continues to be stored in existing facilities. It should also be noted that the potential impacts of construction and operation of the proposed MOX facility (as summarized in

---

[2] A pit is the core element of a nuclear weapon's "primary" or fission component.

Section 4.3) would be avoided by implementation of the continued storage alternative. The impacts of continued storage are presented in the following sections.

The DOE is currently working to close the Rocky Flats Environmental Technology Site (RFETS) by the year 2006. Such a closure entails the shipment of all radioactive waste and special nuclear materials, including the surplus plutonium, to off-site locations. Storage of the RFETS surplus plutonium at other DOE sites currently storing surplus plutonium is expected to result in a long-term reduction of radiological exposure to workers and the public. For example, approximately 6 MT (6.6 tons) of plutonium dioxide is expected to be shipped from the RFETS to the SRS (Roberson 2002). Storage of the additional plutonium material during normal operations was estimated to result in small, if any, impacts to noninvolved workers and the public (DOE 2002c). The eventual removal and return of the shipping containers was estimated to result in a dose of no greater than 1 mrem/yr to a maximally exposed individual (MEI) of the public (DOE 2002c). Thus the cumulative risks from the no-action alternative presented in Table 4.1, which includes the RFETS, are expected to bound the risks that the surplus plutonium will contribute to other DOE storage sites following shipment from the RFETS.

4.2.2 Human Health Risk

4.2.2.1 Radiological Risk

The radiological doses and risks for members of the public are shown in Table 4.1 for all ongoing activities at each of the storage sites; radiological doses and risks from maintaining the surplus plutonium are portions of the totals. The doses are less than 2% of doses associated with natural background (see Section 3.10.3 and Table 3.7 for information on background radiation).

The average annual dose to facility workers maintaining the surplus plutonium inventories at the storage sites is also shown in Table 4.1. The maximum individual worker dose for the sites (3.2 mSv/yr [320 mrem/yr] at Pantex) is 16% of the administrative limit set by DOE (DOE 1999b) and 6% of the radiological limit of 50 mSv/yr (5,000 mrem/yr) as specified in 10 CFR 835, "Occupational Radiation Protection."

4.2.2.2 Chemical Exposure and Risk

Health risks from exposure to hazardous chemicals used in ongoing operations at the storage sites within the DOE complex were estimated in the *Storage and Disposition of Weapons-Usable Fissile Materials Final Programmatic Environmental Impact Statement* (DOE 1996a, Appendix M) (these risks are also summarized in the SPD EIS [DOE 1999a]). The estimated baseline cancer risks for the storage sites include inhalation exposures to all carcinogens measured from site point emission sources. Surplus plutonium storage would account for only a small portion of the total exposures from ongoing operations at the various DOE sites. For members of the public, the estimated increased lifetime cancer risks from continued operations

Table 4.1. Radiological impacts from continued plutonium storage in current locations[a,b]

Site	Annual population dose within 80 km in 2030 [person-Sv (person-rem)]	Expected number of fatal cancers in population from 50 years of storage[c]	Annual dose to the public MEI [mSv (mrem)]	Public MEI 50-year fatal cancer risk[c]	Average worker dose [mSv/yr (mrem/yr)]
Hanford	4.7×10^{-4} (4.7×10^{-2})	1×10^{-3}	4.1×10^{-6} (4.1×10^{-4})	1×10^{-8}	2.5 (250)
INEEL	7.6×10^{-7} (7.6×10^{-5})	2×10^{-6}	1.4×10^{-7} (1.4×10^{-5})	4×10^{-10}	0.26 (26)
Pantex	6.3×10^{-8} (6.3×10^{-6})	2×10^{-7}	1.8×10^{-10} (1.8×10^{-8})	5×10^{-13}	3.2 (320)[d]
SRS	2.9×10^{-6} (2.9×10^{-4})	9×10^{-6}	6.8×10^{-8} (6.8×10^{-6})	2×10^{-10}	2.5 (250)
LLNL	6.7×10^{-5} (6.7×10^{-3})	2×10^{-4}	3.1×10^{-6} (3.1×10^{-4})	9×10^{-9}	2.5 (250)
LANL	0.027 (2.7)	8×10^{-2}	6.5×10^{-2} (6.5)	2×10^{-4}	2.5 (250)
RFETS[e]	1.0×10^{-3} (0.10)	3×10^{-3}	4.8×10^{-3} (0.48)	1×10^{-5}	2.5 (250)

[a]The population doses and cancer risks are from all ongoing activities at each site. The worker doses are for workers involved in surplus plutonium continued storage activities.

[b]MEI = maximally exposed individual, INEEL = Idaho National Engineering and Environmental Laboratory, SRS = Savannah River Site, LLNL = Lawrence Livermore National Laboratory, LANL = Los Alamos National Laboratory, RFETS = Rocky Flats Environmental Technology Site.

[c]Latent cancer fatalities are calculated by multiplying dose by the Federal Guidance Report (FGR) 13 health risk conversion factor of 0.06 fatal cancer per person-Sv (6×10^{-4} fatal cancer per person-rem) (Eckerman et al. 1999).

[d]This is the dose for workers involved in gasket replacement activities projected to occur over a period of 10 years; the dose for other storage workers at Pantex would be 1.16 mSv/yr (116 mrem/yr).

[e]Closure of the RFETS is planned for 2006. As discussed in Section 4.2.2.1, the risks presented here are expected to bound the impacts on storage of the RFETS surplus plutonium at other DOE storage sites.

Source: DOE (1999a, Section 4.2.4, based on data in DOE 1996a).

at all the storage sites were estimated to be lower than or within the risk range of 1×10^{-6} to 1×10^{-4} (the target used by the U.S. Environmental Protection Agency (EPA) to determine whether mitigation actions are needed [EPA 1990; see Section 3.10.4]). Except for Lawrence Livermore National Laboratory (LLNL), the hazard index (HI) for members of the public was also less than 1 in every case (an HI of less than 1 indicates no or small noncancer health risk; see Section 3.10.4). The general public HI for LLNL was estimated as 1.1, narrowly exceeding the noncancer health risk screening criterion. For the site employee populations, the noncancer HI values for all sites except the SRS and LLNL were less than 1; the value for the SRS was 1.2, and the value for LLNL was 2.4. Estimated cancer risks from ongoing operations for employees at several sites (i.e., Idaho National Engineering and Environmental Laboratory [INEEL], SRS, LANL, RFETS) also exceeded EPA's tolerable risk range, although none was greater than 10^{-3}.

The emissions data used as the basis for the HI values and cancer risks from all ongoing operations at the storage sites are several years old. The methods used to estimate the HI values and cancer risks are generally conservative (assuming such things as the public receptor present at the site boundary for 24 hours per day), resulting in overestimates of actual exposure. Furthermore, only a small portion of the total exposures from site emissions would be from plutonium storage activities. Therefore, although it is possible on the basis of the cited data that members of the public (for LLNL) or on-site employees (for several sites) might experience adverse health impacts as a result of exposures from ongoing plutonium storage operations, it is more likely that actual exposures would be less than those that would result in adverse health impacts.

4.2.2.3 Physical Hazards

The number of full-time employees required to maintain continued storage of the excess plutonium at the various sites was not given in the SPD EIS (DOE 1999a). Therefore, it is not possible on the basis of available information to estimate the annual number of fatalities and injuries that would be associated with continued plutonium storage under the no-action alternative.

4.2.2.4 Facility Accidents

The potential for accidental release of plutonium from storage vaults is much lower than for release from MOX fuel fabrication, which involves numerous operations. In the SPD EIS (DOE 1999a), the health risks of beyond-design-basis earthquake events on plutonium storage facilities were reported for the off-site population. Of the DOE sites evaluated, a high value of 0.4 latent cancer fatality (LCF) was reported for the 80-km (50-mi) off-site population at INEEL (see Section 3.10.3 for LCF definition). For a MEI of the public, an explosive airplane crash at Pantex was estimated to result in an LCF probability of 0.04.

There is no known use of hazardous chemicals required for the continued storage of the surplus plutonium at the various storage sites. Therefore, accidental release of hazardous chemicals during continued storage would not be expected.

4.2.3 Air Quality

The SPD EIS (DOE 1999a) summarized ambient concentrations of criteria pollutants (carbon monoxide [CO], nitrogen dioxide [NO_2], particulate matter with a diameter of 10 μm or less [PM_{10}], and sulfur dioxide [SO_2]) at each storage site from total site contributions, including plutonium storage operations. With one exception, the total site contributions were in compliance with applicable standards. At LLNL, however, the estimated maximum 1-hour ambient concentration of NO_2 was 2.5 times higher than the State of California standard. Because plutonium storage operations do not generate appreciable quantities of NO_2, continued storage of the plutonium would not change the impacts of ongoing operations on air quality at LLNL.

4.2.4 Hydrology

The annual water usage and wastewater discharges for all ongoing activities at each of the storage sites are shown in Table 4.2. Water use and wastewater generation for maintaining the surplus plutonium storage are small portions of the totals. No impacts to surface or ground-water resources from continued storage are anticipated beyond those of existing activities.

4.2.5 Waste Management

For all the storage locations, wastes generated by activities required to maintain continued storage of surplus plutonium would be a portion of the existing waste generation rates and are not anticipated to change appreciably. Continued storage should not have a major impact on waste management activities at any of the sites.

4.3 Impacts of the Proposed Action

This section presents the direct impacts of the proposed action. As discussed in Section 2.2, the proposed action is for NRC to authorize DCS to construct and later operate the proposed MOX facility at the SRS to convert 34 MT (37.5 tons) of surplus plutonium to MOX fuel. Section 4.3.1 presents the estimated impacts to human health. Sections 4.3.2 and 4.3.3 cover potential impacts to air and water, respectively. Waste management impacts (Section 4.3.4), potential accident impacts (Section 4.3.5), and environmental justice impacts (Section 4.3.7)

**Table 4.2. Annual water usage and wastewater
discharges for the sites of continued
plutonium storage**

Site	Water requirement (million L/yr)[a]	Wastewater discharge (million L/yr)
Hanford	13,511/195	246
INEEL	0/7,570	540
Pantex	0/249	141
SRS	127,000/13,247	700
LLNL	NA[b]	NA[b]
LANL	0/5,760	693
RFETS	439/0	130

[a]Surface water/groundwater.

[b]NA = not available.

Source: DOE (1996a, Section 4.2).

were also evaluated. The scope of the proposed action includes decommissioning of the proposed facilities (Section 4.3.6).

As discussed in Section 1.4.1, the technology option to substitute sand filters for the proposed high-efficiency particulate air (HEPA) filters was identified during the scoping process. Discussions of the differences in impacts between sand filters and HEPA filters are summarized in Section 4.3.8.

Construction of the proposed MOX facility is assumed to occur over a 5-year period. Construction of the WSB is assumed to occur during the same 5-year period; whereas construction of the PDCF is assumed to begin 2 years after the construction start for the other facilities (DCS 2002c).

If construction of the proposed MOX facility is authorized, DCS plans to submit an application for a 20-year license to possess and use special nuclear material to manufacture MOX fuel. The actual operation period may be 10 to 14 years, with the additional time needed for facility startup, testing, and decommissioning prior to license termination. For purposes of evaluating operational impacts, a 10-year period was assumed for processing the 34 MT (37.5 tons) of surplus plutonium. That period is based on the facility design for a maximum annual throughput of 3.5 MT (3.9 tons) of plutonium. If the actual period of operation is greater than 10 years because the actual throughput is less than the maximum facility design capacity, the annual impacts would be less, but they would occur over a longer time period.

The following sections present potential impacts on human health, air quality, hydrology, waste management, and environmental justice. A discussion of the impacts in other technical areas is presented in Appendix H.

4.3.1 Human Health Risk

4.3.1.1 Radiological Risk

4.3.1.1.1 Construction

The construction workers for the proposed MOX facility, the PDCF, and the WSB, like other workers at the SRS, would be subject to exposure to baseline radiation from other SRS activities. However, no additional radiological impacts to the construction workers, to existing SRS workers, or members of the public off-site are expected from the construction activities because no surface contamination is present.

Although radioactive contamination is present in the groundwater underlying the Old F-Area Seepage Basin and the proposed MOX facility, the primary movement of this contamination is expected to follow the direction of the groundwater flow. This direction is toward the north-northwest, where the groundwater discharges to Upper Three Runs Creek (WSRC 1995), away from the proposed facilities. Another possible source of exposure of the construction workers would be any radioactively contaminated soil in the area disturbed by construction activities. An exploration and sampling program across the project site, however, did not identify any radioactive contaminants (DCS 2000b; Fledderman 2002). As discussed in Section 5.2.8, soil would be further sampled for radioactive contamination before excavation begins at the site. If contamination was found, potential exposures and health impacts to the construction workers would be assessed.

4.3.1.1.2 Operations

Radiological impacts to human health from normal operations would result from releases to the environment and direct exposure of facility workers to sources of radiation (see description in Section 3.10). The impacts were evaluated for three receptor groups (facility workers, SRS employees, and members of the public).

All radiological impacts were assessed in terms of committed dose and associated health effects. The dose calculated was the total effective dose equivalent (TEDE) (10 CFR Part 20), which is the sum of the deep dose equivalent (DDE) from exposure to external radiation and the 50-year committed effective dose equivalent (CEDE) from exposures to internal radiation. Details of the dose calculations are provided in Appendix E. The DDE is the dose equivalent at a tissue depth of 1 cm and applies to external whole-body exposure. The CEDE is the dose equivalent to organs or tissues that is received over a 50-year period following the intake of radioactive material.

For each of the receptor groups, doses were estimated for the group as a whole (population or collective dose) and for an MEI. The MEI was defined as a hypothetical person who — because of proximity, activities, or living habits — could receive the highest possible dose. The MEI for SRS employees and members of the public usually was assumed to be at the location of the highest on-site or off-site air concentrations of contaminants, respectively — even if no individual actually worked or lived there. Under actual conditions, all radiation exposures and releases of radioactive material to the environment are required to be as low as reasonably achievable (ALARA), a practice that has as its objective the attainment of dose levels as far below applicable limits as is practical, taking into account social, technical, economic, and public policy considerations. Annual estimated radiological impacts from normal operations of the proposed MOX facility, the PDCF, and the WSB are provided in Table 4.3.

Facility Workers

MOX facility: Approximately 400 workers are expected to be employed at the MOX facility. Facility workers during normal operations were estimated to receive an annual collective dose of 0.15 person-Sv (15 person-rem). Approximately 0.12 person-Sv (12 person-rem) would be from external exposure and the remaining 0.03 person-Sv (3 person-rem) from internal exposure. The resulting health effects were calculated to be approximately 0.009 LCF/yr. On average, the facility workers' dominant exposure pathway would be external exposure.

Table 4.3. Annual estimated radiological impacts to facility workers, SRS employees, and the public from normal operations at the proposed facilities

	PDCF		MOX facility		WSB	
Receptor	Dose [person-Sv (person-rem)]	Latent cancer fatalities/yr[a]	Dose [person-Sv (person-rem)]	Latent cancer fatalities/yr[a]	Dose [person-Sv (person-rem)]	Latent cancer fatalities/yr[a]
Collective population						
Facility workers	1.97 (197)	0.1	0.15 (15)	0.009	0.50 (50)	0.03
SRS employees (13,295)[b]	0.00031 (0.031)	2×10^{-5}	0.00022 (0.022)	1×10^{-5}	—[c]	—
Public (1,042,000 persons off-site)	0.015(1.5)	0.0009	0.00073 (0.073)	4×10^{-5}	—	—

	PDCF		MOX facility		WSB	
	Dose [Sv (rem)]	LCF risk[d]	Dose [Sv (rem)]	LCF risk[d]	Dose [Sv (rem)]	LCF risk[d]
Maximally exposed individual						
Facility worker	0.020 (2.0)	0.001	0.017 (1.7)	0.001	0.020 (2.0)	0.001
SRS employee (225 m to the ENE)	5.6×10^{-7} (5.6×10^{-5})	3×10^{-8}	4.2×10^{-7} (4.2×10^{-5})	3×10^{-8}	—	—
Public (10,680 m to the N)	3.5×10^{-8} (3.5×10^{-6})	2×10^{-9}	5.1×10^{-9} (5.1×10^{-7})	3×10^{-10}	—	—

[a] Latent cancer fatalities are calculated by multiplying dose by the Federal Guidance Report (FGR) 13 health risk conversion factor of 0.06 fatal cancer per person-Sv (6×10^{-4} fatal cancer per person-rem) (Eckerman et al. 1999).

[b] *Source:* Birch (2001).

[c] Impacts from the WSB are included in the proposed MOX facility results.

[d] For annual individual exposure estimates, number represents the lifetime risk of fatality from a radiologically induced cancer.

However, the MEI dose of approximately 0.017 Sv/yr (1.7 rem/yr) with a fatal cancer risk of 1 chance in 1,000 (0.001) was estimated from inhalation exposure. The facility worker estimates were based on operational experience from a similar facility, as discussed in Appendix E.

PDCF: Average annual worker exposures are expected to remain below 0.005 Sv/yr (0.5 rem/yr), the SRS guideline. For 393 workers, an annual collective dose should not exceed 1.97 person-Sv (197 person-rem) with the potential for 0.1 LCFs/yr of operation. The maximum annual exposure to a single facility worker is expected to be maintained less than the DOE administrative limit of 0.02 Sv/yr (2 rem/yr) (DOE 1994). Such an exposure has an expected lifetime risk of developing a fatal cancer of approximately 0.001 (1 chance in 1,000).

WSB: Average annual worker exposures are expected to remain below 0.005 Sv/yr (0.5 rem/yr), the SRS guideline. For 100 workers, an annual collective dose should not exceed 0.50 person-Sv (50 person-rem) with the potential for 0.03 LCFs/yr of operation. The maximum annual exposure to a single facility worker is expected to be maintained at less than the DOE administrative limit of 0.02 Sv/yr (2 rem/yr). Such an exposure has an expected lifetime risk of developing a fatal cancer of approximately 0.001 (1 chance in 1,000).

SRS Employees

MOX facility and WSB: Normal operations were estimated to result in an annual collective SRS employee dose of 0.00022 person-Sv/yr (0.022 person-rem/yr), which corresponds to approximately 1×10^{-5} LCF/yr. The MEI dose was found to occur at a location 225 m (738 ft) east-northeast of the proposed MOX facility stack location. The MEI was estimated to receive a dose of 4.2×10^{-7} Sv/yr (4.2×10^{-5} rem/yr), which results in an annual fatal cancer risk of 3×10^{-8} (1 chance in 33 million).

PDCF: Normal operations were estimated to result in an annual collective dose of 0.00031 person-Sv (0.031 person-rem) to the SRS employee population, resulting in an estimated 2×10^{-5} LCFs/yr of operation. An MEI located 225 m (738 ft) east-northeast of the facility stack location was estimated to receive an annual dose of 5.6×10^{-7} person-Sv (5.6×10^{-5} person-rem). The resulting lifetime LCF is approximately 3×10^{-8} (1 chance in 33 million).

Members of the Public

Operation of the facilities is considered to have an insignificant impact on members of the public. Maximally exposed individuals of the public were estimated to receive exposures that are about 10,000 times less than that received from the baseline radiological exposures as discussed in Section 3.10.3.

MOX facility and WSB: For members of the public, operations were estimated to result in an annual collective population dose of 0.00073 person-Sv/yr (0.073 person-rem/yr), which is about 3.2% of the estimated dose received by the public from air emissions from the SRS for the year 2000 (0.023 person-Sv [2.3 person-rem]), as discussed in Section 3.10. The number of expected annual LCFs from operations was estimated to be 4×10^{-5}. The MEI location was determined to be at the SRS fenceline, 10,680 m (35,040 ft) north of the proposed MOX facility stack location. An MEI at this location would receive an estimated annual dose of 5.1×10^{-9} Sv/yr (5.1×10^{-7} rem/yr). This dose corresponds to an annual fatal cancer risk of 3×10^{-10} and is 1.3% of the estimated dose received by the public MEI from air emissions from the SRS for the year 2000 (4×10^{-7} Sv [4×10^{-5} rem]), as discussed in Section 3.10.

PDCF: Normal operations were estimated to result in an annual collective population dose of 0.015 person-Sv (1.5 person-rem) that corresponds to approximately 0.0009 LCFs/yr of operation. Thus, the average member of the public would receive a dose of approximately 1.4×10^{-8} Sv (1.4×10^{-6} rem), with an expected lifetime risk of developing a fatal cancer of 9×10^{-10} (1 chance in 1.1 billion). The pubic MEI was estimated to receive an individual dose of 3.5×10^{-8} Sv (3.5×10^{-6} rem) that has an expected lifetime fatal cancer risk of 2×10^{-9} (1 chance in 500 million).

4.3.1.2 Chemical Exposure and Risk

4.3.1.2.1 Construction

The potential airborne emissions of criteria pollutants (a group of air pollutants for which federal ambient standards exist) from construction of the proposed MOX facility and supporting facilities are summarized in Section 4.3.2.1. Emissions of toxic air pollutants during construction would be very low (less than 1 kg/yr (2 lb/yr) [DCS 2000a, 2002a]) and would not result in adverse health impacts. The potential ambient concentrations of criteria pollutants at or beyond the SRS boundary resulting from facility construction emissions were modeled. The estimated incremental criteria pollutant levels varied between 0.01% and 5% of the applicable ambient standard levels (see Table 4.6 in Section 4.3.2.1). Levels of criteria pollutants above the ambient standard levels would not be expected in the vicinity of SRS.

Wastewater generated during construction would be transported to the SRS Central Sanitary Wastewater Treatment Facility for treatment (DCS 2002a). No adverse impacts from human exposure to contaminants in wastewater effluents are expected from the construction of the facilities.

Hazardous wastes generated during construction would be shipped off-site to permitted commercial recycling, treatment, and disposal facilities. Exposure to hazardous materials used during construction (e.g., paints, solvents) would be kept to a minimum by following applicable OSHA regulations and precautions, such as ensuring good ventilation and cleaning up small chemical spills as soon as they occur.

If soil contamination from past site activities exists in the construction area for the proposed facilities, construction workers doing excavation work could be exposed, primarily through inhalation or incidental soil ingestion. The project site is located at the northern boundary of the main processing facility in the F-Area. Historically, the site proposed for facility construction has been used as a disposal area for excavated soil from F-Area construction projects (Wike 2000).

A recent limited investigation of possible contamination in the proposed construction area included 50 shallow soil samples (i.e., cores from 0 to 12 in.) (Fledderman 2002). Data were available for 10 metals (aluminum, beryllium, chromium, copper, gallium, iron, lead, manganese, nickel, and zinc). The concentrations in all samples were lower than the corresponding EPA Region IX health-based screening levels for industrial use properties. These results do not indicate an initial cause for concern regarding potential chemical exposures for excavation workers. However, the number of substances analyzed was low, and past operating history shows extensive contamination at SRS with such substances as trichloroethylene and arsenic, which were not analyzed in the soil samples. Also, if contamination was present at lower soil depths it would not have been detected. Therefore, if indications of possible chemical contamination (e.g., chemical odors, presence of old construction rubble) are observed during excavation activities, further soil testing to evaluate the potential for adverse health impacts to construction workers would be necessary.

4.3.1.2.2 Operations

During operations, the proposed MOX facility would use about 30 chemicals for processing, mostly for aqueous polishing to remove impurities from the plutonium (DCS 2004a; Table 3-2; DCS 2002b; 2004b); the chemicals would include dodecane, hydrazine, hydrogen peroxide, hydroxylamine nitrate, nitric acid, nitrogen, nitrogen tetroxide, and tributyl phosphate. The WSB would use three chemicals for waste processing: aluminum nitrate, nitric acid, and sodium hydroxide (DCS 2004a; Table G-2). Operation of the PDCF would require about 15 processing chemicals, including nitrogen, chlorine, sulfuric acid, phosphoric acid, and aluminum sulfate (DOE 1999a; Table E-7). At all three facilities, the chemicals would generally be stored in liquid or compressed gas form. Accidental releases of the process chemicals are discussed in Section 4.3.5.3 and Appendix E. After the chemicals were used in operations, resulting wastes would be recycled through the systems or disposed of at appropriate licensed facilities for hazardous or radioactive waste. The facilities would not discharge any process liquid directly to the environment.

Facility Workers. For normal operations, inhalation exposures and risks for facility workers (those working at the proposed MOX facility and related facilities) are difficult to estimate. This is due, in part, to the large amount of uncertainty associated with estimating airborne chemical concentrations in various rooms of the facilities. For this reason, quantitative estimates of risks to facility workers from inhalation of substances emitted during facility operations were not developed for this FEIS. However, the workplace environment would be monitored to ensure that airborne chemical concentrations were below applicable occupation exposure limits. In addition, health risks from occupational exposure through all pathways would be minimized by using enclosed operations (e.g., gloveboxes) to the extent possible.

SRS Employees and the Public. SRS employees and members of the public could be exposed to chemicals emitted to air, water, or soil from the proposed MOX facility, the PDCF, and the WSB.

In general, the chemicals involved in processing at the three facilities would be used in small amounts, have low volatilities[3], and/or have low toxicities. On the basis of information that emissions of hazardous chemicals from all three facilities to air and water would be very low (Sections 4.3.2 and 4.3.3), no hazard index or increased cancer risk estimates were made for SRS employees and the public. Adverse impacts to SRS employees and the public from exposure to air or water emissions from the facilities would not be expected. Two process chemicals from the proposed MOX facility requiring special consideration, hydrazine and uranium dioxide, are discussed below.

Hydrazine would be used in the aqueous polishing process to separate plutonium from the solvent. Hydrazine is highly reactive and corrosive; it is a carcinogen and a reproductive hazard. The maximum anticipated on-site inventory of hydrazine would be 480 L (126 gal); annual use would be 2,000 L (530 gal). In the Reagent Storage Building, hydrazine would be kept in sealed containers. Prior to use in the aqueous polishing process, the hydrazine would be blanketed with nitrogen (a process in which the nitrogen gas, which does not mix well with hydrazine, shields the liquid hydrazine from unwanted side reactions). As discussed in Section 3.10.4.2, current SRS sitewide hydrazine emissions do not result in exceedance of the ambient level specified in the South Carolina Department of Health and Environmental Control (SCDHEC) standard. During permitting of the proposed MOX facility, demonstration that operational hydrazine emissions would be limited to levels that would not cause exceedance of the SCDHEC standard would be conducted.

During the fuel fabrication process, purified plutonium dioxide powder would be mixed with depleted uranium dioxide powder. The health risk from plutonium exposure is dominated by the radiological risk, whereas the health risk from uranium exposure is dominated by the chemical risk (i.e., possible damage to the kidney). The radiological health risk from plutonium emissions during operations of the proposed MOX facility and related facilities is addressed above in Section 4.3.1.1.2.

In the proposed MOX facility, uranium powder would be processed in closed containers located in gloveboxes to confine contamination to inaccessible areas and keep occupational exposures within specified guideline and standard levels (DCS 2004a). Air exhaust from gloveboxes would be equipped with HEPA filters to collect particulate emissions. Operation of the facility would generate less than 1 g of uranium emissions annually (see Table E.1). These uranium emissions would result in small exposures and chemical health risks for SRS employees and the public.

---

[3] A chemical with a "low volatility" does not readily change from a liquid to a gas at a relatively low temperature (e.g., near room temperature).

4.3.1.3 Physical Hazards

4.3.1.3.1 Construction

As with any construction project, there would be occupational hazards to construction workers at the proposed MOX facility and related facilities. Occupational hazards were estimated by using the same method as was discussed in Section 3.10.5 for baseline physical hazards. The annual fatality and injury rates for construction activities used were as follows: 13.6 fatalities per 100,000 full-time workers and 4.2 injuries per 100 full-time workers (NSC 2001). On the basis of this methodology, the annual number of fatalities was calculated to be less than 1 for all facilities, assuming peak year employment (see Table 4.4). The estimated annual number of injuries was about 40 per year for each facility. The injuries included in these numbers are those resulting in lost workdays, not including the day of injury.

4.3.1.3.2 Operations

Occupational hazards associated with normal operations at the proposed MOX facility and related facilities were estimated by the same method discussed in Section 3.10.5; impacts are summarized in Table 4.4. Annual fatality and injury rates used were as follows: 3.3 fatalities per 100,000 full-time workers and 4.6 injuries per 100 full-time workers (NSC 2001). Annual fatality and injury rates for the manufacturing sector were used because that sector was assumed to be the most representative for operational work at the proposed facilities. The annual number of fatalities was estimated to be less than 1 for all facilities. The estimated number of injuries was 36 per year collectively for operation of the proposed MOX facility and the PDCF, and 5 per year for the WSB (includes only injuries resulting in lost workdays, not including the day of injury).

4.3.2 Air Quality

This section presents the maximum potential air quality impacts associated with construction and operation of the proposed MOX facility, the PDCF, and the WSB. Air quality impacts associated with construction and operation of the facilities were assessed by determining the concentrations of pollutants in the air caused by emissions associated with the facilities and comparing those concentrations with generally accepted measures of air quality impact, typically standards set by regulatory agencies. Two types of standards exist. Incremental standards set maximum concentrations that cannot be exceeded by emissions from sources associated with a facility or facilities. Total standards set maximum concentrations that cannot be exceeded by total emissions from both sources associated with a facility or facilities and other nearby sources, such as existing SRS sources.

Determining the air quality concentrations involves three steps. First, the emissions of the sources associated with a facility or facilities are calculated. Next, the incremental concentrations caused by these emissions are determined with an air quality model that uses emissions and meteorological data to estimate concentrations at various locations. To

Table 4.4. Annual physical hazard impacts from normal operations[a]

Facility	Peak year construction FTEs[b]	Annual operations FTEs[b]	Projected annual fatalities – construction	Projected annual fatalities – operations	Projected annual injuries – construction	Projected annual injuries – operations
MOX facility	950	400	0.13	0.013	40	18
PDCF	1,024	400	0.14	0.013	40	18
WSB	1,000	100	0.14	0.003	42	5

[a]Fatality estimates of less than 0.5 should be interpreted as "no expected fatalities." Construction of each of the facilities is projected to require 3 to 5 years. The duration of operations is estimated as 10 or more years.

[b]Full-time equivalent employees; the numbers of FTEs were obtained from DCS (2004a) for the proposed MOX facility and the WSB, and from DOE (1999a) for the PDCF.

determine a total concentration, the impacts of other sources not associated with a facility or facilities must be added to the incremental concentrations. The impacts of these other sources are determined either by additional modeling or by selecting a measured background concentration representative of the impacts of the sources not modeled. Finally, the incremental concentrations due to a facility or facilities alone or the total concentrations due to a facility or facilities and other sources are compared against appropriate measures of impact.

In this analysis, incremental impacts of construction activities and operations were determined separately using the Industrial Source Complex Short Term (ISCST3) air quality model (EPA 1995). (Appendix F provides additional detail on the calculations of emissions and the assumptions and data used in the model.) The ISCST3 model is recommended by the EPA for modeling construction activities and operations. The meteorological data used in modeling came from Athens, or Atlanta, Georgia, and Columbia, South Carolina, nearby locations where meteorological data are recorded. The maximum modeled pollutant concentrations were selected to represent the impact of construction activities or operations.

The impacts of other sources were taken into account by adding two additional concentrations to the facility maximum: an SRS maximum concentration for other sources at the SRS (SRS maxima) and a background concentration representing the overall impact of non-SRS sources. The total concentrations were then compared with the applicable ambient standard levels given in Table 3.3. Facility maxima were compared with the incremental PSD standards to provide another measure of impact.

The background concentrations are those used by the State of South Carolina to evaluate air quality impacts. The SRS environmental staff modeled the maxima in support of its air permit process (SCDHEC 2001). These SRS maxima are based on the assumption that all permitted sources operate at their fully permitted limits; thus these values are conservative estimates of SRS impacts. In addition, for a given pollutant and averaging time, maximum values associated with the proposed action and other SRS facilities are unlikely to occur at the same locations. Adding them together for comparison with the corresponding standard level adds additional conservatism to the procedure.

A slightly different procedure was used to evaluate potential impacts of $PM_{2.5}$. Implementation of the $PM_{2.5}$ standard has been delayed, and states have not developed plans for attaining it. SRS maxima and background values were not available for $PM_{2.5}$. Background values were taken as the maximum concentrations measured at background monitors within 80 km (50 mi)[4] of the SRS and were added to the modeled facility maxima for comparison with applicable standard levels. Background concentrations also were not available for air toxics and are generally considered negligible. Therefore, for air toxics, the sum of the facility maximum concentration and the SRS maxima was taken to be the total concentration for comparison with ambient standard levels.

---

[4] $PM_{2.5}$ background values were the 2001 maximum annual average and the maximum 98th percentile concentrations measured at the two rural background monitors within 80 km (50 mi) of the MOX facility. Compliance with the 24-hour $PM_{2.5}$ standard is based on the 98th percentile values being below the standard level.

4.3.2.1 Construction

The earth-moving activities during the construction period for the proposed MOX facility and the WSB will not overlap the earth-moving activity period for the PDCF. The impacts presented below assume simultaneous construction of the proposed MOX facility and the WSB and were found to exceed the impacts from construction of the PDCF. The impacts presented are, therefore, considered to be bounding for construction activities.

During construction, emissions of criteria pollutants (see Section 3.4.2), total suspended particulates (TSP), and volatile organic compounds (VOCs) would include fugitive dust emissions from earthmoving activities, fugitive dust emissions from the concrete batch plant, and exhaust emissions from diesel-powered construction equipment and from worker and delivery vehicles. The emissions associated with constructing the proposed MOX facility and the WSB are listed in Table 4.5. The tabulation does not include emissions of lead, a criteria pollutant. The phaseout of lead in gasoline has led to a significant reduction in lead levels throughout the country. Appendix F summarizes the emission factors and assumptions used in estimating construction emissions.

Fugitive dust emissions would be the emissions of principal concern during construction of the facilities. Dust from construction activities and exhaust from diesel construction equipment would be emitted within the limited area of the construction site. Other vehicles used by construction workers and for deliveries would emit exhaust along various roadways around the site, and this dispersal would reduce the impacts of these emissions relative to emissions from the limited construction area. Therefore, only fugitive dust emissions from construction activities and operation of the concrete batch plant and exhaust emissions from construction equipment were analyzed for the construction phase.

The results of the impact analysis for construction of the proposed MOX facility and the WSB, including the total concentration and its individual components (i.e., the modeled facility maximum, the SRS maximum, and the background concentration) are presented in Table 4.6. As noted above, the totals are conservative in that they overestimate the likely concentrations. Comparison of the total concentrations with applicable ambient standard levels provides a measure of the impact of construction.

Annual maxima would occur 10.7 to 9.5 km (5.9 to 6.7 mi) west northwest of the proposed MOX facility site. Short-term maxima would occur 9.5 to 10.4 km (5.9 to 6.5 mi) west or west northwest of the site except for the 1-hour CO maximum, which would occur 20.6 km (12.8 mi) to the southeast.

The total TSP concentration would be close to, but still less than, the maximum value allowed by the applicable standard. Most of this TSP concentration would be due to existing sources; the TSP concentration from facility construction would be at most only 0.06% of the standard level. Expected PM_{10} ambient levels would not exceed standard levels, and the concentrations from construction of the facilities would be equivalent to, at most, 5.0 and 0.05% of the 24-hour and annual PM_{10} standard levels, respectively.

Table 4.5. MOX facility and WSB construction emissions[a,b,c]

Pollutant	Construction fugitive dust[d]		Concrete batch plant		Construction equipment exhaust	
	Annual (kg/yr)	Hourly (g/h)	Annual (kg/yr)	Hourly (g/h)	Annual (kg/yr)	Hourly (g/h)
TSP	121,000	59,200	5,670	2,730	5,580	2,680
PM_{10}	36,900	17,800	1,640	790	5,580	2,680
$PM_{2.5}$	18,500	8,880	850	409	5,580	2,680
CO	0	0	0	0	25,600	12,300
NO_2	0	0	0	0	67,600	32,500
SO_2	0	0	0	0	6,510	3,130
VOC	0	0	0	0	6,550	3,150

[a]See Appendix F for details on emission calculations.

[b]Hourly values are based on a construction schedule of 8 hours per day, 5 days per week, 52 weeks per year.

[c]The proposed MOX facility and the WSB are assumed to be constructed at the same time. The construction of the PDCF is expected to occur outside the time frame for construction of the other two facilities.

[d]Calculations assume that water is applied to control dust, resulting in a 50% reduction in emissions, and that emissions from earth-moving activities occur over a 9-month period.

Expected $PM_{2.5}$ ambient levels would not exceed standard levels. Construction of the facilities would not exceed 4.3 and 0.070% of the 24-hour annual $PM_{2.5}$ standard levels, respectively.

The CO, SO_2, and NO_2 construction emissions would be from construction equipment exhaust. Concentrations from these emissions would amount to at most 0.29% of any ambient standard level and would not contribute to concentrations in excess of a standard level.

4.3.2.2 Operations

DCS has proposed to treat exhausts from the proposed MOX facility with (at a minimum) a two-stage HEPA filter system to remove radioactive materials before the exhaust is discharged to the atmosphere.

The introduction to Section 4.3.2 provides a short discussion of the method used to assess air quality impacts. Sections 4.3.1 and 4.3.5 discusses the human health impacts of routine and accidental chemical and radiological releases to the air. In addition to the emissions discussed in this section, the facilities also would emit the radionuclides listed in Table E.5.

Table 4.6. Maximum air quality impacts during construction of the facility

Pollutant	Averaging time	Pollutant concentration ($\mu g/m^3$)					Percent of standard		Receptor location[a]	
		Facility maximum[b]	SRS maximum[c,d]	Background[c]	Total[e]	Ambient standard[f]	Total concentration	Facility maximum	Distance [km (mi)]	Direction
TSP	Annual	0.045	46.6	28	74.6	75	99.5	0.061	10.7 (6.7)	WNW
PM$_{10}$	24 hours	7.5	97.0	41	145.5	150	97.0	5.0	10.4 (6.5)	WNW
	Annual	0.023	6.9	19	25.9	50	51.8	0.047	10.7 (6.7)	WNW
PM$_{2.5}$	24 hours	2.8	–g	27	29.8	65	45.8	4.3	10.4 (6.5)	WNW
	Annual	0.011	–g	13.6	13.6	15	90.7	0.070	10.7 (6.7)	WNW
CO	1 hour	40	262.7	10,100	10,400	40,000	26	0.10	20.6 (12.8)	SE
	8 hours	8	67.4	6,800	6,880	10,000	69	0.08	9.5 (5.9)	WNW
SO$_2$	3 hours	3.7	1,171.3	50	1,225	1,300	94	0.29	9.6 (6.0)	W
	24 hours	0.83	337.2	18	356	365	98	0.23	9.5 (5.9)	WNW
	Annual	0.006	27.1	4	31	80	39	0.008	9.5 (5.9)	WNW
NO$_2$	Annual	0.063	17.32	9	26	100	26	0.06	9.5 (5.9)	WNW

[a] Location of facility maximum from center of proposed MOX facility site.

[b] Maximum concentration due to facility construction, modeled with ISCST3 model.

[c] Based on SCDHEC (2001) and EPA (2003).

[d] The SRS maxima are based on maximum permitted emissions from SRS sources and do not necessarily quantify actual air quality impacts.

[e] Sum of facility maximum, SRS maximum, and background.

[f] South Carolina and Georgia standards are the same as NAAQS except for TSP, which is a South Carolina standard.

[g] SRS maxima and background levels are not available for PM$_{2.5}$. Values for background are the 2001 maximum annual average and maximum 98 percentile 24-hour average values measured at the two rural background monitors within 80 km (50 mi) of the MOX facility.

For purposes of this analysis, it was assumed that the proposed MOX facility, PDCF, and WSB would operate at the same time. While this may not always be the case, the combined analysis bounds the air quality impacts from normal operations.

The emissions from operation of the facilities are summarized in Table 4.7. It is expected that all these facilities would use electric boilers; there would be no emissions associated with production of hot water or steam. Air pollutants associated with the MOX process would be emitted from the stack located toward the eastern end of the proposed MOX facility. Nonradiological emissions from this stack would be limited to NO_2 from the aqueous polishing process. There would be no process emissions from the PDCF (DOE 1999a, Table G-59). Particulates from the cementation process in the WSB would be controlled to meet the condition specified in the SCDHEC permit.

Emissions from emergency and standby diesel-powered generators and storage of diesel fuel have been considered. Emergency and standby generators and associated fuel storage facilities would be located at each of the three facilities and would emit criteria pollutants, TSP, VOCs, and air toxics (see Table 4.7). The tabulated process VOCs would result from the storage of diesel fuel and would be small because of the low volatility of diesel fuel.

Air Toxics

Air toxics, also known as hazardous air pollutants, are substances judged to have adverse impacts on human health when present in the ambient air. The EPA and some states have issued lists of substances regulated as air toxics. The specific substances listed and the types of regulations applied differ among jurisdictions.

Parking lots and access roads would be paved to minimize fugitive dust emissions. Vehicle combustion emissions would be released along various roadways around the site, and this dispersal would reduce emission impacts compared with the emissions from the emergency/standby generator diesels. Only the process emissions from the facilities and diesel generators were modeled to evaluate emissions for the operations phase.

The results of the impact analysis for normal operations, including the total concentration and its individual components — the modeled facilities maxima, the SRS maximum, and the background levels — are presented in Table 4.8. As noted above, the totals are conservative in that they overestimate the likely total concentration. Impacts during normal operations were estimated by assuming that all three facilities were operating simultaneously. For short-term concentrations of 24 hours or less, emergency generators were assumed to operate 24 hours per day to simulate an extended power loss. For annual averages, the generators and process sources were modeled with emissions appropriate to their expected schedules (see Appendix F). Comparison of the total modeled concentrations with applicable ambient standard levels provides a measure of the potential impact of normal facility operations on air quality.

The total concentrations are all less than the levels stipulated in the corresponding standards, and the three facilities would contribute concentrations equivalent at most to 1.9% (for 24-hour PM_{10}) of the corresponding standard level. Given the conservative overestimation in the SRS maxima, ambient levels above the standard levels would not be expected.

Table 4.7. MOX, PDCF, and WSB operations emissions[a]

Pollutant[b]	Process		Emergency generators	
	Annual (kg/yr)	Hourly (g/h)	Annual (kg/yr)	Hourly (g/h)
TSP	6.00	463	761	4,222
PM_{10}	3.00	234	692	3,740
$PM_{2.5}$	0.90	70.2	649	3,500
SO_2			1,640	11,800
CO			3,440	25,900
NO_2	13,700	31,100	29,300	217,100
VOCs[c]	1.48	0.169	1,160	8,720
Chlorine	15.0	1.71		
Acetone	2.9	9.75		
Benzene			7.48	48.6
Toluene			2.71	17.6
Xylenes			1.86	12.1
Propylene			26.9	175
Formaldehyde			0.760	4.94
Acetaldehyde			0.243	1.58
Acrolein			0.076	0.493
Naphthalene			1.25	8.14
Total PAHs[d]			2.04	13.3

[a]See Appendix F for details on emission calculations.

[b]Except for PAHs, directly emitted criteria pollutants, their precursors, and federally listed air toxics are included. Naphthalene is both an air toxic and a component of PAH.

[c]Process emissions are from storage of diesel fuel.

[d]PAHs = polycyclic aromatic hydrocarbons.

Sources: DCS (2002a,c,d; 2004a,c); DOE (1999a).

The concentrations of toxic air pollutants and total polycyclic aromatic hydrocarbons (PAHs) associated with emissions from emergency and standby generators are all calculated to be less than 0.03% of the South Carolina standard levels.

Comparing the incremental facility concentrations with Prevention of Significant Deterioration (PSD) increments (see Table 4.9) provides another perspective on operational impacts even when a PSD analysis is not required. As the table shows, maximum concentrations for 3-hour and 24-hour averaging times would all be less than 6.0% of the PSD Class II increments

Prevention of Significant Deterioration (PSD)

The NAAQS establish maximum pollutant levels that should not be exceeded. The PSD program limits the deterioration of existing air quality in areas with air cleaner than the NAAQS. The program establishes a baseline level of air quality and specifies increments that cap the increases in pollutant levels above that baseline. The program applies to sulfur oxides, PM_{10}, and nitrogen dioxide emitted by major new or modified sources. Smaller increments apply in special areas such as national parks (Class I areas) than in other areas (Class II areas).

Table 4.8. Maximum air quality impacts during operation of the proposed facilities

Pollutant	Averaging time	Concentration (µg/m³)				Ambient standard[f]	Percent of standard		Receptor location[a]	
		Facility maximum[b]	SRS maximum[c,d]	Background[c]	Total[e]		Total concentration	Facility increment	Distance (km [mi])	Direction
TSP	Annual	0.0017	46.6	28	74.6	75	99.5	0.002	16.5 (10.2)	NE
PM_{10}	24 hours	1.31	97.0	41	139	150	93.0	0.87	9.6 (6.0)	W
	Annual	0.0015	6.9	19	25.9	50	52	0.003	16.5 (10.2)	NE
$PM_{2.5}$	24 hours	1.21	—[g]	27	28.2	65	43.4	1.9	9.5 (5.9)	WNW
	Annual	0.0014	—[g]	13.6	13.6	15	90.7	0.009	16.5 (10.2)	NE
NO_2	Annual	0.074	17.3	9	26.4	100	26	0.060	16.5 (10.3)	NE
SO_2	3 hours	22	1,171.3	50	1,243	1,300	96	1.7	9.6 (6.0)	W
	24 hours	4.9	337.2	18	360	365	99	1.3	9.5 (5.9)	WNW
	Annual	0.0035	27.1	4	31.1	80	39	0.004	16.8 (10.4)	NE
CO	1 hour	116	262.7	10,100	10,478	40,000	26	0.29	9.7 (6.0)	NW
	8 hours	26	67.4	6,800	6,890	10,000	69	0.26	9.7 (6.0)	NW
Benzene	24 hours	0.019	4.6	NA[h]	4.6	150	3.1	0.01	9.5 (5.9)	WNW
Toluene	24 hours	0.007	14.6	NA	14.6	2,000	0.7	0.0004	9.5 (5.9)	WNW
Xylene	24 hours	0.005	69	NA	69.0	4,350	1.6	0.0001	9.5 (5.9)	WNW
Propylene	24 hours	0.067	NA	NA	NA	NA	NA	NA	9.5 (5.9)	WNW
Formaldehyde	24 hours	0.002	0.15	NA	0.152	7.5	2.0	0.03	9.5 (5.9)	WNW
Acetaldehyde	24 hours	0.0006	<0.01	NA	0.011	1,800	<0.001	<0.0001	9.5 (5.9)	WNW
Acrolein	24 hours	0.0002	<0.01	NA	0.010	1.25	0.82	0.02	9.5 (5.9)	WNW
Naphthalene	24 hours	0.003	<0.01	NA	0.013	1,250	0.001	0.0002	9.5 (5.9)	WNW
Chlorine	24 hours	0.0003	0.04	NA	0.04	75	0.054	0.0004	10.8 (6.7)	N
Acetone	24 hours	0.002	NA	NA	NA	NA	NA	NA	9.8 (6.1)	W
Total PAHs	24 hours	0.005	<0.01	NA	0.015	160	<0.010	0.003	9.5 (5.9)	WNW

[a]Location of facility maximum from center of the proposed MOX facility site.

[b]Maximum concentration due to normal facility operations, modeled using ISCST3 model (EPA 1995).

[c]SCDHEC (2001) and EPA (2003) for criteria pollutants; Hunter (2001) for air toxics.

[d]The SRS maxima are based on maximum permitted emissions from SRS sources and do not necessarily quantify actual air quality impacts.

[e]Sum of facility maximum, SRS maximum, and background.

[f]South Carolina and Georgia standards are same as NAAQS for PM_{10}, $PM_{2.5}$, NO_2, SO_2, and CO. The TSP standard and the air toxic standards are South Carolina standards.

[g]SRS maxima and background levels are not available for $PM_{2.5}$ and acetone. Values for $PM_{2.5}$ background are the 2001 maximum annual average and maximum 98 percentile 24-hour average values measured at the two rural background sites within 80 km (50 mi) of the MOX facility.

[h]NA = not available.

Table 4.9. Comparison of maximum concentration increments and PSD increments[a]

Pollutant	Averaging time	Maximum increment ($\mu g/m^3$)	PSD increment ($\mu g/m^3$) Class I	Class II	Percent PSD II increment
SO_2	3 hours	22	25	512	4.30
	24 hours	4.9	5	91	5.38
	Annual	0.0035	2	20	0.02
NO_2	Annual	0.074	2.5	25	0.30
PM_{10}	24 hours	1.31	8	30	5.33
	Annual	0.0014	4	17	<0.01

[a]Class I increments apply only in Class I areas. An appropriate comparison is made in the text.

for SO_2, NO_2, and PM_{10}. These pollutants are emitted by the emergency generators, not the processes, and the concentration estimates assume all generators at all three facilities operate continuously. For annual averages, the maximum concentrations would all be less than 0.02% of the PSD Class II increments.

Class I PSD increments were compared with the concentrations expected to be experienced at the closest receptor location to the Cape Romain National Wildlife Refuge, the nearest PSD Class I area. This receptor location is 51 km (32 mi) from the site, near the maximum distance at which the ISCST3 model can reliably estimate concentrations. All concentration increments were less than 1% of the Class I increments. Concentration increments attributable to the three facilities would be even lower at Cape Romain, located about 160 km (100 mi) from the site.

Concentrations of lead and ozone were not modeled. Facility operations would not emit lead. Ozone is formed by photochemical reactions of precursors (including NO_2 and VOCs) in the atmosphere. Contributions of individual sources to ozone formation cannot be quantified accurately. As shown in Tables 3.1 and 4.7, ozone precursor emissions from facility operations would be a small percentage of the four-county totals, about 0.3% and 0.02% for NO_2 and VOCs, respectively. The impact of facility operations on ozone concentrations in the area would be negligible.

Under the Clean Air Act (CAA), federal actions in nonattainment and maintenance areas must demonstrate that they conform to the applicable state implementation plan (SIP). The SRS is located in an attainment area for all NAAQS and is not covered by a maintenance plan. Thus, the requirement to demonstrate conformity with the SIP would not apply to the proposed MOX facility, PDCF, and WSB. At some time in the future, EPA will issue conformity regulations for the new NAAQS for ozone and $PM_{2.5}$. Those regulations could impose requirements to demonstrate conformity with the SIP on the proposed MOX facility, PDCF, or WSB.

4.3.3 Hydrology

4.3.3.1 Surface Water

4.3.3.1.1 Construction

The estimated annual average water use for constructing the proposed MOX facility is 125 million L (33 million gal) (DCS 2002a). An additional 12 million L/yr (3.2 million gal/yr) of water would be needed for constructing the PDCF (DOE 1999a), and 2 million L/yr (0.5 million gal/yr) of water would be needed for constructing the WSB. Because surface water would not be used for supplying this water, there would be no impacts to surface water levels or flows. No direct releases of contaminated effluent are planned for construction operations. Sanitary waste would be collected with a combination of portable toilets and semipermanent facilities connected to the SRS Central Sanitary Waste Treatment Facility. All wastewater would be treated in the sitewide treatment system, which has sufficient hydraulic and organic capacity to treat the flows expected from construction activities (DCS 2002a).

During construction, surface water quality could, however, be impacted by contaminated runoff from sources such as accidental oil or diesel fuel spills and sediment from disturbed areas and from construction materials stockpiled in areas that are exposed to precipitation. Two areas of concern identified in the Scoping Comments (see Appendix I) are Upper Three Runs Creek, which would receive runoff water from the affected area via nearby unnamed tributaries, and the Savannah River, which receives water from Upper Three Runs Creek. To comply with South Carolina standards for storm-water management and sediment reduction, detention ponds would be built at strategic locations as part of the SRS construction program. These detention ponds would be designed to control the release of storm-water runoff at a rate equal to or slightly less than that of the predevelopment stage. Good engineering practices, as required by the SCDHEC (see Chapter 6), such as the use of siltation fences or straw bales to control sediment and runoff, would be followed during construction, and a sediment control plan would be developed for areas exceeding 2 ha (5 acres) that are disturbed by construction (DCS 2002a). Therefore, impacts to surface water quality from construction activities are expected to be small. Similarly, impacts from accidental releases of contaminants such as gasoline, oil, diesel fuel, or paint during construction are expected to produce small impacts on surface water quality because cleanup activities would be prompt and thorough, as required in the facility's Spill Prevention Control and Countermeasures Plan. This plan would be developed by DCS to meet EPA regulations (40 CFR Part 112).

4.3.3.1.2 Operations

Normal operations of the proposed MOX facility would utilize 9.1 million L (2.4 million gal) of water per year (DCS 2002a). An additional 48 million L/yr (12.7 million gal/yr) of water would be needed for operating the PDCF, and 19 million L/yr (5 million gal/yr) of water would be needed for operating the WSB, but none of this water would be from surface water resources.

Therefore, there would be no impacts to surface water levels or flows. The nonhazardous wastewater produced by the proposed facilities would be discharged to an existing National Pollutant Discharge Elimination System (NPDES) outfall (H16) in the F-Area under an existing South Carolina Discharge permit, SC0000175. This water flows into Upper Three Runs Creek and ultimately the Savannah River. Because the concentrations of nonhazardous wastes in the discharge would be under the guidelines of the NPDES permit, impacts to water quality in Upper Three Runs Creek and the Savannah River would be small. The uncontaminated heating, ventilation, and air conditioning (HVAC) condensate would be discharged to the stormwater system in accordance with SCDHEC standard stormwater permit conditions. Sanitary wastewater would be sent to the WSRC Central Sanitary Waste Treatment Facility.

Storm-water runoff from the proposed MOX facility, the PDCF, and the WSB would be controlled under existing NPDES storm-water permits. These permits would limit potential contaminants to safe concentrations, and compliance with the permit conditions would ensure that any surface water impacts were small.

4.3.3.2 Groundwater

4.3.3.2.1 Construction

During construction, the groundwater system beneath the SRS would be directly affected by additional pumping from existing wells because groundwater would be the only source of water used for construction activities. Groundwater for constructing the MOX facilities would be obtained from the A-Area loop, which obtains groundwater from wells in the F- and A-Areas. The capacity of the A-Area loop wells in 2000 was about 11,360 L/min (3,000 gal/min) (DCS 2003a). Water use from the loop, including F-Area use, averaged about 2,850 L/min (754 gal/min) in 2000. Construction of the MOX facility, PDCF, and WSB would require about 264 L/min (70 gal/min). This additional groundwater demand would represent an increase of about 10% for the A-Area loop and about 3% of the excess loop capacity. This withdrawal would have a small impact on the groundwater system at SRS.

In addition to impacts from groundwater use, impacts during construction (e.g., grading and excavating) could also occur because groundwater beneath the proposed MOX facility site is contaminated (Section 3.3.2). Impacts from this contamination would not be measurable because the deepest construction activities would occur at least 9.1 m (30 ft) above the zone of groundwater contamination (DCS 2002a). Because direct releases of contaminated effluent to groundwater during construction are not planned, there would be no direct impacts to groundwater quality. Groundwater quality, however, could still be indirectly affected by accidental releases of contaminated effluents and infiltration of contaminated runoff. However, these impacts are expected to be small because appropriate good engineering practices would be implemented during construction, detention basins would be used to control runoff, and any spills would be promptly and thoroughly cleaned up as required under the facility Stormwater Pollution Prevention Plan.

4.3.3.2.2 Operations

During normal operations, groundwater would be the only source of water used for the facilities, and the groundwater system beneath the SRS would be directly impacted by additional pumping that would deplete the resource. Operation of the proposed MOX facility would require 9.1 million L/yr (2.4 million gal/yr), the PDCF would require 48 million L/yr (12.7 million gal/yr), and the WSB would require 19 million L/yr (5 million gal/yr) (DCS 2002a). This water would be obtained from the A-Area loop groundwater wells. Impacts on the SRS groundwater system would be small because the total water use, approximately 145 L/min (38 gal/min), would represent an increase of about 5% of the water demand for the A-Area loop in 2000 and about 2% of the excess A-Area loop capacity.

Groundwater quality would not be affected because there would be no discharges (either shallow or deep) to underlying aquifers. During the scoping process, several commenters expressed concerns about potential contamination of groundwater resources by plutonium. Because no direct releases of contaminated effluent to the groundwater are planned during normal operations of the proposed facilities and because the facilities would not use settling or holding basins as part of the wastewater treatment system, there would be no direct impacts to groundwater quality (DCS 2002a).

Indirect impacts to groundwater could also occur during normal operations. These impacts would result from discharges to the NPDES outfall and surface spills. The impacts of such spills are expected to be small because appropriate good engineering practices would be implemented during the operational period, discharges would comply with NPDES guidelines, and any spills would be promptly and thoroughly cleaned up as required under the facility Spill Prevention Control and Countermeasures Plan.

4.3.4 Waste Management

This section presents the waste management impacts associated with the construction and operation of the proposed MOX facility, the PDCF, and the WSB. Waste management impacts relate to the types and quantities of radioactive, hazardous, and nonhazardous wastes generated and how these wastes are handled. Wastes generated by the three facilities would be managed similarly to wastes generated by other SRS facilities. The NRC conducted an evaluation to determine if existing and proposed facilities and capacities at SRS and within the DOE complex (e.g., the Waste Isolation Pilot Plant [WIPP]) would be adequate for handling and disposing of the generated waste. Because the types of wastes generated by the proposed MOX facility, the PDCF, and the WSB would be similar to the types of wastes already generated by existing SRS facilities and the volumes would be relatively small compared to the overall existing or projected volumes, the human health impacts discussed in Section 3.10 for current activities at SRS are expected to bound the human health impacts, if any, resulting from the waste generated by the proposed action. Also, the human health impacts discussed in Section 3.10 are not anticipated to change significantly as a result of the waste generated from the proposed action.

The WSB would process waste from both the proposed MOX facility and the PDCF. The waste volumes presented in the tables in this section are based on where the particular waste type is generated (e.g., solid TRU waste generated at the WSB as a result of processing the liquid high-alpha-activity waste transferred from the proposed MOX facility is presented as TRU waste volume for the WSB). The waste types that would be generated include TRU waste, liquid and solid LLW, hazardous/mixed waste, and liquid and solid nonhazardous waste.

4.3.4.1 Construction

The construction of the proposed MOX facility and the WSB is expected to take 5 years; the construction of the PDCF is expected to take 3 years. Waste generated from construction activities would be similar to that from construction of any industrial building and would include liquid and solid waste (nonhazardous) and hazardous wastes. Such solid wastes would be managed consistently with SRS waste management practices (see Section 3.9). No high-level (radioactive) (HLW) waste, TRU waste, low-level (radioactive) (LLW) waste, or mixed LLW would be expected to be generated during construction. No hazardous or radiologically contaminated soil is expected to be generated (DCS 2002a).

Hazardous wastes that would be generated would be similar to those expected during the construction of any industrial facility. Examples of these wastes include liquids (such as motor oil), batteries, and other machinery-related products, cleaning products, and other chemicals (such as insecticides and pesticides). These wastes would be managed in accordance with the hazardous waste management practices in place at the SRS. The current practice includes accumulating the waste at the generating facility (which in this case would be in the F-Area) for a maximum of 90 days as necessary, and packaging such wastes in U.S. Department of Transportation (DOT)-approved containers to ship off-site to permitted commercial recycling, treatment, or disposal facilities.

As shown in Table 4.10, the following waste types and estimated volumes would be generated during construction of the three facilities:

- For the proposed MOX facility: 77 m³/yr (100 yd³/yr) of hazardous wastes; 36 million L/yr (9.5 million gal/yr) of nonhazardous liquid waste and 8,410 m³/yr (11,000 yd³/yr) of nonhazardous solid waste;

- For the PDCF: 50 m³/yr (65 yd³/yr) of hazardous waste, 5.3 million L/yr (1.4 million gal/yr) of nonhazardous liquid waste and 120 m³/yr (157 yd³/yr) of nonhazardous solid waste; and

- For the WSB: 35 m³/yr (46 yd³/yr) of hazardous waste, 21 million L/yr (6.3 million gal/yr) of nonhazardous liquid waste and 2,200 m³/yr (2,880 yd³/yr) of nonhazardous solid waste.

Table 4.10. Annual waste volumes from the construction of the facilities compared with waste management capacities at the SRS

Waste type	Estimated MOX facility construction waste[b]	Estimated PDCF construction waste[c]	Estimated WSB construction waste[b]	SRS capacity[a]		
				Characterization or treatment (annual capacity)	Storage (total capacity in m³)	Disposal (total capacity in m³ unless specified)
TRU (m³/yr)	—[d]	—[d]	—[d]	1,720	34,400	168,500[e]
LLW						
Liquid (L/yr)	—[d]	—[d]	—[d]	17,830,000	NA[f]	594,000,000 L
Solid (m³/yr)	—[d]	—[d]	—[d]	17,830	NA	30,500
Hazardous[g] (m³/yr)	77	50	35	17,830	5,170	NA
Nonhazardous						
Liquid[h] (L/yr)	36,000,000	5,300,000	21,000,000	1,033,000,000	NA	NA
Solid (m³/yr)	8,410	120	2,200	NA	NA	24,900,000[i]

[a] Storage and disposal capacity estimates presented represent total capacity at the SRS. Sources of estimates: DOE (1999a).

[b] The construction period for the proposed MOX facility is assumed to be 5 years; the construction period of the PDCF is assumed to be 3 years. The construction period of the WSB is assumed to be 5 years. Source of estimates: DCS (2003a).

[c] Source of estimates: DOE (1999a).

[d] No radioactive waste would be generated by facility construction.

[e] Value represents limit for TRU waste at the WIPP.

[f] NA = not applicable.

Footnotes continued on next page.

Table 4.10. Continued

[g]Hazardous waste that would be generated is less than 4% of the treatment and about 3% of the storage capacity at the SRS. For estimating impact on the storage capacity, the annual generation rates for the three facilities were summed and the total divided by the storage capacity at the SRS. Hazardous wastes are generally not stored on-site for more than 90 days, consistent with permit requirements. Hazardous wastes are sent off-site for disposal.

[h]Nonhazardous liquid waste generated during construction of the facilities is equivalent to about 6% of the treatment capacity at SRS.

[i]The disposal capacity presented for nonhazardous solid waste is for a privatized landfill (Three Rivers Landfill) that is located on site. The combined volume of nonhazardous solid waste that would be generated from the construction of the three facilities constitutes less than 1% of the disposal capacity at the landfill.

The impact of the facilities construction waste on SRS waste management capacities would be small. The hazardous waste that would be generated would be shipped off-site to permitted facilities. The impacts at these permitted facilities from the proposed MOX facility, PDCF, and WSB wastes are expected to be within the bounds of the evaluations performed for the waste facilities. The nonhazardous liquid waste generated by the facilities would constitute a small percentage of the SRS's capacity for treatment (about 6%). Nonhazardous solid wastes are packaged in conformance with standard industrial practice and shipped to commercial or municipal facilities for recycling or disposal. Estimates for waste volumes that would be generated during construction of the facilities are presented in Table 4.10.

4.3.4.2 Operations

This section describes the waste management impacts of operating the proposed MOX facility, the PDCF, and the WSB. A discussion of radioactive effluents and wastes for each facility is provided in Sections 2.2.2.3, 2.2.3.3, and 2.2.4.3. The WSB would process some waste streams from the proposed MOX facility and PDCF. Other wastes would be managed by existing SRS waste management facilities. This section is divided into two parts. The first part describes where the waste is generated at each facility. A more detailed description of the processes that generate waste is provided in Chapter 2. The second part describes how those wastes would be handled and describes the potential waste management impacts. Consistent with waste management practices at the SRS, all wastes generated from operations of the facilities would be transferred to the WSB or to the appropriate facilities or areas elsewhere within the SRS or outside of the SRS for subsequent treatment, storage, shipment off site, or disposal. The period of operation for the proposed MOX facility is expected to be about 10 years.

Wastes that would be generated and the impacts from such wastes were identified as concerns during scoping. The waste types that would be generated from the three facilities include the following: solid TRU waste, liquid and solid LLW, hazardous/mixed waste, and nonhazardous liquid and solid waste. The estimated waste generation rates from the operation of each of the facilities are discussed in Sections 4.3.4.2.1 and 4.3.4.2.2 and are summarized in Table 4.11. Overall, the operation of the facilities would have a small impact on the SRS waste management system. The DOE has concluded (DOE 2003) that impacts are bounded by its SPD EIS (DOE 1999a).

4.3.4.2.1 Operating Facility Description

MOX Facility. The proposed fabrication of MOX fuel consists primarily of two steps: the aqueous polishing process and the fuel fabrication process. These two processes generate several types of waste that are discussed below. The aqueous polishing step removes impurities from the plutonium. The fuel fabrication process involves the blending of the purified plutonium with the depleted uranium dioxide to form pellets. The pellets would be incorporated

Table 4.11. Waste volumes from the 10-year operational period of the facilities compared with waste management capacities at the SRS

Waste type	Estimated MOX facility operational waste[b]	Estimated PDCF operational waste[b]	Estimated WSB operational waste[b]	SRS capacity[a]		
				Characterization or treatment (annual capacity)	Storage (total capacity)	Disposal (total capacity)
TRU (m³)[c]	2,340	180	1,911	1,720	34,400	168,500[d]
LLW[e]						
Liquid (L)	10,800,000	416,000	11,570,000	17,830,000	NA[f]	594,000,000
Solid (m³)	1,760	184	4,108	17,830	NA	30,500
Hazardous[g] (m³)	110	10	0	17,830	5,170	NA
Nonhazardous[h]						
Liquid (L)	333,000,000	250,000,000	19,000,000	1,033,000,000	NA	NA
Solid (m³)	13,400	18,000	10,000	NA	NA	NA

[a]Storage and disposal capacity estimates presented represent total capacity at the SRS. Sources of estimates: DOE (1999a).

[b]The facilities are assumed to be in operation for a 10-year period. Sources for estimates: MOX facility (DCS 2004a); PDCF (DOE 1999a and DCS 2004a); WSB (DCS 2004a and DOE 1999a). The volumes presented for WSB TRU and solid LLW represent that generated from processing the high-alpha liquid and stripped uranium waste streams from the MOX facility. Liquid LLW volume was obtained from DCS 2004a (presented on page 5-23 of DCS 2004a as 890 m³ annually and multiplied by 13 years of approximate operation for the WSB). Hazardous and nonhazardous waste volumes obtained by subtracting PDCF volumes presented in this table from values presented in DCS 2004a Table 5-15c.

[c]The combined values of TRU waste that would be generated from the three facilities is estimated to be approximately 26% and 13% of the treatment and storage capacity, respectively, at the SRS. The generated TRU waste is approximately 2.6% of the disposal capacity at WIPP.

[d]Value represents limit for TRU waste at the WIPP.

Footnotes continued on next page.

Table 4.11. Continued

[e]The volume reported for PDCF (in DOE 1999a) is 60 m^3/yr (or 600 m^3/10 yrs), but the liquid versus solid amounts were not specified. The volume of 41,600 L/yr or 416,000 L (416 m^3) over 10 years, as reported in DCS 2004a, was subtracted from the volume reported in DOE 1999a to obtain the volume for solid LLW (i.e., 600 m^3 – 416 m^3 = 184 m^3). The liquid LLW generated by the three facilities constitutes 4% of the discharge capacity at SRS. The solid LLW generated constitutes about 21% of the disposal capacity at SRS (if disposed of entirely at the SRS). Disposal of solid LLW will either be at the SRS or at another approved facility.

[f]NA = Not applicable.

[g]Hazardous waste that would be generated is less than 1% of the treatment and less than 2% of the storage capacity at the SRS.

[h]The nonhazardous liquid waste generated constitutes about 6% of the treatment capacity at SRS.

into the fuel rods, which would then be placed in fuel assemblies. Figure 4.1 depicts the waste streams and volumes generated and the final disposition for each.

The aqueous polishing process would generate approximately 33,300 L/yr (8,800 gal/yr) of liquid high-alpha waste, 174,000 L/yr (46,000 gal/yr) of stripped uranium waste, 1,078,000 L/yr (285,000 gal/yr) of chloride removal waste, and 10,600 L/yr (2,800 gal/yr) of excess solvent waste. The liquid high-alpha waste consists of three waste streams (liquid americium waste stream, excess acid waste stream, alkaline wash waste stream). The liquid high-alpha waste and the stripped uranium waste stream would be sent to the WSB via separate pipelines for further treatment. Because the liquid high alpha waste and stripped uranium waste would be processed at the WSB, the final waste volumes following processing are included in the discussion of the WSB. The chloride removal waste would be collected in tanks and transferred to the WSB. The excess solvent waste would be sent to SRS facilities or to a commercial facility for treatment and disposal as a contaminated solvent waste.

The fuel fabrication process and maintenance activities would generate approximately 1,340 m^3/yr (1,750 yd^3/yr) of solid nonhazardous waste, 176 m^3/yr (230 yd^3/yr) of solid LLW, and 234 m^3/yr (306 yd^3/yr) of solid TRU waste. The solid non-hazardous waste consists of sanitary waste (e.g., garbage, machine shop waste, and other industrial waste) and non-sanitary waste (e.g, paper, metal cans, plastic and glass bottles).

The MOX facility would also generate approximately 33.3 million L/yr (8.8 million gal/yr) of nonhazardous liquid waste. This waste includes uncontaminated HVAC condensate, rinse water, and sanitary waste from sinks, showers, urinals, and water closets from the inactive area. The uncontaminated HVAC condensate (94,600 L/yr [25,000 gal/yr]) would be discharged to the stormwater system. The remaining nonhazardous liquid waste would be sent to SRS for processing at the CSWTF.

PDCF. The PDCF would be used to recover the plutonium metal from the pits of disassembled weapons and would convert the weapons-grade plutonium to plutonium dioxide powder. The PDCF would accommodate the following surplus plutonium-processing activities: pit receipt, storage, and preparation; pit disassembly; plutonium conversion; oxide blending and sampling; nondestructive assay; product canning; product storage; product inspection and sampling for international inspection; product shipping; declassification of parts not made from special nuclear material (SNM); highly enriched uranium (HEU) decontamination, packaging, storage, and shipping; tritium capture, packaging, and storage; and waste packaging, sampling and certification.

Aside from the 41,600 L/yr (11,000 gal/yr) of laboratory radioactive liquid waste that would be transferred to the WSB for further processing, the operations at the PDCF would also generate about 18 m^3/yr (24 yd^3/yr) of solid TRU waste. TRU waste generated during operations would include spent filters, contaminated beryllium pieces and cuttings, used containers and equipment, paper and cloth wipes, analytical and quality control samples, and solidified inorganic solutions. Liquid TRU wastes would be evaporated or solidified before being packaged for storage. About 60 m^3/yr (78 yd^3/yr) of LLW (assumed to be all solid) would also be generated. LLW generated during operations would originate from activities in the

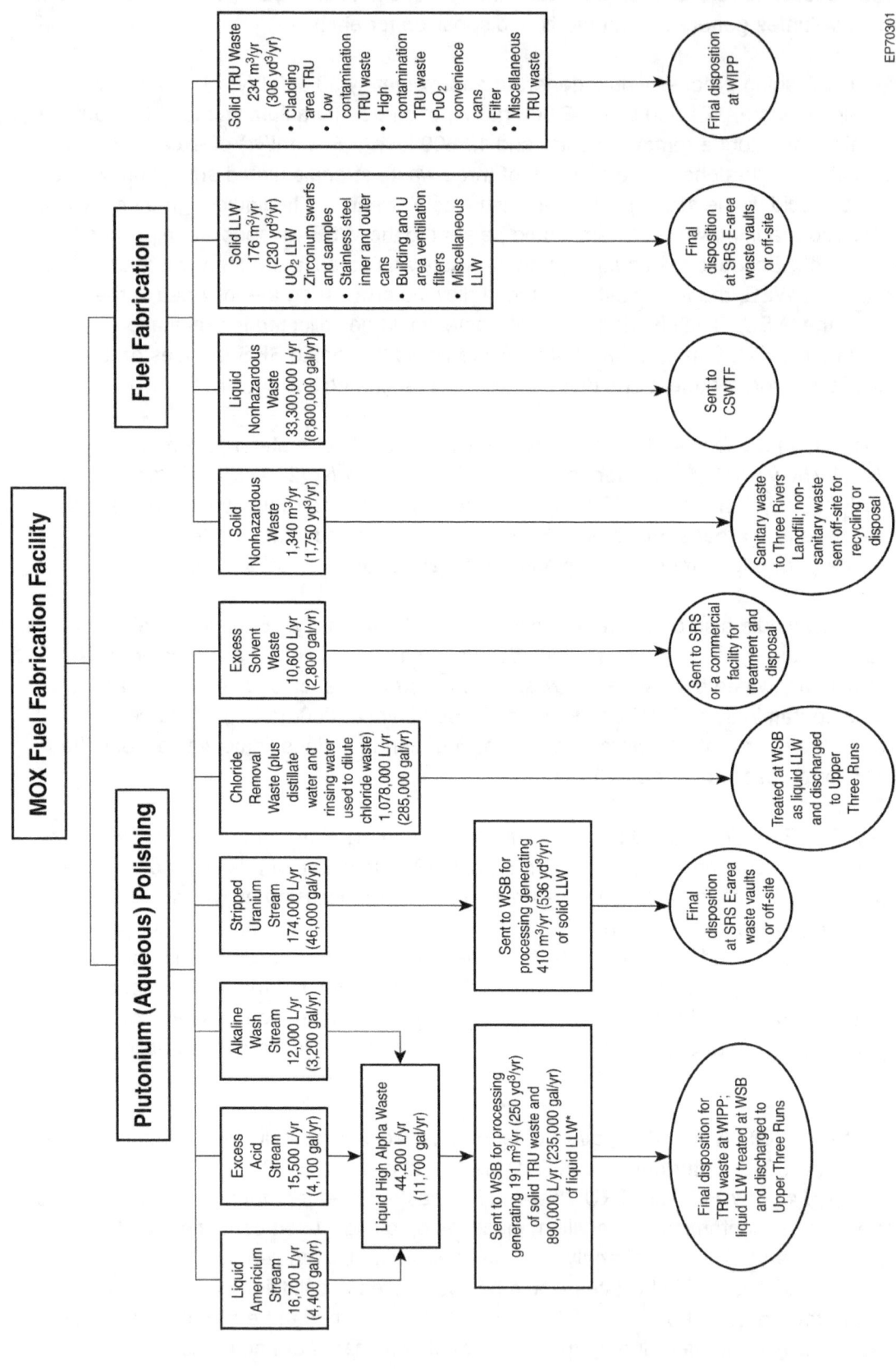

Figure 4.1. Waste streams generated by the proposed MOX facility (*Source:* Modified from DCS 2003a; 2004a).

processing areas. LLW would include equipment, wipes, protective clothing, solidified inorganic solutions, and tritium. Liquid LLW would be evaporated or solidified before being packaged for accumulation. About 1 m³/yr (1.3 yd³/yr) of hazardous/mixed waste generated during operations would include spent cleaning solutions, vacuum pump oils, film processing fluids, hydraulic fluids, antifreeze solutions, paints, chemicals, lead packaging, and contaminated rags or wipes. Hazardous waste would be packaged for treatment and disposal at off-site permitted commercial facilities.

Two types of nonhazardous waste would be generated; 25 million L/yr (6.6 million gal/yr) liquid waste and 1,800 m³/yr (2,350 yd³/yr) of solid waste. Nonhazardous solid waste would include office garbage, machine shop waste, and other industrial wastes from utility and maintenance operations. Recyclable solid waste would be sent off the site for recycling. Nonhazardous liquid waste would include sanitary waste from sinks, showers, urinals, and water closets and process wastewater from lab sinks and drains, mop water, and cooling tower blowdown.

Waste Solidification Building. The WSB would process three waste streams from the proposed MOX facility (i.e., liquid high-alpha waste, stripped uranium waste, and liquid LLW) and two waste streams from the PDCF (i.e., PDCF laboratory liquid stream and liquid LLW). The WSB would be expected to generate about 191 m³/yr (250 yd³/yr) of solid TRU waste from the processing of the liquid high-alpha-activity waste resulting from the aqueous polishing step conducted at the proposed MOX facility. About 890,000 L/yr (235,000 gal/yr) of liquid LLW would also be generated from the processing of the liquid high-alpha-activity waste and the stripped uranium waste from the aqueous polishing step, and the laboratory liquid waste from the PDCF. The waste streams would be batch-transferred as a separate waste to the WSB through separate double-walled stainless steel pipelines. The wastes would be collected in the waste receipt area of the WSB. This area would be equipped with separate collection tanks for each waste type, with capacities to hold waste volumes generated for a period of 6-8 weeks at a time.

Following receipt at the WSB, the high-alpha-activity waste would be reduced in volume by evaporation, and the still bottoms would be neutralized with sodium hydroxide. The distillate would be subjected to further treatment at the WSB and discharged to a permitted outfall. The neutralized bottoms would be blended with cement to produce a solid TRU waste matrix suitable for disposal at WIPP. The high-activity waste overheads (materials that evaporate and are collected) would be transferred to the low-activity waste head tank for a second evaporator process.

The stripped uranium waste and the PDCF laboratory liquids would also be evaporated at the WSB to reduce the volume. As noted above, the high-activity waste overheads would be further evaporated in the low-activity waste evaporator. The process is similar to what would be used for the liquid high-alpha waste. About 410 m³/yr (536 yd³/yr) of solid LLW is expected to be generated at the WSB from processing the stripped uranium waste transferred from the proposed MOX facility (DCS 2004c).

4.3.4.2.2 Waste Management Impacts from Operation

This section describes how the TRU, liquid and solid LLW, mixed LLW, hazardous, and nonhazardous wastes would be managed. It also describes the potential waste management impacts for a 10-year period. As discussed above, approximately, 4,431 m³ (5,796 yd³) of TRU waste would be generated each 10-year period during the operation of the three facilities. The DOE has a national program for the management and disposal of defense-related TRU waste. Subsequently, waste acceptance criteria (WAC) for receipt of TRU waste at WIPP have been established for contact-handled TRU (CH-TRU) waste. The TRU wastes generated from the proposed MOX facility, the PDCF, and the WSB are expected to be in this category. The WAC that must be met for CH-TRU waste to be transported to, managed at, and disposed of at WIPP address container properties, radiological properties, physical properties, chemical properties, and data package contents. The generator facilities are required to transmit characterization, certification, and shipping data to WIPP before shipping waste.

The liquid LLW generated (22,786,000 L [6.0 million gal]/10 yr) from the three facilities would be transferred to the WSB for treatment and then discharged to the Upper Three Runs Creek consistent with permit discharge limitations. The liquid LLW from the three facilities would be about 4% of the discharge capacities at SRS. Solid LLW generated (6,052 m³ [7,916 yd³]/10 yr) would be packaged, certified, and accumulated at the F-Area before transfer to the appropriate facilities for treatment and disposal (at the SRS E-Area waste vaults or at an approved off-site facility). The solid LLW from the three facilities would constitute about 21% of the disposal capacity at SRS (if disposed of entirely at SRS).

Hazardous wastes (120 m³ [157 yd³]/10 yr) generated from the three facilities would either be transferred to the SRS for treatment and storage at either on-site or off-site facilities and disposal at off-site, permitted facilities or shipped off site for treatment and disposition at permitted facilities. If the treatment and disposal are assumed to be on-site, the expected wastes volumes from the facilities would represent less than 2% of the capacities at the SRS. Therefore, the facilities' waste should not affect the SRS hazardous waste management system.

Nonhazardous solid waste (41,400 m³ [54,149 yd³]/10 yr) generated from the three facilities would be packaged and transported in accordance with standard industrial practices. Recyclable waste would be sent off-site, with the remaining waste (primarily solid sanitary waste) sent to the Three Rivers Landfill for disposal. The nonsanitary waste would be sent off-site for recycling or disposal.

Nonhazardous liquid wastes (602,000,000 L [159 million gal]/10 yr) from the three facilities would be treated before being discharged to the F-Area sanitary sewer system, which connects to the SRS Central Sanitary Wastewater Treatment Facility. The wastes of this type expected to be generated by operations of the facilities are estimated to be about 4% of the capacity of the Central Sanitary Wastewater Treatment Facility. These additional wastes would constitute a small contribution and should not affect the nonhazardous liquid waste management system at the SRS.

Although the current plans call for treating all liquid LLW generated at the proposed MOX facility, the PDCF, and the WSB at the WSB and discharging the treated effluents to a permitted outfall on the SRS site following the NPDES permit guidelines, it is possible that at some future date liquid LLW streams generated at these facilities may be sent to the Effluent Treatment Facility (ETF) on the SRS. If that should happen, the waste management impacts discussed in this EIS would still be comparable to or would bound the impacts that would occur during the management of wastes resulting from the operation of the three facilities, namely the proposed MOX facility, the PDCF, and the WSB.

4.3.5 Accident Impacts

This section discusses hypothetical accidents that could occur at the proposed facilities (the MOX facility, the PDCF, and the WSB), and the estimated maximum impacts that such accidents could produce. Table 4.12 lists the various accidents considered, and Tables 4.13, 4.14, and 4.15 list the estimated radiological impacts on SRS employees, the collective off-site public, and the maximally exposed member of the public, respectively. The potential impacts of accidental chemical releases from the proposed facilities are discussed in Section 4.3.5.3. This section describes the potential accident impacts in more detail and includes a discussion of impacts on local groundwater quality that could result from accidental releases.

4.3.5.1 Accidents Considered

4.3.5.1.1 Proposed MOX Facility

To obtain a possession and use license, DCS is required under 10 CFR Part 70, Subpart H, to perform an integrated safety analysis (ISA) to identify the hazards of the proposed MOX facility in a systematic and comprehensive manner. As an initial part of that process, DCS has completed a safety assessment that identified the following types of events that could lead to releases to the environment — natural phenomena, loss of confinement, internal fire, explosion, load handling events, external man-made events, criticality, direct radiation exposure, and chemical releases (DCS 2002a).

With respect to natural phenomena, DCS has shown that flooding does not pose a credible threat to the proposed MOX facility. For the remainder of the credible natural phenomena events, which include extreme winds, earthquakes, tornadoes, external fires, rain, snow, ice, and lightning, the applicant has committed to design criteria and standards that would prevent accidents associated with these hazards. For this reason, the effects of accidents caused by these phenomena are not described in this EIS.

External man-made events were also considered in DCS's hazard evaluation. These events include hazards from nearby facilities or vehicles. These hazards may include industrial facilities, military facilities, chemical facilities, nearby SRS facilities, pipelines, automobiles, and aircraft. A screening evaluation by DCS determined that credible external man-made events

Table 4.12. Accidents evaluated for the proposed facilities

Facility/ accident	Description
Proposed MOX facility	
Internal fire	A fire was postulated to occur in a storage location for polished plutonium dioxide powder (the PuO_2 Final Dosing Unit). The frequency of this event is considered to be unlikely or lower because multiple failures are required for this event to occur.
Explosion	A hypothetical explosion event was postulated to occur in an aqueous polishing process cell and involved the maximum material at risk in any process cell. Simultaneous failure of the design features and administrative controls resulting in an explosion and the subsequent release of radioactive materials is highly unlikely.
Load handling	The load-handling event postulated to produce the largest radiological consequences was a drop event involving the glovebox in the Jar Storage and Handling Unit. This glovebox would contain jars of plutonium powder. The frequency associated with this event is estimated to be unlikely or lower since multiple failures would be required for this event to occur.
Criticality	A criticality hazard arises whenever fissionable materials (e.g., uranium-235 or plutonium-239) are present in sufficient quantities to attain a self-sustaining fission chain reaction under optimal conditions. Thus, a generic hypothetical criticality event was evaluated.
Chemical releases	Chemical releases were modeled by assuming that the largest container for each chemical in storage at the facility was punctured. Chemical-specific characteristics were used to determine the amount of material released.
PDCF	
Fire	The bounding fire accident was assumed to occur in a plutonium glovebox. Against procedure, a flammable cleaning liquid was assumed to be taken into the glovebox used for blending plutonium powder. The liquid is inadvertently spilled and ignited, involving all of the gloves.
Explosion	Multiple equipment failures and operator errors were postulated to result in the ignition of a hydrogen and oxygen gas mixture in an inert-atmosphere glovebox. The resulting explosive pressure was assumed to damage the glovebox windows but would be insufficient to compromise the building HEPA filtration system.
Leak/spill	A forklift or other heavy vehicle running over a package of plutonium dioxide was postulated as the most catastrophic leak or spill. A portion of the released oxide becomes airborne and is filtered by the HEPA filtration system before entering the environment.

Table 4.12. Continued

Facility/ accident	Description
Criticality	A criticality involving plutonium dioxide powder was postulated because the PDCF handles amounts in excess of that required for such an accident. However, facility design and procedures are intended to preclude such an occurrence. No specific scenario was identified other than multiple failures due to human error.
Earthquake	During an earthquake event, the PDCF was expected to maintain its structural integrity, and the major safety systems, including building confinement and HEPA filtration, were assumed to continue to function. It was conservatively assumed that loose plutonium powder in gloveboxes would be resuspended and result in some minor spills.
Tritium release	Tritium contamination of parts in a glovebox was assumed to be released during a major glovebox fire. The formation of tritiated water vapor is postulated to occur, and the resulting vapor is released through the building ventilation system.
Chemical releases	Chemical releases were modeled by assuming that the largest container for each chemical in storage at the facility was punctured. Chemical-specific characteristics were used to determine the amount of material released.
WSB	
Loss of confinement	A facility-wide spill of all material in the low-activity process area was considered due to natural phenomena or an external event. The high-activity waste in this area is in hardened structures that are designed to withstand such an event (DCS 2003b).
Fire	The bounding fire accident was postulated to be an area fire in the low-activity processing section of the WSB. As a result of structural damage to the facility, thousands of gallons of unprocessed low-activity waste, low-activity bottoms, low-activity overheads, effluent bottoms, and effluent overheads are released.
Earthquake	An earthquake event was assumed to cause a spill of all material in the low-activity process area. A fire was then assumed to occur throughout the entire facility except for within the hardened structure that contains the high-activity cells. The potential impacts are taken to be the sum of the loss of confinement and fire events evaluated for the WSB.
Chemical releases	Chemical releases were modeled by assuming that the largest container for each chemical in storage at the facility was punctured. Chemical-specific characteristics were used to determine the amount of material released.

Table 4.13. Estimated human health radiological impacts to SRS employees from hypothetical facility accidents

Facility/accident	SRS employee MEI			SRS employee population		
	Dose [Sv (rem)]	Likelihood of LCF[a]	Major exposure pathway	Dose [person-Sv (person-rem)]	Fatalities (LCFs)[a]	Major exposure pathway
Pit Disassembly and Conversion Facility						
Criticality	0.00070 (0.070)	4×10^{-5}	External	0.062 (6.2)	0.004	External
Earthquake	4.0×10^{-5} (0.0040)	2×10^{-6}	Inhalation	0.023 (2.3)	0.001	Inhalation
Explosion	0.00033 (0.033)	2×10^{-5}	Inhalation	0.19 (19)	0.01	Inhalation
Fire	1.2×10^{-6} (0.00012)	7×10^{-8}	Inhalation	0.00071 (0.071)	4×10^{-5}	Inhalation
Leak/spill	4.0×10^{-7} (4.0×10^{-5})	2×10^{-8}	Inhalation	0.00023 (0.023)	1×10^{-5}	Inhalation
Tritium release	0.026 (2.6)	0.002	Inhalation	18 (1,800)	1	Inhalation
Proposed MOX Facility						
Criticality	0.023 (2.3)	0.001	External	3.0 (300)	0.2	External
Explosion	0.0068 (0.68)	0.0004	Inhalation	3.9 (390)	0.2	Inhalation
Internal fire	0.00025 (0.025)	2×10^{-5}	Inhalation	0.15 (15)	0.009	Inhalation
Load handling	0.0010 (0.10)	6×10^{-5}	Inhalation	0.60 (60)	0.04	Inhalation
Waste Solidification Building						
Loss of confinement	0.00030 (0.030)	2×10^{-5}	Inhalation	0.16 (16)	0.01	Inhalation
Fire	0.0058 (0.58)	0.0003	Inhalation	3.2 (320)	0.2	Inhalation
Earthquake	0.0061 (0.61)	0.0004	Inhalation	3.4 (340)	0.2	Inhalation

[a] Latent cancer fatalities are calculated by multiplying dose by the FGR 13 health risk conversion factor of 0.06 fatal cancer per person-Sv (6×10^{-4} fatal cancer per person-rem) (Eckerman et al. 1999). Values are rounded to one significant figure.

Table 4.14. Estimated human health radiological impacts to the collective off-site public from hypothetical facility accidents

Facility/accident	Dose [person-Sv (person-rem)]	Fatalities (LCFs)[a]	Major exposure pathway
Short-Term Exposure			
Pit Disassembly and Conversion Facility			
Criticality	0.048 (4.8)	0.003	External
Earthquake	0.054 (5.4)	0.003	Inhalation
Explosion	0.44 (44)	0.03	Inhalation
Fire	0.0017 (0.17)	0.0001	Inhalation
Leak/spill	0.00053 (0.053)	3×10^{-5}	Inhalation
Tritium release	42 (4,200)	3	Inhalation
Proposed MOX Facility			
Criticality	1.3 (130)	0.08	Inhalation
Explosion	9.1 (910)	0.5	Inhalation
Internal fire	0.35 (35)	0.02	Inhalation
Load handling	1.4 (140)	0.08	Inhalation
Waste Solidification Building			
Loss of confinement	0.38 (38)	0.02	Inhalation
Fire	7.3 (730)	0.4	Inhalation
Earthquake	7.7 (770)	0.5	Inhalation
1-Year Exposure without Ingestion			
Pit Disassembly and Conversion Facility			
Criticality	0.052 (5.2)	0.003	External
Earthquake	0.054 (5.4)	0.003	Inhalation
Explosion	0.44 (44)	0.03	Inhalation
Fire	0.0017 (0.17)	0.0001	Inhalation
Leak/spill	0.00053 (0.053)	3×10^{-5}	Inhalation
Tritium release	42 (4,200)	3	Inhalation
Proposed MOX Facility			
Criticality	1.5 (150)	0.09	Inhalation
Explosion	9.1 (910)	0.5	Inhalation
Internal fire	0.35 (35)	0.02	Inhalation
Load handling	1.4 (140)	0.08	Inhalation
Waste Solidification Building			
Loss of confinement	0.38 (38)	0.02	Inhalation
Fire	7.3 (730)	0.4	Inhalation
Earthquake	7.7 (770)	0.5	Inhalation

Table 4.14. Continued

Facility/accident	Dose [person-Sv (person-rem)]	Fatalities (LCFs)[a]	Major exposure pathway
1-Year Exposure with Ingestion			
Pit Disassembly and Conversion Facility			
Criticality	0.13 (13)	0.008	Ingestion
Earthquake	0.16 (16)	0.01	Ingestion
Explosion	1.3 (130)	0.08	Ingestion
Fire	0.0049 (0.49)	0.0003	Ingestion
Leak/spill	0.0016 (0.16)	0.0001	Ingestion
Tritium release	1,800 (180,000)	100	Ingestion
Proposed MOX Facility			
Criticality	9.6 (960)	0.6	Ingestion
Explosion	27 (2,700)	2	Ingestion
Internal fire	1.1 (110)	0.07	Ingestion
Load handling	4.1 (410)	0.2	Ingestion
Waste Solidification Building			
Loss of confinement	0.65 (65)	0.04	Ingestion
Fire	13 (1,300)	0.8	Ingestion
Earthquake	14 (1,400)	0.8	Ingestion

[a]Latent cancer fatalities are calculated by multiplying dose by the FGR 13 health risk conversion factor of 0.06 fatal cancer per person-Sv (6×10^{-4} fatal cancer per person-rem) (Eckerman et al. 1999). Values are rounded to one significant figure.

will not significantly impact facility operations (DCS 2002a). For this reason, the effects of accidents caused by such events are not described in this FEIS.

Direct radiation hazards generally arise from radioactive material or other sources that emit penetrating gamma or neutron radiation. The radioactive material that would be used in the proposed MOX facility produces mostly alpha radiation, which is not as penetrating and is a less significant direct radiation hazard, but could cause adverse health effects when inhaled. As a result, there would be no accidents at the proposed MOX facility that would produce a direct radiation hazard to the public. In addition, other than a criticality event, there would be no accidents that would produce a direct radiation exposure hazard for an SRS employee.

The events for which accident consequences were evaluated in this FEIS are internal fire, explosion, load handling event, criticality, and chemical releases. The methods employed to analyze accident consequences were based on conservative assumptions and were intended to provide a comprehensive, bounding analysis for all potential events up to and including design basis accidents.

**Table 4.15. Estimated human health radiological impacts
to the maximally exposed member of the public
from hypothetical facility accidents**

Facility/accident	Dose Dose [mSv (mrem)]	Likelihood of LCF[a]	Major exposure pathway
Short-Term Exposure			
Pit Disassembly and Conversion Facility			
Criticality	0.0038 (0.38)	2×10^{-7}	External
Earthquake	0.0011 (0.11)	7×10^{-8}	Inhalation
Explosion	0.0094 (0.94)	6×10^{-7}	Inhalation
Fire	3.5×10^{-5} (0.0035)	2×10^{-9}	Inhalation
Leak/spill	1.2×10^{-5} (0.0012)	7×10^{-10}	Inhalation
Tritium release	0.90 (90)	5×10^{-5}	Inhalation
Proposed MOX Facility			
Criticality	0.098 (9.8)	6×10^{-6}	External
Explosion	0.2 (20)	1×10^{-5}	Inhalation
Internal fire	0.0077 (0.77)	5×10^{-7}	Inhalation
Load handling	0.030 (3.0)	2×10^{-6}	Inhalation
Waste Solidification Building			
Loss of confinement	0.0081 (0.81)	5×10^{-7}	Inhalation
Fire	0.16 (16)	1×10^{-5}	Inhalation
Earthquake	0.17 (17)	1×10^{-5}	Inhalation
1-Year Exposure without Ingestion			
Pit Disassembly and Conversion Facility			
Criticality	0.0042 (0.42)	3×10^{-7}	External
Earthquake	0.0011 (0.11)	7×10^{-8}	Inhalation
Explosion	0.0094 (0.94)	6×10^{-7}	Inhalation
Fire	3.5×10^{-5} (0.0035)	2×10^{-9}	Inhalation
Leak/spill	1.2×10^{-5} (0.0012)	7×10^{-10}	Inhalation
Tritium release	0.90 (90)	5×10^{-5}	Inhalation
Proposed MOX Facility			
Criticality	0.11 (11)	7×10^{-6}	External
Explosion	0.2 (20)	1×10^{-5}	Inhalation
Internal fire	0.0077 (0.77)	5×10^{-7}	Inhalation
Load handling	0.030 (3.0)	2×10^{-6}	Inhalation
Waste Solidification Building			
Loss of confinement	0.0081 (0.81)	5×10^{-7}	Inhalation
Fire	0.16 (16)	1×10^{-5}	Inhalation
Earthquake	0.17 (17)	1×10^{-5}	Inhalation

Table 4.15. Continued

Facility/accident	Dose Dose [mSv (mrem)]	Likelihood of LCF[a]	Major exposure pathway
1-Year Exposure with Ingestion			
Pit Disassembly and Conversion Facility			
Criticality	0.012 (1.2)	7×10^{-7}	Ingestion
Earthquake	0.0016 (0.16)	1×10^{-7}	Inhalation
Explosion	0.013 (1.3)	8×10^{-7}	Inhalation
Fire	4.9×10^{-5} (0.0049)	3×10^{-9}	Inhalation
Leak/spill	1.3×10^{-5} (0.0013)	8×10^{-10}	Inhalation
Tritium release	39 (3,900)	0.002	Ingestion
Proposed MOX Facility			
Criticality	0.6 (60)	4×10^{-5}	Ingestion
Explosion	0.23 (23)	1×10^{-5}	Inhalation
Internal fire	0.012 (1.2)	7×10^{-7}	Inhalation
Load handling	0.045 (4.5)	3×10^{-6}	Inhalation
Waste Solidification Building			
Loss of confinement	0.010 (1.0)	6×10^{-7}	Inhalation
Fire	0.20 (20)	1×10^{-5}	Inhalation
Earthquake	0.21 (21)	1×10^{-5}	Inhalation

[a]Latent cancer fatalities are calculated by multiplying dose by the FGR 13 health risk conversion factor of 0.06 fatal cancer per person-Sv (6×10^{-4} fatal cancer per person-rem) (Eckerman et al. 1999). Values are rounded to one significant figure.

Radiological release accidents were classified into likelihood categories on the basis of qualitative estimates (DCS 2001, 2002a). The likelihood categories were defined as follows:

- Not Unlikely – Event may occur during the facility's lifetime.

- Unlikely – Event is not expected to occur during the facility's lifetime, but may be considered credible.

- Highly Unlikely – Event originally classified as "not unlikely" or "unlikely" to which sufficient controls have been applied to further reduce its likelihood to an acceptable level.

DCS did not classify the likelihood of chemical release accidents. An assessment was conducted that assumed the largest container for each chemical in storage was punctured, although safety precautions are exercised to avoid such occurrences.

A short description of each event evaluated for the accident risk assessment is given in Table 4.12. Additional details of the assessment methodology are provided in Appendix E.

4.3.5.1.2 Pit Disassembly and Conversion Facility

A wide range of accident scenarios was considered previously for the PDCF (DOE 1999a). Potential accidents from both man-made and natural phenomena were considered. The potential accidents evaluated for this FEIS were taken from DOE (1999a) and are listed in Table 4.12.

4.3.5.1.3 Waste Solidification Building

A procedure similar to those used for the proposed MOX facility and the PDCF was used to identify potential accidents at the WSB. Those accidents considered to be credible were evaluated (DCS 2003b). A description of the accidents is presented in Table 4.12.

4.3.5.2 Radiological Human Health Risk

For exposures to depleted uranium, the health impacts would be expected to be dominated by the chemical toxicity of the compounds rather than by their radiological effects (see Section 4.3.5.3). A lethal exposure from the chemical toxicity of uranium (resulting from kidney failure), would occur with an internal radiation dose of about 0.01 Sv (1 rem) (over a lifetime), a dose that is not considered to have any significant radiation health effects.

Receptors: Radiation doses and health risk effects were calculated for SRS employees and the public. General definitions of these receptor groups are given in Section 3.10.2.

For radiological hazards, the dose consequences to facility workers and SRS employees following an accident would generally be dominated by the 50-year committed effective dose equivalent from radioactive material inhaled immediately following the event. For the purposes of analyses in this FEIS, this period of inhalation is assumed to last 8 hours. This exposure pathway would dominate the dose (except in the case of criticality accidents) because it is assumed that direct exposure to contaminated areas following an accident can be effectively limited. In addition, no food is grown on the SRS, so the consumption of contaminated food is not included in the dose for facility workers or SRS employees. Criticality accidents involve radionuclides, other than uranium or plutonium, that pose a higher direct radiation hazard than do inhalation or ingestion.

Unlike SRS employees, members of the public could reasonably be expected to be exposed to both contaminated soil and food for some time beyond the early phase of an accident if no protective action is taken. Initial food contamination occurs through the direct deposition of airborne radioactive material onto crops. A lower level of contamination occurs through crop root uptake of radioactive material from contaminated soil. Thus, the largest ingestion exposure would occur if crops were ready for harvest immediately following an accidental release. Many stakeholders want to know what could happen if no interdiction of crops occurred. Whether an individual would be exposed to contaminated soil and food would depend on the specific protective actions that the applicant and government agencies might take following an accident. The NRC recognizes that some interdiction would likely occur following a significant accident, even if contamination levels were below the protective action guides. Therefore, three separate sets of impacts to members of the public were assessed for accidents. The first set of impacts is for the early phase (short-term period) of an accident similar to the exposure pathways evaluated for the SRS employees. The second and third sets of impacts are for the intermediate/long-term period (1 year) following an accident. The second set presents the impacts without the ingestion pathway (if interdiction occurred). The third set presents the ingestion pathway included in the impacts (if interdiction did not occur) with crops assumed to be ready for harvest immediately following an accidental release (a bounding analysis). Thus, a range of impacts to the public are presented to provide perspective on the potential exposures associated with the consumption of contaminated crops for the 1-year exposure period.

Population doses were calculated for up to a distance of 80 km (50 mi) from the release point for 10 downwind distances and 16 wind directions. Radiation doses were calculated for the following receptors for accident conditions:

- *SRS employee MEI:* For the purposes of the accident consequence assessment, an employee on the SRS at the point of maximum air concentration located close to, but outside, the facility's protected area fence (at least 100 m [330 ft] or more from the accident location). Exposure pathways assessed were inhalation exposure and direct radiation from the passing cloud of airborne radioactive material (cloudshine) released by the accident. A period of 8 hours of direct radiation exposure from deposited radioactive material on the ground (groundshine) following the accident was also considered.

- *SRS employee population:* All employees on the site located more than 100 m (330 ft) from the accident location outside the facility. The same exposure pathways as evaluated for the SRS employee MEI were evaluated for the collective SRS employee population.

- *Off-site MEI:* A hypothetical individual member of the public living off-site and receiving the maximum exposure from accidental releases. For the purposes of the accident consequence assessment, this individual was assumed to be located at the SRS boundary. A short-term exposure period, involving the same exposure pathways assessed for the SRS employees, and a 1-year exposure period were evaluated. The 1-year exposure evaluation included the short-term exposures, but it also included a 1-year exposure, not 8 hours, to groundshine and a 1-year ingestion exposure to contaminated food grown locally. Contaminated crops were not assumed to be condemned; all locally grown food was assumed to have been consumed.

- *General population:* All members of the public within an 80-km (50-mi) radius of the site where the accident might occur. Short-term and 1-year impacts to the general population were assessed on the basis of the same exposure pathways as for the public, or off-site, MEI.

During an accident, facility workers might be subject to severe physical and thermal (fire) forces and could be exposed to releases of chemicals and radiation. The risk to the facility workers would be very sensitive to the specific circumstances of each accident and would depend on how rapidly the accident developed, the exact location and response of the workers, the direction and amount of the release, the physical and thermal forces causing or caused by the accident, meteorological conditions, and characteristics of the room or building if the accident occurred indoors. Quantitative facility worker accident impacts are not provided in this FEIS. For most events, the applicant has conservatively assumed that consequences to the facility worker MEI would exceed the applicable performance requirements in 10 CFR 70.61 and has identified preventive or mitigative features in the facility's design basis in order to meet the performance requirements. However, it is recognized that worker injuries and fatalities would be possible from chemical, radiological, thermal, and physical forces if an accident did occur.

Impacts: Estimated radiological impacts from the four hypothetical accident scenarios considered are presented in Tables 4.13, 4.14, and 4.15 and are discussed below. While the consequences of many of these accidents are significant, the likelihood of significant accidents will be very low (highly unlikely) through the use of safety systems discussed in DCS's Construction Authorization Request. Thus, the overall risk of significant accidents is considered to be low.

SRS employee population: SRS employees were assumed to be unshielded from the passing plume of airborne radioactivity released during an accident. The impacts for the collective SRS employee population given in Table 4.13 were estimated for inhalation and external radiation exposure. External radiation exposure consisted of cloudshine and groundshine. Groundshine exposure was evaluated for 8 hours following an accident and was negligible, less than approximately 0.02% of the total dose, in all cases. The impacts presented in Table 4.13 are the highest potential impacts to the SRS employee population and were found to occur in the direction of the major F-Area facilities, toward the south-southwest. The dominant exposure pathway was inhalation for all accidents except for the hypothetical criticality events. For those hypothetical criticality events, exposure to cloudshine was estimated to account for approximately 70% of the collective dose; the remaining dose was estimated to result from inhalation.

The SRS employee MEI was estimated to receive a maximum dose, 0.026 Sv (2.6 rem), from the tritium release at the PDCF. This dose was from the inhalation pathway. For this dose, the likelihood of developing a latent fatal cancer was estimated to be 0.002 (about 1 chance in 500). SRS employee MEI impacts for all accidents considered are presented in Table 4.13.

Members of the public: As discussed above, impacts to the public were assessed for a short-term period immediately following the accident and for a 1-year exposure period following the accident that includes the short-term exposures. With the exception of nuclear criticality accident events, inhalation was the dominant exposure pathway for the public in the short term and 1-year exposure without ingestion. Maximum inhalation doses would occur to the west-northwest of the SRS and would be more than 100 million times any external exposure. For the 1-year exposure to the public with ingestion, the ingestion pathway was the dominant exposure pathway. The highest potential 1-year ingestion dose would be to the southwest of the SRS.

Inhalation would account for the remainder of the dose except in the case of the criticality accidents where external exposure and inhalation make up the balance of the dose. Further details of the accident risk analysis are given in Appendix E.

The tritium release accident at the proposed PDCF was estimated to result in the largest short-term exposure. An estimated collective dose of 42 person-Sv (4,200 person-rem) was projected to be received by a population of approximately 309,900 persons extending out to 80 km (50 mi) to the west-northwest of the proposed MOX facility. The average individual dose was projected to be approximately 0.14 mSv (14 mrem), about 4% of the value an individual would receive on an annual basis from existing natural and man-made sources in the SRS vicinity. However, persons living closer to the accident location would receive a higher dose on average as discussed below for the hypothetical public MEI. The collective population dose received from this accident is estimated to have a risk of an additional 3 LCFs in the affected population.

The tritium release accident at the PDCF also produced the largest 1-year collective population doses. For the case without ingestion, the results were the same as discussed above for the short-term impacts because inhalation of the passing airborne emissions was the dominant exposure pathway. For the case with ingestion, the largest impact was calculated for winds blowing toward the southwest, where 18,010 people reside. The estimated collective population dose was 1,800 person-Sv (180,000 person-rem). This dose corresponds to a human health effect of up to 100 LCFs. However, for the purposes of this EIS, all contaminated food that would be grown in an affected area is assumed to be eaten. Because the amount of contaminated food exceeds the amount that would be consumed by persons living within the affected area, it is further assumed that some of the affected food would be shipped out of the region and consumed by persons living outside the region. Excluding ingestion, the dose received by the people residing in the southwest sector was 1.7 person-Sv (170 person-rem). The remainder of the dose was attributed to the ingestion of all contaminated crops in the southwest sector. Therefore, the collective dose of 1,800 person-Sv includes doses to persons both within the affected area and outside the region. As shown in Table 4.15, the public MEI was estimated to receive a dose of 0.039 Sv (3.9 rem) for this hypothetical accident, on the basis of individual consumption rates in Appendix E. Assuming that all 18,010 persons received the MEI dose, which would be an overestimate of the dose, the corresponding collective population dose would be about 40% of the total collective dose estimated above for the case including ingestion. Therefore, the people living within the affected area would receive less than 40% of the collective dose estimated.

The potential 100 LCFs among members of the public estimated from the PDCF tritium release accident is intended to be an upper bound for such an accident when the ingestion of contaminated food is considered. The GENII code used for the accident analysis provides impacts for the four seasons of the year (winter, spring, summer, and autumn), which correspond to various phases of crop growth. Ingestion impacts increase from winter (from radionuclide deposition on soil only) through autumn (from radionuclide deposition on plants immediately prior to harvest). As discussed earlier in this section, when impacts were estimated, crops were assumed to be ready for harvest (autumn) at the time of an accidental release. This assumption was made to place an upper bound on any expected impacts resulting from the ingestion of contaminated food. In addition, ingestion pathway impacts estimated with GENII typically display a steady increase upon progressing from winter through spring, summer, and autumn, resulting from an increase in direct deposition on crops due to increased crop growth. However, in the case of tritium contamination, an ingestion dose of 0 person-Sv was estimated for winter, spring, or summer, and an ingestion dose of 1,800 person-Sv (180,000 person-rem) was estimated for autumn.

GENII incorporates a tritium-specific model that recognizes that tritium, in the form of water vapor, is an integral part of the environment and human metabolism and exchanges readily with other water in the environment. As modeled, the deposited tritium has a chance to dissipate in the environment prior to crop harvest (i.e., winter, spring, and summer impacts), but if deposited immediately prior to harvest (autumn impacts), the tritium is assumed to remain in the crops. Thus, the 100 LCFs calculated from the collective population dose of 1,800 person-Sv (180,000 person-rem) from the PDCF tritium release accident is a high upper-bound estimate because further dissipation of the tritium after crop harvest would be likely to occur before ingestion.

Impacts were assessed for an MEI living at the SRS boundary for short-term, 1-year without ingestion, and 1-year with ingestion exposures. In all three cases, maximum impacts were found to occur to a hypothetical individual located 9,070 m (5.6 mi) northwest of the facilities as a result of the PDCF tritium release accident. As shown in Table 4.15, the highest estimated dose to the public MEI was 0.90 mSv (90 mrem) in the short term from inhalation exposure. The potential maximum 1-year exposure without ingestion accident impact was estimated to be the same as the short-term exposure impact because both are dominated by inhalation exposure to the passing airborne contaminant plume immediately following an accidental release. If ingestion of contaminated crops is considered, a total exposure of 39 mSv (3,900 mrem) was estimated for the MEI. The resulting health effects were estimated to be a chance of contracting a latent fatal cancer over their lifetime of 5×10^{-5} (1 chance in 20,000) and 0.002 (about 1 chance in 500) as a result of the short-term or 1-year without ingestion exposures and the 1-year with ingestion exposure, respectively.

No mitigative actions were considered in the above analysis for the 1-year MEI exposure with ingestion. However, current Food and Drug Administration (FDA) recommendations (FDA 1998) include a protective action guide (PAG) of 5 mSv (500 mrem) CEDE and 50 mSv (5,000 mrem) committed dose equivalent to an individual tissue or organ, whichever is more limiting. These intervention levels of dose are radiation doses at which protective actions should be considered. The maximum public MEI ingestion dose of 39 mSv (3,900 mrem) would exceed the FDA PAG of 5 mSv (500 mrem) CEDE.

The impacts presented here are intended to provide a comprehensive bounding analysis for all potential events up to and including design basis accidents as discussed in Section 4.3.5.1. While non-credible "worst-case" accidents were not evaluated, a number of conservative assumptions were used to ensure that potential future impacts are bounded. Should an accident occur, potential nearby receptors would be the most vulnerable immediately after the event because they might not be aware of the accident and might not receive notification in time to take protective actions. However, those individuals farther from an accident would be more likely to receive notification in time and would be in a position to reduce doses by taking protective actions. The consequences reported here provide a range of impacts including the assumption that no protective actions are taken. Protective actions include sheltering or evacuation in the short-term and the banning of locally grown food in the long-term. Further, the 1-year results with ingestion presented here are based on the assumption that an accident occurs immediately before harvest. This is a bounding assumption because the direct deposition of radioactivity on crops would cause the highest ingestion exposures. However, long-term exposure without ingestion was also included for perspective. In addition, this analysis assumes that individuals are not sheltered during the accident and passing of the radioactive plume. Thus, the estimated accident impacts presented in this EIS are considered to bound future possible outcomes.

The radiological risks of accidents described in this FEIS are considered to be low because either the likelihood of these accidents would be significantly diminished, or sufficient controls would be applied to ensure the dose consequences are much lower than those presented here. The requirements to reduce the risk of accidents that could result in high consequences are contained in the NRC's regulations in 10 CFR Part 70, "Domestic Licensing of Special Nuclear Material," and the DOE's 10 CFR Part 830 "Nuclear Safety Management." In order to obtain a license to possess and use special nuclear material from the NRC, for example, the applicant must show that the risk of each credible high-consequence event is limited through the use of engineered controls, administrative controls, or both. Pursuant to this and other performance requirements, mitigation measures identified in Chapter 5 of this FEIS include those controls identified by the applicant to reduce the risks of potential accidents.

4.3.5.3 Chemical Human Health Risk

An analysis of potential impacts from accidental chemical releases was conducted. The analysis considered maximum inventories of stored chemicals at the proposed facilities and each chemical's physical characteristics (e.g., volatility) and its toxic concentration levels. Liquid storage containers with the largest chemical inventories were assumed to be punctured (e.g., by a forklift), resulting in a spill of the entire chemical contents of the container on an outdoor concrete surface. In general, it was assumed that the spill would occur onto an impervious surface from which evaporation could occur, rather than onto a soil surface where absorption would limit evaporation. (Two chemical releases were modeled as pressurized releases; see below.) Evaporation from the chemical pool was assumed to be of limited duration, not more than an hour, because of rapid mitigative response. The Areal Locations of Hazardous Atmospheres (ALOHA, Version 5.2.3) model (Reynolds 1992) was used with the aid of a liquid pool evaporation algorithm to assess the downwind consequences of such bounding-case spills. An assessment of the accidental release of uranium dioxide powder was also included.

For each release, potential impacts to two populations were evaluated — the off-site general public and SRS employees. For the SRS employee evaluation, a wind speed of 2.2 m/s (4.9 mph), F atmospheric stability class, and a temperature of 25.8°C (78.5°F), was determined to represent the site-specific 95th percentile concentration. This was established on the basis of the ARCON96 model chi/Q value (ratio of concentration to emissions) estimated at a distance of 100 m (330 ft) from the release. For the off-site general public evaluation, the bounding conditions were determined to be a wind speed of 1.3 m/s (3.0 mph), F atmospheric stability class, and a temperature of 25.8°C (78.5°F), representing site-specific, 95th percentile nighttime bounding meteorology. The 95th percentile meteorology was assumed to be a reasonable approximation of conditions that would produce the 95th percentile concentration consistent with the ARCON96 estimate at 100 m (330 ft). Details on the modeling assumptions are provided in Appendix E.

The criteria levels used to assess potential exposures were temporary emergency exposure limits (TEELs) adopted by the DOE Subcommittee on Consequence Assessment and Protective Action (SCAPA) (Craig 2002). TEEL values are available for about 2000 substances; they are derived by using a hierarchy of other available criteria values (Craig et al. 2000). If Emergency Response Planning Guidelines (ERPGs) developed by panels of toxicologists for the American Conference of Governmental Industrial Hygienists (ACGIH) are available, these are used for the TEEL values. If ERPGs are not available, TEELs usually are based on emergency planning and other guideline levels developed for the protection of workers (Craig 2002).

Several TEEL concentration values are available for each chemical (see text box on next page). For the purposes of this analysis, modeled exposures of SRS employees (assumed to be located 100 m [330 ft] from the release location) to levels greater than TEEL-3 for any chemical were defined as large consequence, and levels less than TEEL-3 but greater than TEEL-2 were defined as moderate consequence. The assessment for the off-site general public differed slightly, as discussed below.

The distance from the release location to the SRS boundary (the nearest location for potential exposures of the general public) is 8.2 km (5.1 mi). Since the ALOHA model restricts release durations to 1 hour, the ambient air concentration at that location could not be readily obtained (the concentrations for downwind distances at times exceeding 1 hour are not directly provided in the ALOHA model). Because plume travel time exceeded 1 hour (i.e., the ALOHA limit) for all of the evaporative spill scenarios considered, the estimated site boundary concentration was obtained by extrapolation methods (see Appendix E). To assess impacts to the general public, site boundary concentrations greater than TEEL-2 levels for any chemical were defined as large consequence, and levels less than TEEL-2 but greater than TEEL-1 were defined as moderate consequence. In addition, the maximum distances from the release point to which chemical TEEL-1 and TEEL-2 air concentrations could extend were estimated using the ALOHA model.

Two release scenarios, one involving nitrogen tetroxide and the other involving chlorine, were modeled as pressurized releases. The HGSYSTEM model (Post 1994a,b; Hanna et al. 1997) was used to simulate pressurized jet releases for punctured containers and the downwind dispersion of the released material. As was done with the ALOHA model for the evaporative dispersion cases, all model runs accounted for the influence of dense vapor cloud behavior on downwind dispersion in releases determined to exhibit this behavior.

Temporary Emergency Exposure Limits (TEELs)

TEEL-1: The maximum concentration in air below which it is believed that nearly all individuals could be exposed for up to one hour *without experiencing other than mild transient adverse health effects or perceiving a clearly defined objectionable odor.*

TEEL-2: The maximum concentration in air below which it is believed that nearly all individuals could be exposed for up to one hour *without experiencing or developing irreversible or other serious health effects or symptoms that could impair their abilities to take protective action.*

TEEL-3: The maximum concentration in air below which it is believed that nearly all individuals could be exposed for up to one hour *without experiencing or developing life-threatening health effects.*

The results of the assessment are summarized in Table 4.16. No accidental releases would result in concentrations exceeding TEEL-1 levels beyond the site boundary. Impacts from these spills on the general public would be small. For all spills, impacts could be minimized with rapid emergency response actions by nearby workers. This response would include quick mitigative action to cover the spill and to minimize evaporation and downwind transport. For SRS employees, impacts could be moderate or large for spills involving chlorine or nitrogen tetroxide. Specific response actions covered under the existing SRS Emergency Response Plan (SRS 2001), including remaining indoors (i.e., sheltering in place) and evacuating (e.g., including rapid evacuation of all nonemergency workers to an upwind location and into designated buildings), would be implemented to minimize worker exposures to spills involving hazardous chemicals of this type. The SRS Emergency Response Plan may be revised to address specific hazards that are not covered in the existing plan subsequent to safety analysis reviews required under DOE chemical safety standards or orders (e.g., DOE-STD-3009-94, DOE Order 420.1).

4.3.5.4 Hydrology

During the scoping process, a concern was raised about groundwater contamination through existing deep boreholes. There are 11 deep boreholes at the SRS. The closest deep borehole is located north of the unnamed tributary that is just north of the proposed MOX facility (see Figure 3.3). Impacts to the groundwater from the proposed facilities have been evaluated. The deep boreholes were determined not to be a credible path by which materials from the proposed facilities could contaminate groundwater, and there would be no discharges to groundwater. Surface spills from the facilities that might travel toward the deep boreholes would be intercepted by the unnamed tributary. Accidental releases that might possibly reach the groundwater would flow in the shallow groundwater aquifer and discharge to Upper Three Runs Creek.

This page is being withheld pursuant to 10 CFR 2.390(a).

Because accidental releases to surface water would be quickly remediated as required by the facility's Spill Prevention Control and Countermeasures Plan, impacts would be negligible. Materials released by leaks or ruptures of vessels and piping used to store and transfer process chemicals and liquid radioactive waste could affect surface water and groundwater. Bulk process chemicals would be stored and chemical mixtures would be prepared in the Reagent Processing Building. DCS has identified a number of chemical process safety controls to prevent significant spills or other accidents that would have the potential to significantly affect the human environment. These measures include administrative controls over segregation and separation of incompatible chemicals, concentration controls on specific reagents, and a process safety instrumentation and control system to measure and control process conditions to ensure safety limits are not exceeded.

Groundwater quality could be indirectly impacted by accidental releases of contaminated effluents or hazardous stored liquids and infiltration of contaminated runoff. Such impacts, however, are expected to be negligible because of adherence to guidelines established in existing NPDES permits and prompt cleanup of any spills as required under the facility's Spill Prevention Control and Countermeasures Plan. Storage vessels for liquid wastes would be located in the Aqueous Polishing Building.

A rupture of the low-level liquid radioactive waste transfer line could release wastewater containing radioactivity at concentrations up to the ETF waste-acceptance criteria levels. DCS, however, has committed to liquid containment features, including containment basins below storage tanks that hold contaminated liquids (stainless-steel-lined floors and portions of walls would be used to create basins in the tank room of the Aqueous Polishing Building) and double-wall pipe and a leak detection system for the transfer line.

The WSB would be connected to the proposed MOX facility and PDCF by stainless steel double-walled pipelines for transfer of stripped uranium wastes and the high-alpha-activity wastes. The waste streams that constitute the high-alpha-activity waste stream include the americium stream, the alkaline wash stream, and the excess acid stream. The combined volumes of these streams would be about 44,200 L/yr (11,700 gal/yr) (DCS 2002a, 2004a). The stripped uranium stream would average about 174,000 L/yr (46,000 gal/yr) during normal operations. The stripped uranium stream would contain only 1% uranium-235 to avoid issues of criticality. To minimize the probability of a pipe failure, both of these waste streams would be transported in double-walled stainless steel pipes. In addition, the pipes would be designed to withstand the effects of a design-basis earthquake and other natural phenomena. If either of these lines ruptured, impacts to surface water or groundwater would be small because of the small quantities of waste involved in the transfer and prompt and thorough cleanup required under the SRS Spill Prevention Control and Countermeasures Plan.

4.3.5.5 Waste Management

Wastes that may be generated from the accident scenarios discussed in this FEIS are expected to be similar in type and of volumes that would be within the bounds of the capacities at the

SRS for waste management. Potential impact to the waste management system at the SRS is expected to be minimal.

4.3.6 Deactivation and Decommissioning

4.3.6.1 Introduction

License termination is considered the final stage of the licensing process for an NRC-licensed facility. License termination entails deactivation and decommissioning of the facility as part of the termination process. Decommissioning involves the removal of the facility safely from service and reduction of residual radioactivity to a level that permits release of the property for unrestricted or restricted use. Termination of the MOX facility license would be governed by 10 CFR 70.38. Decommissioning of the proposed MOX facility would be conducted in accordance with criteria of 10 CFR 20 Subpart E (Radiological Criteria for License Termination). The PDCF and WSB may not be decommissioned after completion of MOX facility operations, but they are included in this evaluation to bound the analysis.

DCS plans to deactivate the proposed MOX facility and request NRC to terminate the license once the facility's mission for disposition of excess plutonium is completed (DCS 2002a). This plan is based on the contract between DOE and DCS that calls for DCS to deactivate the proposed MOX facility and place it in a safe-shutdown condition once operations have ended. In addition, the supporting DOE-owned and -operated support facilities, the PDCF and the WSB, would also require decommissioning once the surplus plutonium mission was completed. The ultimate fate of the facilities would then become the responsibility of DOE.

Deactivation
Deactivation is the process of removing a facility from operation and placing it in safe-shutdown condition for an extended period of time. Deactivation would involve: • Removal of unused plutonium and uranium feedstock, process chemicals, and loose surface contamination; • Depressurization of all facility systems; and • Sealing of gloveboxes and ventilation systems.

DOE may choose to reuse or decommission the facilities once the surplus plutonium mission has been completed. DOE will make a decision on when and how to decommission the facilities.

Currently, it is difficult to determine the possible final disposition of the facilities following the completion of their intended mission. The proposed MOX facility would be owned by DOE and operated by DCS under the terms of the DOE-DCS contract and scope of work. The course of decommissioning and future use of all three facilities would depend largely on DOE decisions that would be made at some future date as the facilities approached the end of their operating lives. Since the scoping process identified decommissioning as a significant issue, the potential impacts of decommissioning the facilities are presented below.

4.3.6.2 Decommissioning Process

Options for decommissioning nuclear facilities are discussed generically in NRC's *Final Generic Environmental Impact Statement on Decommissioning of Nuclear Facilities* (NUREG/CR-0586 [NRC 1988]). As stated in that document, it is the objective of the NRC to conduct decommissioning as an end point of the license termination process.

Other options, such as safe storage, deferred decommissioning, or restricted release, could have been evaluated. However, for safe storage and deferred decommissioning, the doses to workers during decommissioning would be greater because of the decay of transuranic radionuclides (e.g., plutonium-241

Decommissioning

Decommissioning is the process of decontaminating and dismantling the facilities following deactivation and returning the site to an end state that meets the prescribed regulatory criteria. Decommissioning would involve:

- Chemical decontamination,

- Physical decontamination of equipment, structures, and materials (e.g., disassembly of equipment and enclosures and removal of materials), and

- Removal of structures and restoration of the site to a prescribed end state.

---

and plutonium-238). That is, the radioactivity in a facility would increase because of the ingrowth of daughter products. Restricted release was not considered at this time because the "base case" for decommissioning under 10 CFR Part 20, Subpart E, would be unrestricted release. DCS would need to provide additional justification to support a request for restricted release, which at this point in the project would be speculative.

On the basis of the EIS on decommissioning of nuclear facilities (NRC 1988), it is assumed that the decommissioning process for the facilities would include 2 years of preparation and planning, followed by actual decommissioning activities. In general, decommissioning planning would be conducted during the last 2 years of normal plant operation. During that time, detailed plans and procedures would be prepared, a decommissioning staff would be trained, safety and environmental reports would be prepared (if necessary), and effluent control system modifications would be started.

Work would begin immediately following facility shutdown. Chemical decontamination would be followed by physical decontamination of most plant areas, including disassembly of equipment and enclosures and removal of resulting materials, such as structural components. These materials would be packaged and transported off-site as waste. The main facility and service system (e.g., decommissioning equipment and accessories) would be removed last. Some buildings, such as the Administration Building at the proposed MOX facility might not require any decommissioning prior to release for unrestricted use.

4.3.6.3 Decommissioning Impacts

4.3.6.3.1 Radiological Impacts

Because of the uncertainties involved in future operation of the facilities, most of the specific information needed to assess actual decommissioning impacts would depend on the actual operating history of the facilities. Because of the lack of a full-scale MOX facility, PDCF, and WSB, the analysis conducted for this FEIS has been extrapolated from the generic information provided in NRC's final generic EIS for a small mixed oxide fuel fabrication plant (NUREG/CR-0129; NRC 1979) and from NUREG/CR-0586 (NRC 1988). The extrapolation is based on a comparison of the size of the facilities as represented by the total area covered (square meters or square feet) by the MOX Fuel Fabrication Building plus the PDCF and the WSB. The objective of this analysis is to obtain baseline information pertaining to the radiological impact associated with decommissioning activities. Thus, the radiological impact from the proposed MOX facility was estimated to be about 28 times that in the NRC's generic EIS. Given the uncertainties in the decommissioning activities that would be undertaken at the proposed facilities in the future, this assumption provides a reasonable estimate of the decommissioning impacts. The radiological impacts associated with decommissioning are presented in Table 4.17.

Table 4.17. Summary of radiological impacts from routine facility decommissioning

Exposure	Dose[a] [person-Sv (person-rem)]
Occupational	
Deactivation[b]	6.3 (630)
Decommissioning	19 (1,900)
Transportation[c]	0.99 (99)
Total	27 (2,700)
Public	
Deactivation	8.2×10^{-9} (8.2×10^{-7})
Decommissioning	1.8×10^{-7} (1.8×10^{-5})
Transportation[c]	1.2 (120)
Total	1.2 (120)
Grand total	28 (2,800)

[a]Doses are rounded to two significant figures.

[b]Assumed to follow the same preparation process for long-term custodial care (NRC 1998).

[c]Assumes 686 shipments. Estimated from single shipment risks for TRU waste shipments from the SRS to WIPP presented in Monette et al. (1996).

4.3.6.3.2 Nonradiological Impacts

Geology and Soils. Soils covered by buildings and paved surfaces would be reclaimed to support the designated vegetation type after decommissioning. Soil treatments, including grading, disking, and fertilizer applications, would be used following removal of concrete foundations of structures and asphalt from paved parking areas. The movements of trucks and other vehicles involved in removing concrete and major facility components during decommissioning might result in soil compaction in localized areas. The use of chisel plows or other equipment might be required to loosen the soil in areas where compaction was severe. Depending on the final engineering design for the facility sites, some earth moving might be needed. Soils in the storm-water retention area might be moved and/or graded to prevent erosion and to enhance establishment of plant species on areas to be revegetated. Attempts would be made to grade the area to fit with the existing topography of this portion of F-Area at the time of decommissioning.

Hydrology. The types of impacts to surface and groundwater during decommissioning of the facilities would be similar to those occurring during construction. Water would be used for dust suppression when necessary and might be needed during planting until vegetation becomes established. Runoff from areas being graded after the removal of concrete or asphalt would be minimized through use of silt fences or straw bales to control erosion. No impacts are anticipated to groundwater during decommissioning activities. Impacts to surface water during decommissioning would be small because of the measures employed to control runoff.

Air Quality and Noise. The types of air quality impacts expected during decommissioning of the facilities would be similar to those anticipated during facility construction. Vehicles used during decommissioning might create fugitive dust during dry conditions at the SRS. Fugitive dust would be controlled by watering during these periods. As described in Section 4.3.2.1, impacts to air quality would be small.

Noise associated with dismantling and removal of facility structures from F-Area and the SRS would be localized and temporary. Impacts of noise would be similar to those generated by initial construction of the facility (see Section H.2.1 in Appendix H) and would be small.

Ecology. Assuming that full decommissioning occurs and DCS removes the facilities and allows restricted use of the facility areas on the SRS, the following ecological impacts could occur. Although decommissioning plans may call for removal of facility structures, other areas designed to support operations may not be changed. The 4.5 ha (11.0 acres) occupied by the relocated 115-kV power line would remain in use as the power line continued to provide electricity to other F-Area facilities. Also, the 2.0 ha (5.0 acres) of new roads and road upgrades would remain. The 0.6 ha (1.5 acres) occupied by the storm-water basin might also be retained for that use. If storm-water control was not necessary, this area could provide wetland and pond habitats. The remaining areas located within the fenced boundaries of the facilities and along the pipeline rights-of-way could be revegetated. Revegetation goals could include establishing landscaped lawn around buildings, grass and forb species (e.g., similar to the vegetated conditions on the existing spoils pile area within the proposed location for the proposed MOX facility area), or evergreen and mixed forest habitats. The choice of treatment would depend upon the restricted use planned for the area in the future.

During decommissioning activities, wildlife would be affected in a manner similar to what would occur during construction (see Section H.3.1.1.2 in Appendix H). Impacts would primarily be disturbance and displacement caused by noise and human presence. Following decommissioning, a potentially diverse wildlife community could reoccupy the facility areas. Reforestation of the areas would be the most productive for wildlife, while use of the area for new facilities would be least productive for wildlife.

On the basis of the assessment of impacts to ecological resources during construction of the proposed facilities (Section H.3.1, Appendix H), the impacts of decommissioning are expected to be minor.

Land Use. The F-Area is classified as developed/industrial land. Construction of the proposed facilities is consistent with this classification and the SRS Long Range Comprehensive Plan (DOE 2000b). Decommissioning of the facility site for unrestricted use at SRS would not interfere with current uses or anticipated future uses of the F-Area. Lands in adjacent areas on the SRS managed by the U.S. Forest Service would not be adversely affected by decommissioning activities.

Cultural and Paleontological Resources. Decommissioning is not likely to affect any archaeological sites, historic structures, or traditional cultural properties at the proposed project site. Mitigation measures to avoid impacts during construction of the facility at one prehistoric archaeological site that is eligible for listing on the *National Register of Historic Places* (NRHP) are described in Section H.5.1.1 (Appendix H). Prior to decommissioning, a plan would be developed by DOE describing actions that would be taken to avoid or protect any known or new archaeological sites discovered in areas likely to experience surface disturbance or impacts from runoff because of decommissioning activities. The plan would also address other impacts of decommissioning workers such as unauthorized pedestrian traffic or vehicular activity in the vicinity of known sites or eligible sites. If the mitigation measures described in Section H.5.1.1 are implemented during decommissioning, the impacts to cultural resources could be avoided or minimized.

Nonradiological Impacts of Transportation. Decommissioning would require the transport of demolished structures and components to on-site or off-site disposal areas. The transport of structural materials and components would be along existing SRS roads and local South Carolina highways and would not require new roadway construction. Vehicular traffic on the SRS and local roadways related to decommissioning activities is not expected to affect traffic volume or traffic flow patterns on local roads.

Waste Management. The demolition of the facilities would generate solid waste in the form of structural materials such as concrete and steel and contaminated facility components. The exact quantities and classification of waste types cannot be determined at this time; the information presented here on waste types and volumes is based only on projections. The handling and disposal of wastes produced during decommissioning would comply with all regulatory requirements.

Socioeconomics. The types of impacts to socioeconomic and community resources during the decommissioning of the facilities would be similar to those occurring during their construction. The number of workers expected to be needed for decommissioning is about the same as for construction. Socioeconomic impacts from construction are described in Section H.7.1 (Appendix H). No adverse impacts are anticipated to local communities relative to housing demand for workers or community services from decommissioning activities. Assuming that they would have sufficient notice of the completion of decommissioning impacts, local communities should be able to plan for the loss of revenue generated by the work force. The projected costs of decommissioning are discussed below.

Decontamination and Decommissioning Costs. Uncertainties surrounding the precise nature of activities and, consequently, the magnitude of the cost associated with decommissioning of the proposed MOX facility have meant that no direct estimates of these costs have been made to date. However, estimates have been made on the basis of the costs of decommissioning efforts for a similar facility at the RFETS in Colorado (DCS 2001). Facilities currently being decommissioned at the RFETS have supported activities that are broadly similar to those likely to take place in a MOX fuel fabrication facility and in the associated aqueous polishing facility. These activities at the RFETS have included the manufacture of plutonium weapons components, including casting and machining in dry gloveboxes, and the recovery of plutonium from plutonium residue in "canyon" rooms. On the basis of the volume and types of wastes generated during the decommissioning of those buildings, estimates of the direct costs of decommissioning of the proposed MOX facility and related facilities are about $377 million (FY 2003 dollars).

In addition to the direct costs of the facilities, a number of indirect costs would also be incurred. These costs include site security, residue and fuel deactivation and removal, environmental programs, project management, and costs associated with borrowing funds to finance the project (DCS 2001). Significant contingency allowances would also have to be included.

On the basis of data gathered from other, similarly large nuclear fuel cycle-related projects, it can be concluded that the indirect costs are likely to be roughly approximate to the direct costs of construction and operation. It has also been estimated that decommissioning costs of similar projects are equivalent to about 80% of project capital cost (DOE 1995). Design and construction costs for the MOX, PDCF, and WSB facilities, including contingency, are estimated to be $1,929 million (NNSA 2002). Using both approaches, the total decommissioning cost for the three facilities would, therefore, lie in the range of $758 million to $1,543 million (2003 dollars).

4.3.7 Environmental Justice

4.3.7.1 Introduction

Executive Order 12898, *Federal Actions to Address Environmental Justice in Minority Populations and Low-Income Populations* (Volume 59, page 7629 of the *Federal Register* [59 FR 7629]), issued by President Clinton on February 11, 1994, requires federal agencies to incorporate environmental justice as part of their missions. Specifically, it directs executive branch agencies to address, as appropriate, any disproportionately high and adverse human health or environmental effects of their actions, programs, or policies on minority and low-income populations. Although independent agencies, such as the NRC, were only requested to comply with Executive Order 12898, the NRC, in a letter dated March 31, 1994, stated it would endeavor to carry out the measures set forth in the Executive Order and accompanying memorandum as part of its efforts to comply with the requirements of NEPA. The NRC has developed guidelines for environmental justice analyses described in *Environmental Review Guidance for Licensing Actions Associated with NMSS Programs* (NRC 2001, NRC 2003).

The analysis of the potential impacts of the no-action and proposed action alternatives on environmental justice communities near the SRS uses demographic data from the 2000 census to describe the distribution of minority and low-income populations in the vicinity of the SRS. The definitions of minority and low-income population groups as used in this analysis are as follows:

- **Minority.** Beginning with the 2000 census, where appropriate, the census form allows individuals to designate multiple population group categories to reflect their ethnic or racial origin. Persons are included in the minority category if they classify themselves as belonging to any of the following racial groups: Hispanic, Black or African American, American Indian or Alaska Native, Asian, Native Hawaiian or Other Pacific Islander. In addition, persons who classify themselves as being of multiple racial origin may choose up to six racial groups as the basis of their racial origins. The "minority population" therefore incorporates all persons, including those classifying themselves in multiple racial categories, except those who classify themselves as not of Hispanic origin and as White or "Other Race" (U.S. Bureau of the Census 2002).

- **Low-Income.** Individuals who fall below the poverty line are classified as low-income. The poverty line takes into account family size and age of individuals in the family. In 1999, for example, the poverty line for a family of five with three children below the age of 18 was $19,882 in annual income. For any given family below the poverty line, all family members are considered as being below the poverty line for the purposes of analysis (U.S. Bureau of the Census 2002).

Data on minority and low-income populations are available at the county, census tract, block group, and block level. To fully evaluate the potential environmental justice impacts of the proposed action alternative, the distribution of minority and low-income populations was analyzed at the census block group level. The analysis was based on guidelines for environmental justice analyses described in *Environmental Review Guidance for Licensing Actions Associated with NMSS Programs* (NRC 2001). An 80-km (50-mi)-diameter buffer zone around F-Area at the SRS was used as the basis for the analysis so as to include potential adverse human health or socioeconomic impacts related to the construction and operation at the SRS. Accidental chemical and radiological releases, for example, have the potential to affect minority and low-income population groups located some distance from the site, depending on the size and nature of potential releases and on meteorological conditions. The actual extent of any such effects would depend on the magnitude and nature of any release at the site.

In addition to demographic data, the NRC solicited comments and information regarding the potential for the proposed action to cause disproportionate impacts to environmental justice communities at the public scoping meetings (see Section 1.4.1). The comments received at these meetings are presented in Appendix I, Section 2.2.13. In summary, environmental justice impacts were a concern to many stakeholders. It was stated that contamination could affect fishing resources that might be used for subsistence by low-income and minority population groups some distance downstream of the site. This information further supported NRC's decision to use a larger assessment area for environmental justice impacts. It was also stated that many low-income people rely to a greater extent on food produced from gardens.

Guidelines for performing environmental justice reviews are described in NRC's NUREG-1748 (NRC 2001). The analysis method is multistep and consists of first determining if a site has a potential environmental justice concern based on the identification of low-income and minority populations that could be affected by the proposed action. Next, a determination is made as to whether possible impacts would disproportionately impact low-income or minority populations. In cases where the low-income and minority populations are located next to the site, potential impacts could be disproportionate. In other cases, specific behavior of low-income and minority populations, such as the consumption of a greater portion of homegrown crops and other food items, for example, may result in a disproportionate impact. Finally, if it is determined that there would be a potential impact, an assessment would be made as to whether the impact of any aspect of construction and operation of the proposed facilities, including accidents, on low-income or minority populations would be both "high and adverse."

Block group level data for minority and low-income populations for all block groups within 80 km (50 mi) of F-Area are shown in Tables 4.18 and 4.19. Data for each population group are compared with the state and county minority and low-income totals. The environmental justice impacts of the transportation of MOX fuel were not considered because of the uncertainty surrounding the routes that would be selected and the timing and quantity of MOX fuel shipments. NRC guidelines suggest that disproportionate effects on minority and low-income populations should be considered if the minority or low-income populations in block groups are more than 20 percentage points higher than the state and county levels, or where the local minority or low-income population exceeds 50%. Using data in Table 4.18, adding 20 percentage points to the state average would mean that disproportionate effects on minority populations should be considered if the percentage of minorities in a block group is greater than 57.2% in Georgia and 53.8% in South Carolina. Disproportionate effects on low-income populations should be considered if the percentage of the low-income persons in a block group is greater than 34.7% in Georgia and 35.4% in South Carolina (Table 4.19). Minority and low-income percentages in each block group were also compared with the county minority and low-income averages by adding 20 percentage points to the corresponding county minority and low-income percentages. This analysis considered block groups with minority and low-income populations more than 20 percentage points above the state or county value as block groups that have environmental justice populations. Any block group where minority and low-income populations exceeded 50% of the block group population was also considered in the analysis.

Figures 4.2 and 4.3 show the census block groups for the 80-km (50-mi) buffer zone area. The shaded areas are those block groups where minority and low-income individuals are 20 percentage points higher than the state or county averages, or greater than 50% of the total population in the block group.

Table 4.18. Minority population characteristics in the vicinity of the SRS

County	White	Hispanic	Black	American Indian or Alaskan Native	Asian	Native Hawaiian or other Pacific Islander	Other	Two or more races	Total minority	Percent minority
Georgia										
Bulloch	2,850	138	1,152	3	9	1	25	22	1,212	29.8
Burke	10,433	316	11,343	51	57	3	141	215	11,810	53.1
Columbia	72,862	2,297	9,952	276	2,997	80	703	1,376	15,384	17.4
Emanuel	674	17	274	1	0	0	0	6	281	29.4
Jefferson	3,041	101	2,713	7	15	1	56	52	2,844	48.3
Jenkins	4,827	287	3,472	13	18	8	177	60	3,748	43.7
Lincoln	571	3	129	1	3	0	1	7	141	19.8
McDuffie	3,862	100	1,115	18	17	3	28	50	1,231	24.2
Richmond	91,006	5,545	99,391	552	3,000	249	2,024	3,553	108,769	54.4
Screven	8,234	147	6,963	22	40	8	31	76	7,140	46.4
Warren	579	14	324	3	1	0	0	3	331	36.4
Within 80-km buffer	198,939	8,965	136,828	947	6,157	353	3,186	5,420	152,891	43.5
State	5,327,281	435,227	2,349,542	21,737	173,170	4,246	196,289	114,188	2,859,172	34.9
South Carolina										
Aiken	101,745	3,025	36,442	566	905	36	1,181	1,677	40,807	28.6
Allendale	3,068	181	7,960	10	14	7	95	57	8,143	72.6
Bamberg	6,075	118	10,411	27	32	1	23	89	10,583	63.5
Barnwell	12,956	327	9,990	81	91	8	182	170	10,522	44.8
Colleton	605	102	261	0	0	0	64	20	345	36.3
Edgefield	13,962	503	10,209	81	59	8	107	169	10,633	43.2
Hampton	6,259	482	8,276	28	22	1	102	69	8,498	57.6
Lexington	40,976	957	6,085	186	117	10	517	477	7,392	15.3
McCormick	1,312	21	1,736	2	2	1	3	13	1,757	57.2
Orangeburg	9,888	127	7,983	121	26	4	44	199	8,377	45.9
Saluda	9,679	1,159	5,011	37	4	0	511	111	5,674	37.0
Within 80-km buffer	206,525	7,002	104,364	1,139	1,272	76	2,829	3,051	112,731	35.3
State	2,695,560	95,076	1,185,215	13,718	36,014	1,628	39,926	39,950	1,316,452	32.8

Table 4.19. Low-income population characteristics in the vicinity of the SRS

County	Low-income population	Percent low-income
Georgia		
Bulloch	711	17.3
Burke	6,348	28.7
Columbia	4,462	5.1
Emanuel	214	22.9
Jefferson	1,155	19.6
Jenkins	2,419	28.4
Lincoln	128	18.8
McDuffie	796	15.6
Richmond	37,522	19.5
Screven	3,043	20.1
Warren	142	15.6
Within 80-km buffer	56,940	16.6
State	1,033,793	12.6
South Carolina		
Aiken	19,388	13.9
Allendale	3,466	34.5
Bamberg	4,403	27.8
Barnwell	4,834	20.9
Colleton	212	21.5
Edgefield	3,407	15.5
Hampton	2,747	22.8
Lexington	5,517	11.4
McCormick	492	16.3
Orangeburg	3,260	17.9
Saluda	2,374	15.7
Within 80-km buffer	50,100	16.2
State	547,869	13.7

4.3.7.2 Impacts of the No-Action Alternative

For all the storage sites, radiological and nonradiological risks from continued storage of surplus plutonium would be small regardless of the racial and ethnic composition of the populations surrounding the sites, and independent of the economic status of individuals constituting the populations. Continued storage would have no disproportionately high and adverse effects on minority or low-income populations.

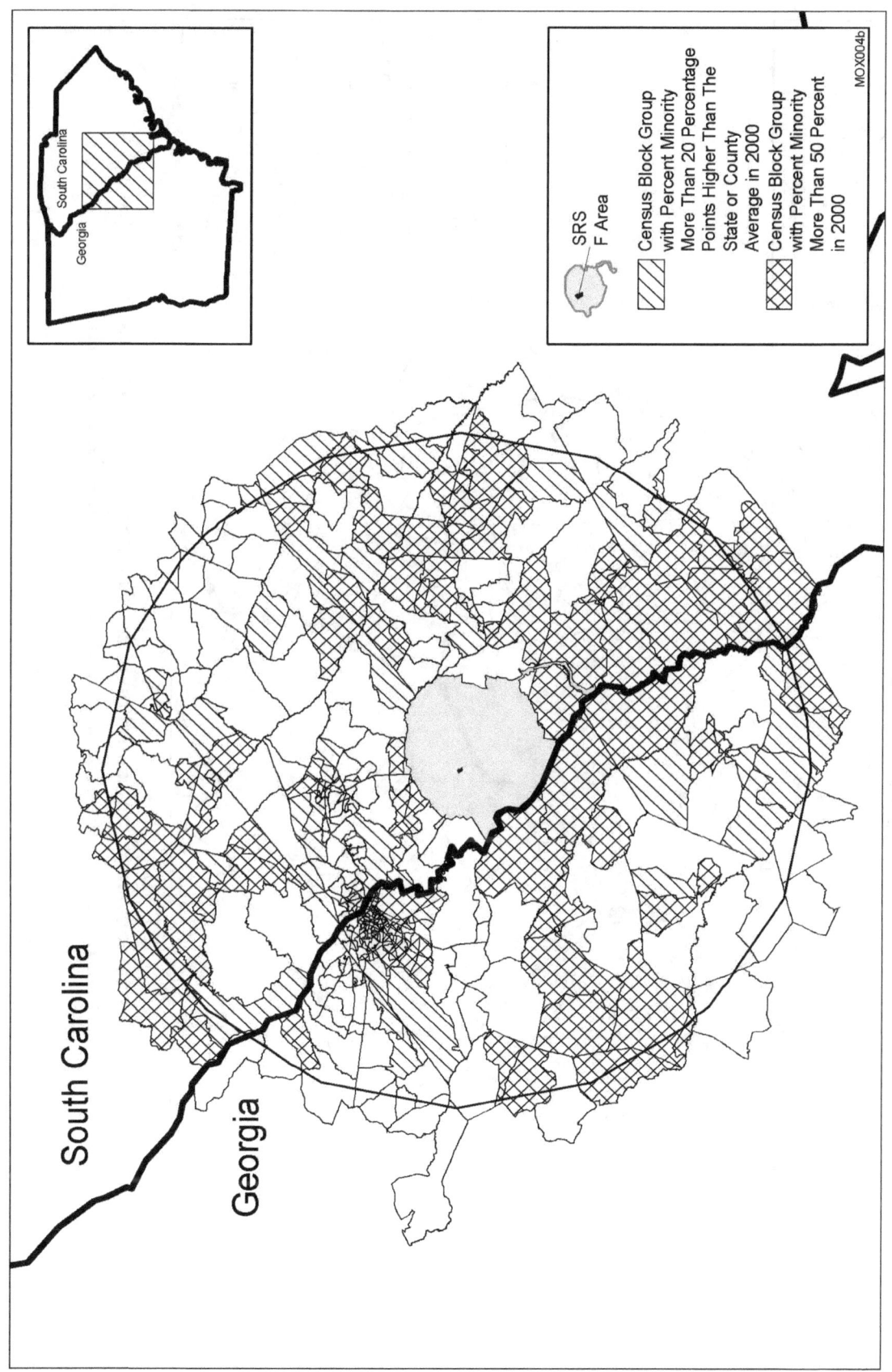

Figure 4.2. Minority population concentration in census block groups within an 80-km (50-mi) radius of the SRS F-Area (*Source:* U.S. Bureau of the Census 2002).

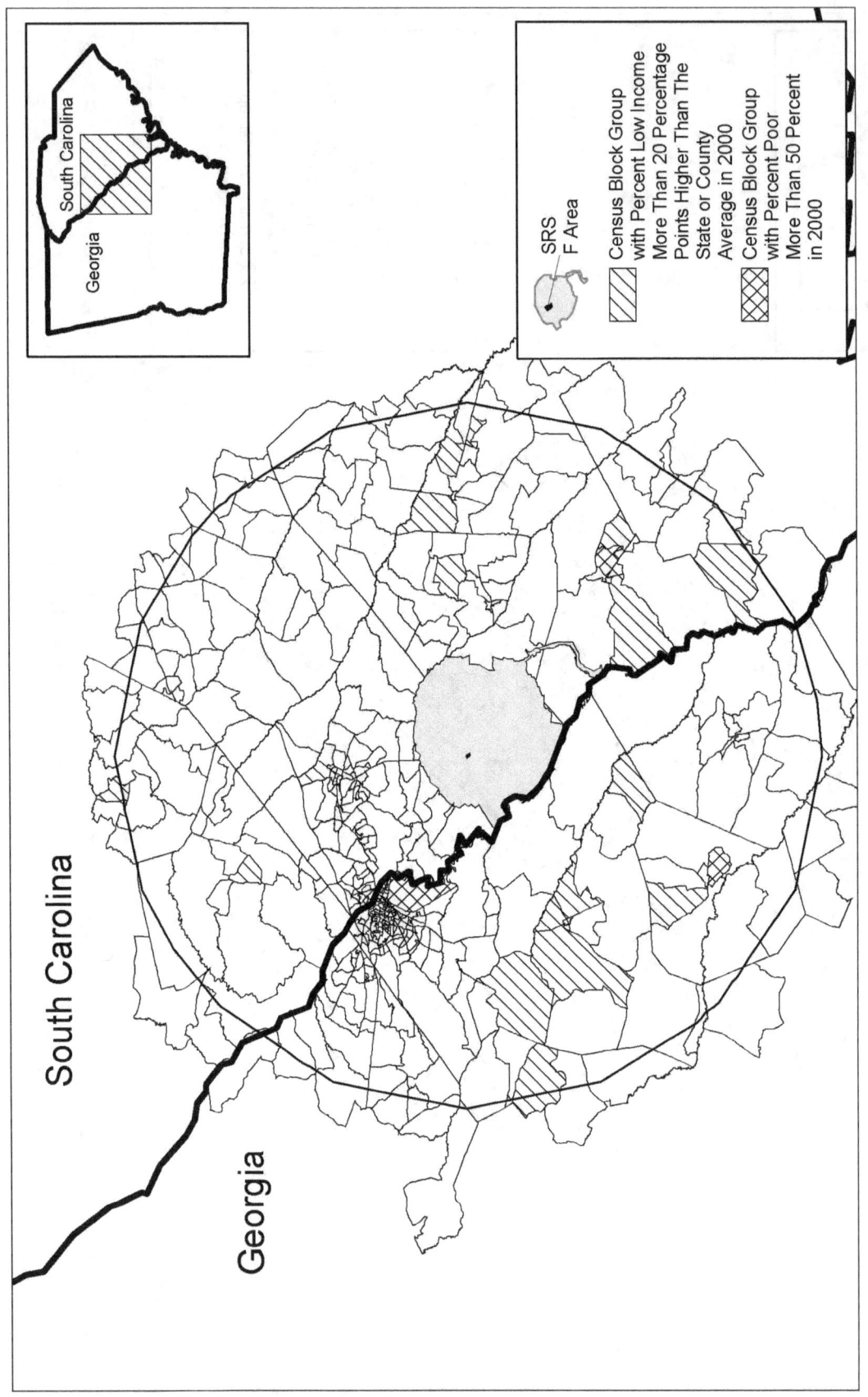

Figure 4.3. Low-income population concentration in census block groups within an 80-km (50-mi) radius of the SRS F-Area (*Source*: U.S. Bureau of the Census 2002).

4.3.7.3 Impacts of the Proposed Action

As discussed above, the analysis of environmental justice impacts is a multistep process. As depicted by the shaded areas in Figures 4.2 and 4.3, low-income and minority populations meeting the definition of environmental justice populations are present within the 80-km (50-mi) assessment area. The next step is to determine whether any impacts would be disproportionate to the low-income or minority populations. Generally, impacts are larger the closer a person is to the source of the impact. Therefore, low-income and minority populations could be disproportionately impacted if they were located closer to the source of the impact than the general population. As depicted in Figures 4.2 and 4.3, the majority of the border of the SRS is populated by predominately minority populations. In addition, specific behavior may result in disproportionate impacts. For example, during the scoping meetings and public meetings on the DEIS, commenters noted that some low-income and minority people relied heavily on homegrown foods and fish from the Savannah River. In addition, it was reported that some in the environmental justice community did not understand the impacts discussed in the DEIS. On the basis of the location of the low-income and minority populations and specific behavior, the NRC concludes that impacts to low-income and minority populations could be disproportionate. The following sections discuss whether the impact of any aspect of construction and operation of the proposed facilities, including accidents, on low-income or minority populations would be both "high and adverse."

4.3.7.3.1 Construction

No radiological risks and only very low chemical exposure and risk are expected during construction. Chemical exposure would be limited to toxic air pollutants released at levels below applicable standards and would not result in any high adverse health impacts. Because the health impacts on the general population within the 80-km (50-mi) assessment area during construction would be negligible, impacts on the minority and low-income population would be small.

4.3.7.3.2 Routine Operations

Radiological impacts to the general public during routine operation of the proposed facilities would be minimal and would not cause any adverse health impacts. The facilities are expected to produce an annual latent cancer risk of approximately 2×10^{-9} for the MEI member of the public. The annual collective dose to members of the public living and working within 80 km (50 mi) of SRS associated with the facilities is expected to produce an LCF risk of approximately 0.0009 or less. In addition, no surface releases that might enter local streams or interfere with subsistence activities by low-income or minority populations are expected to occur. Because the health impacts of routine operations on the general public would be small and there would be no releases that would affect any water or food used for subsistence, there would be no disproportionately high adverse impact on low-income or minority population groups within the 80-km (50-mi) assessment area.

4.3.7.3.3 Accidents

An airborne release following an accident at the proposed facilities has the potential for causing up to 3 LCFs in the area surrounding SRS in the short term because of inhalation exposure. Up to 100 LCFs could occur following the ingestion of contaminated crops. These estimated latent cancer fatalities apply to the entire population within a given sector, which would include both environmental justice populations and non-environmental justice populations. (See discussion in Section 4.3.5 on the accident assessment methodology). If an accident producing such an airborne release were to occur, people living closer to SRS would be impacted to a greater degree than those living farther away from SRS. In the unlikely event of such an accident at the proposed facilities, many of the communities most likely affected would be minority or low income, given the demographics within the 80-km (50-mi) assessment area (see Figures 4.2 and 4.3). In addition, following a hypothetical accident severe enough to produce such a significant airborne release, impacts would be larger if contaminated crops were ingested. In the long-term, the impacts to low-income and minority groups could be higher because of the reliance on homegrown foods. On the basis of the above estimate of accident impacts and considering that low-income and minority populations would be more likely to rely on homegrown foods, the NRC concludes that the impacts to low-income and minority populations could be high and adverse in the event of an accident as described above. However, it is highly unlikely that such an accident would occur. Therefore, the risk to any population, including low-income and minority communities, is considered to be low.

In the event that accidents producing significant contamination occurred as described above, appropriate measures are expected to be taken to ensure that the impacts to all populations, including low-income and minority populations, would be minimized (see Section 5.2.12). The extent to which low-income or minority population groups would be affected would depend on the amount of material released and the direction and speed at which airborne material was dispersed from the facility by the wind. Although the overall risk would be very small, the greatest short-term risk of exposure following an airborne release would be to the population located to the west-northwest of SRS. The greatest 1-year exposure risk would be to population groups residing to the southwest of the site following the ingestion of contaminated crops. With no ingestion, the greatest 1-year risk would still be to the west-northwest. Airborne releases following an accident would likely have a larger impact area than would an accident that released contaminants directly onto the soil surface. A surface release entering local streams could temporarily interfere with subsistence activities by low-income and minority populations located within a few kilometers downstream of SRS.

Monitoring of contaminant levels in soil and surface water following an accident would provide the public with information on the extent of any contaminated areas. Analysis of contaminated areas to decide how to control use of high health risk areas would reduce the potential impact to local residents.

4.3.7.3.4 Decommissioning

Impacts of decommissioning are not expected to disproportionally affect low income or minority populations in the SRS vicinity. A detailed analysis of impacts would be prepared by DOE in a NEPA document specifically on decommissioning and site closure if plans call for full decommissioning of the facilities. Important elements of the environmental analysis in the DOE NEPA document would likely address the disposal process and locations of disposal sites for structural materials and facility components resulting from decommissioning.

4.3.8 Sand Filter Technology Option

Sand filters are air filtration systems used to prevent the release of radioactive material from nuclear facilities to the atmosphere. In a sand filter, the airborne radioactive material is forced through large beds of stone, gravel, and sand that capture and retain radioactive material. Filtered air is discharged to the atmosphere from a nearby stack.

As discussed in Sections 1.4.1 and 2.2.5, the use of sand filters was identified during the EIS scoping process as a potential substitute for final HEPA filters. Differences in impacts between sand filters and HEPA filters are discussed below. Specifically, this section presents the impacts to human health, air quality, hydrology, waste management, potential accident impacts, and facility decommissioning.

Relative to radiological impacts during routine operations, those human receptors who would be affected by such a change would be the proposed MOX facility workers, SRS employees, and the public. However, the differences in emissions between the two filter types is not significant. Thus, the impacts presented in Section 4.3.2.2 on routine operational impacts from the proposed MOX facility to SRS employees and the public would hold for both sand filters and the proposed HEPA filters. In the case of the proposed MOX facility workers, exposure would occur from maintenance activities during normal operations. Monitoring to ensure adequate performance would be required for both filter types. However, HEPA filters, unlike sand filters, would require periodic replacement in addition to monitoring (Orr 2001). The additional exposure in the case of HEPA filters would be minimized with the use of a bag-in/bag-out system (one that isolates the filters from personnel and the environment during replacement) and the maintenance of practices to limit releases of radioactivity to levels ALARA (Orr 2001).

With regard to chemical risks, the difference in chemical removal efficiency between HEPA filters and sand filters is small. Therefore, the impacts presented in Section 4.3.2.2 would be representative for either filter type.

Because air quality impacts associated with the proposed MOX facility would be dominated by the emission of gaseous chemical compounds, and neither HEPA filters nor sand filters are effective for gases, sand filters do not present a clear advantage over HEPA filters. Air quality impacts would be mitigated by other off-gas treatment systems associated with the proposed action.

If sand filters were chosen over HEPA filters at the proposed MOX facility, excavation would be needed for the filter foundations. Excavation is not expected to extend to a depth likely to encounter groundwater. The depth of the sand filter would depend on spatial configuration and topography at the specific site selected for the filter. A surface area of 3,162 m^2 (33,650 ft^2) would be required for the sand filter (Orr 2001). Operation of a sand filter at the proposed MOX facility would not impact groundwater resources. The filter would be covered to prevent precipitation from enhancing recharge of the underlying aquifers and would have a concrete wall and bottom.

The impact to waste management practices was also evaluated with regards to the type of air filters that could be used during proposed MOX facility operations. The waste volume and associated disposal costs from routine operations using HEPA filters versus use of sand filters are compared in Table 4.20. TRU waste and LLW would be generated if HEPA filters were used, and primarily TRU waste would be generated if sand filters were used.

Relative to radiological impacts resulting from accidents, sand filters may provide a larger margin of safety for SRS employees and the public. Two of the four accidents evaluated, the internal fire event and the explosion event, have the potential to damage HEPA filters. If the major vent duct work itself remained intact for these accidents, filter efficiency would not be lost if sand filters were used, and the impacts for the internal fire event and the explosion event could be approximately 100 times lower than the impacts presented for HEPA filters in Section 4.3.5. (Appendix E presents more information on the amount of radioactivity released from each accident considered.) DCS has committed to a strategy of making explosions highly unlikely if they could result in high consequences to SRS employees and members of the public. By preventing explosions, DCS would prevent impaired function of the facility HEPA filters. Further, DCS would maintain safety controls in the proposed MOX facility that would either prevent fires, or for some areas, ensure that fires are contained to single fire areas that would limit the amount of radioactive material involved a fire. Where fires are limited to fire areas, DCS would ensure that the facility HEPA filters would continue to function in the high temperature and soot environment created by the bounding fire.

The decommissioning impacts described in Section 4.3.6 were based on the proposed use of HEPA filters. However, if a sand filter was used, there is the possibility that it could be left in

Table 4.20. Comparison of waste volume and disposal cost for HEPA and sand filters

Parameter	HEPA filter	Sand filter
Waste amount	2,178 filters	9,543 m^3
Disposal cost[a]	$9,333,000	$8,411,750
Type of waste	TRU, LLW	TRU

[a]Estimated disposal cost for HEPA filters is based on the number of filters required, while the cost for the sand filters is based on total volume of sand and rock requiring disposal.

place, incurring little additional decommissioning work. Otherwise, there could be significant impacts, such as economic costs and human health risks, from excavating the contaminated material and possibly transporting and disposing of significant amounts of low-level or transuranic waste, depending on the level of contamination (Orr 2001).

In conclusion, the technology option to install sand filters would not clearly result in lower net environmental impacts than the use of HEPA filters. By selecting sand filters, DCS could reduce environmental impacts in the areas of human health risk to facility workers and accident mitigation. However, controls on HEPA filter change-out and a DCS safety strategy to prevent accidents that would challenge HEPA filter function provide an equivalent reduction of impacts.

4.4 Indirect Impacts

4.4.1 Transportation

This assessment is based on the transportation assessment presented in the NRC's NUREG-0170 report (NRC 1977). Since that assessment was conducted, computer models and basic assumptions have been refined, but the overall approach to estimating transportation impacts has remained the same.

4.4.1.1 Scope of the Analysis

The technical approach for estimating transportation risks involves use of several computer models and databases. For assessment of normal transport, risks were calculated for the collective populations of all potentially exposed individuals, as well as for an MEI receptor. Potentially exposed populations include those persons living and working along the transport route, those present at vehicle stops, and those on the road near the shipment. The accident assessment included consideration of the probabilities and consequences of a range of possible transportation-related accidents, including low-probability accidents that have high consequences, and high-probability accidents that have low consequences. The details of the transportation analysis are provided in Appendix C. Transportation impacts are presented in Section 4.4.1.2.

Transportation concerns raised during the scoping process for this EIS (see Appendix I) included the impacts of transporting MOX feed materials (depleted uranium hexafluoride [UF_6] and the surplus plutonium metal) transport. As discussed below, impacts from the transportation of depleted uranium and surplus plutonium metal (pit material) feed materials were analyzed. Impacts of transporting the plutonium dioxide from the proposed PDCF to the proposed MOX facility are not considered because of the short distance involved and the absence of public roads in this area (DCS 2002a). The NRC intended to evaluate truck and rail transportation impacts of shipping fresh MOX fuel from the SRS (see Appendix I). However, this FEIS evaluated only truck shipments of such fuel because of the added security provided through the use of the Safeguards Transporter, as described in Appendix C, Section C.2.3.

The transportation risk assessment conducted for operation of the proposed MOX facility involved estimating the potential human health risks during transport of feed and waste materials associated with the MOX fuel fabrication process. The risk assessment also considered the risks associated with the transport of the MOX fuel following fabrication.

Transport of the depleted uranium feed materials analyzed included shipment of depleted UF_6 from Portsmouth, Ohio, to Wilmington, North Carolina, and depleted uranium dioxide (UO_2) from Wilmington to the proposed MOX facility at the SRS. Assessment of the transport of plutonium pit material considered shipments from existing storage sites to the SRS. Of the 34 MT (37.5 tons) of plutonium expected to be processed into MOX fuel, 7.3 MT (8.0 tons) would be initially available at the SRS site. Under a separate action (DOE 2002a), approximately 6 MT (6.6 tons) of surplus plutonium is to be shipped from RFETS to SRS (Roberson 2002), which currently has 1.3 MT (1.4 tons) (DOE 1996a). The proposed action would therefore require the shipment of another 26.7 MT (29.4 tons) of plutonium, approximately 21.3 MT (23.4 tons) of which is expected to come from the Pantex Plant in Texas. This FEIS analyses the transportation impacts of the Pantex shipments and the remaining 5.4 MT (5.9 tons) of plutonium whose origins are not yet determined. However, the remaining plutonium would come from storage at other DOE sites. For the purposes of this FEIS, the analysis assumed that the remaining 5.4 MT (5.9 tons) of plutonium would come from the Hanford Site, the plutonium storage site farthest from the SRS. Thus, the actual transportation impacts are expected to be lower than those presented here because some plutonium from closer storage sites is expected to be used. Impacts of shipping TRU waste from the WSB to the Waste Isolation Pilot Plant (WIPP) in New Mexico were evaluated for two cases that bound the potential number of shipments. No volume reduction of the TRU waste is analyzed for the first option, resulting in approximately 2,300 truck shipments over the life of the project. The second option analyzes a case involving a volume reduction of TRU waste by a 3:1 ratio, shipments being constrained by a wattage limit.

Additionally, the FEIS evaluates the impacts of shipping all the fresh MOX fuel from the SRS to a surrogate commercial nuclear plant. The fresh MOX fuel is expected to be used in reactors in the eastern to midwestern portion of the United States. For purposes of impact assessment, a midwestern site was chosen for the surrogate nuclear plant because such a location maximizes the distances necessary to transport the fuel, thus providing conservative estimates of potential impacts. A surrogate nuclear power plant was chosen because no licensed nuclear plant has applied to NRC for authority to use MOX fuel. Thus, the impacts presented here are expected to bound the impacts for future shipments of fresh MOX fuel.

For all shipments, risks were estimated for truck transport for both normal (incident-free) and accident conditions. In both cases, "vehicle-related" and "cargo-related" impacts were evaluated.

Vehicle-related risks result simply from moving any material from one location to another, independent of the characteristics of the cargo. For example, increased levels of pollution from vehicular emissions during normal conditions may affect human health. Similarly, accidents during transportation may cause fatalities from physical trauma.

Cargo-related risk, on the other hand, refers to risk attributable to the characteristics of the cargo being shipped. The radiological cargo-related risks from the transportation of depleted uranium, surplus plutonium, fresh MOX fuel, and TRU waste would be caused by exposure to ionizing radiation. Exposures to radiation occur during both normal transportation and during accident conditions. In the case of the depleted uranium materials considered, cargo-related risks also include chemical hazards during accident conditions.

The risks from exposure to hazardous chemicals during transportation-related accidents can be either acute (result in immediate injury or fatality) or latent (result in cancer that would present itself after a latency period of several years). The acute health end point — potential irreversible adverse effects — was evaluated for the assessment of cargo-related population impacts from transportation accidents. Accidental releases during transport of the uranium compounds (UF_6 and UO_2) were evaluated quantitatively. The analysis of UF_6 effects included consideration of the formation of hydrogen fluoride (HF) from the reaction of UF_6 with moisture in the air. Chemical health effects from transportation of plutonium compounds were not assessed because the radiological impacts are far greater than any chemical impacts.

Unlike the case for radiological exposure, the acute chemical effects evaluated were assumed to exhibit a threshold nonlinear relationship with exposure; that is, some low level of exposure can be tolerated without inducing a health effect. To estimate risks, chemical-specific concentrations were developed for potential irreversible adverse effects. All individuals exposed at these levels or higher following an accident were included in the transportation risk estimates. In addition to acute health effects, the cargo-related risk of excess cases of latent cancer from accidental chemical exposures could be evaluated. However, none of the chemicals that might be released in any of the transportation accidents involving UF_6, UO_2, plutonium, or the MOX fuel would be carcinogenic. As a result, no predictions for excess chemically induced latent cancers are presented in this assessment for accidental chemical releases.

4.4.1.2 Transportation Impacts

The estimated exposures and the associated human health effects are discussed in this section and summarized in Table 4.21.

4.4.1.2.1 Routine Transportation

Radiological risks during routine transportation would result from the potential exposure of people to low levels of external radiation near a loaded shipment. DOT and NRC regulations — 49 CFR Part 173.441 (*Radiation Level Limitations*) and 10 CFR Part 71.47 (*External Radiation Standards for All Packages*) — were set to maintain these external radiation levels at a value considered to be protective of the public. The maximum allowable external dose rate is 0.1 mSv/h (10 mrem/h) at 2 m (6.5 ft) from the outer lateral sides of the transport vehicle. In this analysis, the external dose rates expected are approximately 0.0024 mSv/h (0.24 mrem/h), 0.0076 mSv/h (0.76 mrem/h), 0.048 mSv/h (4.8 mrem/h), and 0.040 mSv/h (4.0 mrem/h) at 1 m

(3.3 ft) for the UF$_6$, UO$_2$, MOX fuel, and TRU waste shipments, respectively (Biwer et al. 1997; DCS 2001; DOE 1997b). Since the regulatory maximum is approximately 0.14 mSv/h (14 mrem/h) at a distance of 1 m (3.3 ft), the external dose rates from the depleted uranium shipments, the MOX fuel shipments, and the TRU waste shipments are expected to be less than 6%, 35%, and 30% respectively, of that regulatory maximum. For this analysis, the external dose rate for the shipments of plutonium metal were set to the regulatory maximum, but it is expected that the dose rate from these shipments would actually be similar to those for the fresh MOX fuel and TRU waste.

Combined total exposures of 3.1 to 5.6 person-Sv (310 to 560 person-rem) and 2.1 to 5.3 person-Sv (210 to 530 person-rem) were estimated for the public and the transportation crews, respectively, from all shipments. The resulting expected LCFs were 0.2 to 0.4 and 0.1 to 0.3, respectively (see Table 4.21). These impacts to the public would be insignificant because the exposure would be spread out over several years among all the people along the transportation routes. If no TRU waste volume reduction occurs, TRU waste shipments from the WSB to WIPP would have the highest average individual dose to the public, 0.0025 mSv (0.53 mrem), estimated from a total collective dose of 3.0 person-Sv (300 person-rem) spread over 566,000 persons along the route. Thus, the routine radiological impacts to the public for the entire shipping campaign would be negligible, an average member of the public would receive only 0.15% or less of the value for exposure to background radiation in one year.

For an MEI member of the public (defined as being located 30 m [98 ft] away from a shipment passing at a speed of 24 km/h [15 mph] [Neuhauser and Kanipe 1992]), the greatest radiological risk would be from the plutonium metal shipments, as shown in Table 4.22. In this case, a risk of 6 × 10^{-10} (a chance of less than 1 in 1 billion) of contracting a fatal cancer is 0.0003% of the value for an annual exposure to background radiation. However, the value for potential exposure to multiple shipments would be correspondingly higher. For example, if the same MEI were present for three shipments of depleted UO$_2$, that individual would receive a dose of approximately 1.1 × 10^{-6} mSv [3 × (3.7 ×10^{-7} mSv)].

For transportation crew members, the largest estimated single shipment dose to one transportation crew member was 0.0013 Sv (0.13 rem) for shipments of plutonium from the Hanford Site to the PDCF. In this case, the risk of contracting a fatal cancer is 1 in 13,000.

A total of up to 2 latent fatalities were estimated from vehicle emissions for the entire shipping campaign. Thus, approximately 2 fatalities or less might be expected from vehicle emissions. This vehicle-related impact is insignificant because the proposed action truck travel on U.S. highways for the high end of the entire shipping campaign, 8,200,000 km (5,090,000 mi) as shown in Table 4.21, is only 0.0038% of similar truck travel on an annual basis in the United States, 217,550,000,000 km (135,179,000,000 mi) (BTS 2002).

Table 4.21. Total collective population transportation risks

Parameter	Depleted UF$_6$	Depleted UO$_2$	Pu metal	TRU waste[a]	MOX fuel	Total campaign[a]
Origin site	Portsmouth, OH	Wilmington, NC	Storage sites	WSB	MOX facility	
Destination site	Wilmington, NC	MOX facility	PDCF	WIPP	surrogate reactor	
Shipment summary						
Shipments	110	60	430	299-2,314	598	1,497-3,512
Distance (km)[b]	103,000	26,500	1,130,000	730,000-5,650,000	1,280,000	3,280,000-8,200,000
Population impacts						
Cargo-related[c]						
Radiological impacts						
Dose risk (person-Sv)[d]						
Routine crew	0.0061	0.0045	0.72	0.46-3.6	0.93	2.1-5.3
Routine public						
Off-link	0.00044	0.00013	0.12	0.019-0.15	0.038	0.18-0.30
On-link	0.0011	0.00035	0.35	0.058-0.45	0.094	0.50-0.89
Stops	0.0045	0.0018	1.7	0.31-2.4	0.34	2.4-4.4
Total	0.0060	0.0022	2.2	0.39-3.0	0.48	3.1-5.6
Accident[e]	0.0025	0.00049	0.00063	0.063	0.16	0.23
Latent cancer fatalities[f]						
Crew	0.0004	0.0003	0.04	0.03-0.2	0.06	0.1-0.3
Public	0.0005	0.0002	0.1	0.03-0.2	0.04	0.2-0.4
Chemical impacts						
Irreversible adverse effects[g]	1.3×10^{-7}	0	NA[h]	NA	NA	1.3×10^{-7}
Vehicle-related[i]						
Emission fatalities	0.04	0.008	0.3	0.2-1	0.6	1-2
Accident fatalities	0.003	0.0012	0.028	0.017-0.13	0.029	0.078-0.20

Table 4.21. Continued

[a]The number of TRU waste shipments will depend on the waste treatment process used (DCS 2004a). The largest volume reduction estimated would result in the fewest number of shipments. The largest number of shipments corresponds to the minimum amount of TRU waste treatment necessary for shipment.

[b]To convert km to mi, multiply by 1.609.

[c]Cargo-related impacts are impacts attributable to the radioactive or chemical nature of the waste material.

[d]To convert person-Sv to person-rem, multiply by 100.

[e]Accident dose risk is a societal risk and is the product of accident probability and accident consequence.

[f]Latent cancer fatalities are calculated by multiplying dose by the FGR 13 health risk conversion factor of 0.06 fatal cancer per person-Sv (6×10^{-4} fatal cancer per person-rem) (Eckerman et al. 1999).

[g]Potential for irreversible adverse effects from chemical exposures. Exposure to HF or uranium compounds is estimated to result in fatality of approximately 1% or less of those persons experiencing irreversible adverse effects (Policastro et al. 1997).

[h]NA = not applicable.

[i]Vehicle-related impacts are impacts independent of the cargo in the shipment.

Table 4.22. Routine single-shipment impacts to a maximally exposed individual[a]

Shipment type	Dose [mSv (mrem)]	Risk of developing a latent fatal cancer
Depleted UF_6	2.3×10^{-7} (2.3×10^{-5})	1×10^{-11}
Depleted UO_2	3.7×10^{-7} (3.7×10^{-5})	2×10^{-11}
Pu metal	1×10^{-5} (1×10^{-3})	6×10^{-10}
MOX fuel	1.5×10^{-6} (1.5×10^{-4})	9×10^{-11}
TRU waste	2.4×10^{-6} (2.4×10^{-4})	1×10^{-10}

[a]Individual is located 30 m (98 ft) from a passing shipment traveling at 24 km/h (15 mph).

4.4.1.2.2 Accident Impacts

The total radiological collective population accident dose risk to the public from all shipments was estimated to be 0.23 person-Sv (23 person-rem). The resulting estimated LCFs are 0.01 for the entire shipping campaign.

Chemical impacts would be negligible; only 1.3×10^{-7} irreversible adverse effect from depleted UF_6 shipments is expected for the entire shipping campaign. As discussed in Appendix C (Section C.2.6), this value corresponds to approximately 1×10^{-9} fatality.

Total fatalities from direct physical trauma as a result of accidents were estimated to be up to 0.20. Thus, no fatalities are expected from accidents for the entire shipping campaign.

4.4.1.3 Highly Enriched Uranium

As described in Section 2.2.2.2, HEU is a by-product of the plutonium pit disassembly process. This recovered HEU from the PDCF would be shipped to the Y-12 facility at the Oak Ridge Reservation for declassification, storage, and eventual disposition. The transportation risks for these shipments were analyzed and included in estimates presented in the SPD EIS for transport of all radioactive material associated with the conversion of 33 MT (36.4 tons) of plutonium to MOX fuel as part of Alternative 3 (see Table L-6 in DOE 1999a). The total radiological transportation risks for Alternative 3 were 0.024 and 0.038 LCFs expected for transportation workers and the public, respectively. Thus, the transportation risks for the HEU shipments are considered to be insignificant because they represent only a small portion of an insignificant impact.

4.4.1.4 Spent MOX Fuel

Transportation of the spent MOX fuel to a final disposal site would be required after irradiation in a commercial nuclear reactor. The types of transportation risks posed would be the same as those considered above for the uranium and plutonium feed materials, the fresh MOX fuel, and the TRU waste. These risks include the radiological cargo-related risks from routine transport and hypothetical accidents and the vehicle-related risks, such as traffic accident fatalities and potential latent fatalities from vehicle emissions.

Estimating specific transportation risks for the spent MOX fuel is premature at this time because of the uncertainty in the actual location of both the commercial reactors that would be used for irradiation of the fresh MOX fuel and the final disposal site. As discussed in Section 4.4.1.1, the actual commercial reactors that would be used to irradiate the fresh MOX fuel are not yet known. The only disposal site currently under consideration in the United States is the proposed geologic repository at Yucca Mountain in Nye County, Nevada (DOE 2002d). For purposes of complying with NEPA requirements, it is assumed that spent MOX fuel would eventually be shipped to the proposed Yucca Mountain repository. However, the DOE's application for a license to operate the Yucca Mountain repository has not yet been submitted to the NRC. There is no assurance that the DOE's application, if submitted, would be approved.

On a per kilometer traveled basis, the routine radiological and vehicle-related transportation risks for spent MOX fuel would be similar to those estimated in this FEIS for fresh MOX fuel, plutonium metal, or TRU waste. The transportation risks of commercial spent nuclear fuel (SNF) and spent MOX fuel transport in particular were estimated in DOE's EIS concerning disposal of SNF and high-level waste at Yucca Mountain (DOE 2002d). In the mostly legal-weight truck scenario, approximately 53,000 truck shipments were estimated to result in approximately 12 LCFs to workers, 3 LCFs to the public, and 5 traffic fatalities. A rough estimate of the transportation risks of the spent MOX fuel can be obtained based on average shipment risks calculated from these results to show that no fatalities would be expected. Shipment of all the spent MOX fuel, approximately 598 shipments assuming three assemblies per cask, might be expected to result in approximately 0.1 worker LCFs, 0.03 public LCFs, and 0.056 transportation fatalities. Actual impacts would be lower or higher depending on the actual shipment distances relative to the average in the Yucca Mountain EIS (DOE 2002d). Thus, no significant impacts would be expected because the estimated risks are only a very small fraction of the radiological and vehicular risks to which the public are exposed to on a routine basis as discussed in Section 4.4.1.2.1.

4.4.2 Conversion of Uranium Hexafluoride to Uranium Dioxide

As discussed in Section 1.2.2, it is assumed that the conversion of uranium hexafluoride to uranium dioxide would take place at the Global Nuclear Fuel-Americas, LLC facility in Wilmington, North Carolina. The impacts of the general conversion process are described in the environmental assessment for the last license renewal of that facility (NRC 1997). At that time, the Wilmington facility was using the ammonium diuranate (ADU) process and was planning to begin using a new dry conversion process (DCP). The ADU process is a "wet" process that has higher impacts than the DCP. The GE facility currently uses the DCP. The environmental assessment includes a discussion of the impacts from both the ADU process and DCP. Therefore, it is believed that the impacts summarized below would bound impacts from the conversion process if another facility was ultimately selected.

No measurable impacts have been observed to the air, surface water, or vegetation due to releases from the Wilmington facility. Impacts to the shallow groundwater aquifer have occurred. The Wilmington facility produces gaseous, liquid, and solid effluent streams. Gaseous effluents are controlled by the use of HEPA filters and scrubbers permitted by the State of North Carolina, as necessary. Liquid effluents are controlled by the use of treatment systems and wastewater retention basins designed to reduce the concentration of contaminants prior to discharge. Solid wastes are managed through segregation, recycling, off-site disposal, and incineration. Discharges are permitted and are monitored to ensure compliance with permit requirements. Impacts to a hypothetical MEI and to the collective population are summarized in Table 4.23.

4.4.3 MOX Fuel Use

This section evaluates on a generic basis the impacts of using MOX fuel in reactors by summarizing analyses performed by the DOE in the SPD EIS (DOE 1999a).

Table 4.23. Comparison of human exposure for ammonium diuranate (ADU) and dry conversion processes (DCPs)

Pathway/receptor	ADU dose	DCP dose
Air		
Maximally exposed individual [mSv/yr (mrem/yr)]	0.001 (0.1)	0.0005 (0.05)
Collective population [person-Sv (person-rem)]	0.0009 (0.09)	0.00045 (0.045)
Liquid		
Maximally exposed individual [mSv/yr (mrem/yr)]	0.007 (0.7)	0.001 (0.1)
Collective population [person-Sv (person-rem)]	NA[a]	NA
Total		
Maximally exposed individual [mSv/yr (mrem/yr)]	0.008 (0.8)	0.00015 (0.15)
Collective population [person-Sv (person-rem)]	0.0009 (0.09)	0.00045 (0.045)

[a]Not applicable because liquid effluent in the river quickly dilutes to background levels; therefore, the collective dose impact is negligible.

The DOE's analysis is provided in Section 4.28 and Appendix K.7 of the SPD EIS. Impacts resulting from both normal operations and postulated accidents were evaluated for six reactors, two each at the Catawba, McGuire and North Anna nuclear stations. The range of impacts at each of these reactors were considered to reasonably bound the impacts of reactors that could use MOX fuel. Therefore, the range impacts is considered to represent a generic analysis. This range includes impacts from both ice condenser-type reactors (i.e., Catawba and McGurie) and non-ice condenser-type reactors. It was assumed that up to 40% of the fuel assemblies in a generic reactor would contain MOX fuel and that the remaining assemblies would contain the type of low-enriched uranium (LEU) fuel now used by commercial reactors. The impacts resulting from the use of MOX fuel in such a hybrid reactor core were estimated and compared with the impacts that would result from the use of a reactor core containing only LEU fuel.

The impacts from normal operations would be the same whether the reactor core contained 40% MOX fuel or 100% LEU fuel. The public surrounding such a generic reactor was estimated to receive a collective dose in the range of 0.057 person-Sv/yr (5.7 person-rem/yr) to 0.203 person-Sv/yr (20.3 person-rem/yr). The estimated number of annual LCFs produced by such a dose would be less than 0.01. No individual would be expected to receive more than 0.0073 mSv/yr (0.73 mrem/yr) due to reactor operations under normal conditions.

Some of the beyond-design-basis accidents were estimated to cause prompt fatalities in the highly unlikely event that they occurred. The change in the number of prompt fatalities due to the use of MOX fuel was estimated to range from 0 to 28 additional fatalities (815 versus 843 in the worst accident).

These doses are a small fraction of the annual average background dose. For comparison, as discussed in Section 3.10, the average annual natural background radiation dose to an individual in the United States is 3.6 mSv (360 mrem).

The SPD EIS (DOE 1999a) also analyzed potential MOX fuel use impacts from both postulated design-basis and beyond-design-basis accidents. The impacts were estimated in terms of both the consequences (the impacts that would result if the accident occurred) and risks (taken to be the consequences multiplied by the probability of occurrence of the accident). The risk was estimated over a 16-year campaign. The risk, over the entire 16-year period, of a LCF associated with design-basis accidents to the public surrounding a reactor using all LEU fuel ranged from 2.19×10^{-4} to 8.98×10^{-4}. The change in risk of a LCF associated with a reactor using 40% MOX fuel ranged from about 6% lower to 3% greater. For beyond-design-basis accidents, the campaign risk of a LCF to the public surrounding a reactor using all LEU fuel ranged from 0.144 to 5.25×10^{-5}. The change in risk of a LCF associated with a reactor using 40% MOX fuel ranged from about 7% lower to 14% greater.

The analysis in this EIS does not specifically consider impacts from the use of the lead test assembly (LTA) program. The LTA program consists of fabricating, transporting, using in a reactor, and analyzing a limited number of fuel assemblies. The DOE estimated the impact of the LTA program in the SPD EIS. The LTA program is considered to be independent of the proposed action. That is, the NRC decision regarding the proposed MOX facility is not affected by the DOE's decision on how to make and test the LTAs.

On February 27, 2003 (as amended September 23, 2003), Duke Power submitted a license amendment request to irradiate four MOX fuel lead test assemblies in the spring of 2005 in its Catawba Nuclear Station Units 1 & 2 (Docket Nos. 50-413, 50-414). The NRC is currently reviewing this license amendment request. In addition, in order for any specific commercial reactor to use MOX fuel on a production scale, an amendment to a 10 CFR Part 50 license, issued by the NRC, would be required. The NRC would perform its own site-specific NEPA analyses in evaluating any license amendment application it may later receive seeking authorization to use MOX fuel.

Impacts of transporting fresh MOX fuel to reactors is presented in Section 4.4.1.2.1, and impacts of transporting spent MOX fuel to a geologic repository is presented in Section 4.4.1.4. The impacts of disposing of the MOX fuel is included in the FEIS for Yucca Mountain (DOE 2002d).

4.5 Cumulative Impacts

This section assesses potential cumulative impacts of construction and operation of the proposed MOX, PDCF, and WSB facilities. Cumulative impacts are distinguished from the direct and indirect impacts of these facilities, which are discussed in Sections 4.3 and 4.4 and Appendix H. Direct effects are caused by the proposed action and occur at the same time and place. Indirect effects are caused by the proposed action and occur later in time or are farther removed in distance but are still reasonably foreseeable.

> **Cumulative Impacts**
>
> *Cumulative impacts* are potential impacts when the proposed action is added to other past, present, and reasonably foreseeable future actions.

Cumulative impacts were determined by adding the expected impacts of past, present, and reasonably foreseeable future actions to the projected direct and indirect impacts of the proposed MOX, PDCF, and WSB facilities. The impacts of construction and normal operations of the proposed facilities were evaluated for each impact area and are presented in Section 4.3. The impacts of past and present actions were determined from site environmental reports and other available documents (e.g., recent EISs). Reasonably foreseeable future actions include among others, those that would occur if the proposed MOX facility is built and operated, and include actions to be undertaken by the DOE as part of its surplus plutonium disposition program. The impacts of reasonably foreseeable future actions were taken from recently published NEPA analyses. Although the cumulative impact analysis focused on impacts at the SRS and vicinity (Section 4.5.1), an evaluation of cumulative impacts of off-site transportation activities is also included (Section 4.5.2).

4.5.1 Cumulative Impacts at the SRS

A review was conducted of past, present, and reasonably foreseeable future activities on the SRS. Past impacts were included in the cumulative impact assessment only if the residual effects of past actions are still in existence (e.g., past land use changes that are still in effect). Past impacts that have come and gone (e.g., operational impacts of decommissioned facilities) were not included in the cumulative impact assessment. The impacts of present activities and residual past activities at the SRS were determined from annual environmental reports that document the results of ongoing monitoring activities (e.g., Arnett and Mamatey 2001), as well as descriptions of the SRS baseline conditions in various recent DOE EISs. The impacts of past and present activities at the SRS are described qualitatively for each impact area in Chapter 3.

Nuclear facilities within an 80-km (50-mi) radius of the SRS include Georgia Power's Vogtle Electric Generating Plant across the river from the SRS; Chem-Nuclear Inc., a commercial low-level waste burial site just east of the SRS; and Starmet CMI, Inc. (formerly Carolina Metals), located southeast of the SRS, which processes uranium-contaminated metals. Radiological impacts from the operations of the Vogtle Electric Generation Plant, a two-unit commercial nuclear power plant, are small, but they are included in this cumulative impact analysis. The South Carolina Department of Health and Environmental Control Annual Report (SCDHEC 1995) indicates that operation of the Chem-Nuclear Services facility and the Starmet CMI facility do not noticeably affect radiation levels in air or liquid pathways in the vicinity of the SRS.

The counties surrounding the SRS host numerous industrial facilities (e.g., Bridgestone Tire, textile mills, paper product mills, and manufacturing facilities) with permitted air emissions that cumulatively affect regional air quality. South Carolina Electric and Gas Company's Urquhart Station, a three-unit, 250-megawatt, coal- and natural-gas-fired steam electric plant, is located near the SRS in Beech Island, South Carolina. All of these facilities contribute to ambient air quality at the SRS and thus are included within the SRS baseline used in the analysis of cumulative air quality impacts.

A number of construction and operating permits for industrial facilities in Aiken, Barnwell, Allendale, and Edgefield Counties have recently been filed with the South Carolina Department of Health and Environmental Control Bureau of Air Quality. No new permits have been applied for in Augusta-Richmond, Columbia, and Burke Counties in Georgia. In addition, a number of road projects are planned in the area. These include relatively minor improvements in the Aiken and North Augusta, South Carolina, areas that are part of the Augusta Regional Transportation Study and would take place in 2003 through 2007. Additional road projects in the area include improvements to a 13-km (8-mi) portion of US 78 from Montmorenci, South Carolina, to Windsor, South Carolina (to the east of Aiken), and the extension of I-520 across the Savannah River into North Augusta. This latter project would take place in 2006 through 2009.

Construction of new facilities and roads would result in short-term air quality impacts and would only contribute to the cumulative impact of MOX facilities if the construction period of facilities overlapped with the MOX construction or operational period. Impacts to air quality resulting from operations of new facilities and roads would result in changes to regional air quality. It is difficult to adequately predict the contribution of these facilities and roads to cumulative air quality impacts with the information available. All facilities would require permitting, and this permit process would take into consideration regional air quality NAAQS compliance.

Reasonably foreseeable future actions at the SRS were identified by reviewing recent NEPA documents for the site. A brief synopsis of future projects at the SRS that are considered in the cumulative impact analysis is presented in the following paragraphs:

- *Final Defense Waste Processing Facility Supplemental Environmental Impact Statement,* DOE/EIS–0082–S (DOE 1994). The Defense Waste Processing Facility (DWPF) has been constructed at the SRS and is currently processing sludge from SRS HLW tanks. However, SRS baseline data do not include the impacts of all planned DWPF operations, including the processing of salt solution from these tanks. Therefore, the cumulative impact analysis includes some effects of DWPF in the impacts of past and present activities and some in the impacts of reasonably foreseeable future actions.

- *Disposition of Surplus Highly Enriched Uranium Final Environmental Impact Statement,* DOE/EIS-0240 (DOE 1996b). The cumulative impact analysis incorporates an alternative at the SRS that would blend highly enriched uranium to 4% low-enriched uranium as uranyl nitrate hexahydrate (61 FR 40619; August 5, 1996).

- *Final Environmental Impact Statement on Management of Certain Plutonium Residues and Scrub Alloy at the Rocky Flats Environmental Technology Site,* DOE/EIS-0277 (DOE 1998). DOE plans to process certain plutonium-bearing materials currently being stored at the RFETS (64 FR 8068; February 18, 1999, and 66 FR 4803; January 18, 2001). These materials are plutonium residues and scrub alloy remaining from nuclear weapons manufacturing operations. DOE has decided to ship certain residues from the RFETS to the SRS for plutonium separation and stabilization. The separated plutonium would be stored at the SRS pending disposition decisions. Environmental impacts from using the F-Canyon to chemically separate the plutonium from the remaining materials at the SRS are included in the cumulative impact analysis.

- *Final Environmental Impact Statement for the Construction and Operation of a Tritium Extraction Facility at the Savannah River Site,* DOE/EIS-0271 (DOE 1999c). DOE plans to construct and operate a facility at the SRS to extract the tritium from commercial light-water reactor targets and targets of similar design (64 FR 26369; May 14, 1999). The proposed action and alternatives would provide tritium extraction capability to support either reactor or accelerator tritium production. Environmental impacts from the maximum processing option in the EIS are included in the cumulative impact analysis.

- *Surplus Plutonium Disposition Final Environmental Impact Statement,* DOE/EIS-0283 (DOE 1999a). The SPD EIS analyzed implementation of DOE's disposition strategy for surplus plutonium. The decision to site the facilities to implement this strategy at the SRS (as described in 65 FR 1608, January 11, 2000) is the basis for the proposed action analyzed in this EIS. The SPD EIS was used in some cases to determine the impacts of the Pit Disassembly and Conversion Facility for inclusion in the cumulative impact analysis.

- *Savannah River Site Spent Nuclear Fuel Management Final Environmental Impact Statement,* DOE/EIS-0279 (DOE 2000c). The selected alternative in the Record of Decision (ROD) for the Spent Nuclear Fuel Management EIS is to prepare for disposal of about 97% by volume (about 60% by mass) of the aluminum-based fuel considered in the EIS (48 MT [53 tons] heavy metal), using a melt and dilute treatment process (65 FR 48224; August 7, 2000). The impacts of this process are included in the cumulative impact analysis. The remaining 3% by volume (about 40% by mass) would be managed using conventional processing in existing SRS chemical separation facilities. As part of the preferred alternative, DOE will develop and demonstrate the melt and dilute technology. Following development and demonstration of that technology, DOE will begin detailed design, construction testing, and startup of a new treatment and storage facility to combine with a new dry storage facility. The SNF will remain in existing wet storage until treated and will then be placed in dry storage.

- *Savannah River Site High-Level Waste Tank Closure Final Environmental Impact Statement,* DOE/EIS-0303 (DOE 2002b). DOE evaluated three alternatives for tank closure. All of these alternatives would start after bulk waste removal. DOE decided (as described in 67 FR 53784; August 19, 2002) to implement the preferred alternative identified in the EIS (i.e., stabilize tanks and fill with grout). The impacts of this alternative are presented in this cumulative impact analysis.

- *Savannah River Site Waste Management Final Environmental Impact Statement,* DOE/EIS-0217 (DOE 1995). This EIS provides a basis for the selection of a sitewide approach to managing present and future (through 2024) wastes generated at the SRS. These wastes would come from ongoing operations and potential actions, new missions, environmental restoration, and decontamination and decommissioning programs. The EIS evaluated the treatment of wastewater discharges in the Effluent Treatment Facility, F- and H-Area Tank Farm operations and waste removal, and construction and operation of an HLW evaporator in the H-Area Tank Farm. In addition, it evaluated the Consolidated Incineration Facility (CIF) for the treatment of mixed waste, including incineration of benzene waste from the in-tank precipitation (ITP) process. (The CIF has suspended operations and the ITP process is to be replaced by an alternative evaluated in DOE 2001.) The first ROD stated that DOE would configure its waste management systems according to the moderate treatment alternatives described in the EIS (60 FR 55249; October 30, 1995). The second ROD (62 FR 27241; May 9, 1997) was deferred regarding treatment of mixed waste to ensure consistency with the *Approved Site Treatment Plan* (WSRC 2000). The Waste Management EIS is relevant to the assessment of cumulative impacts because it provides the baseline forecast of waste generation from operations, environmental restoration, and decontamination and decommissioning. This forecast was updated in 1999 (Halverson 1999).

- *Final Environmental Impact Statement for the Treatment and Management of Sodium-Bonded Spent Nuclear Fuel,* DOE/EIS-0306 (DOE 2000d). DOE plans to treat all spent nuclear fuel from the Experimental Breeder Reactor-II (EBR-II) and sodium-bonded spent nuclear fuel at Argonne National Laboratory-West (ANL-W) (located at INEEL) (65 FR 56565, September 19, 2000). Fermi-1 sodium-bonded spent nuclear fuel will be stored pending a decision on alternative treatments. DOE does not plan to implement any of the alternatives proposed for the SRS. However, some of the impact projections from other EISs (e.g., cumulative waste generation from the High-Level Waste Tank Closure EIS [DOE 2000a]) include impacts at the SRS from sodium-bonded spent nuclear fuel, and these impacts were excluded from the cumulative impact analysis.

• *Savannah River Site Salt Processing Alternatives Final Supplemental Environmental Impact Statement,* DOE/EIS-0082-S2 (DOE 2001). A process to separate the high-activity and low-activity waste fractions in high-level waste solutions is planned to replace the in-tank precipitation process assessed in the Defense Waste Processing Facility EIS (DOE 1994). The Salt Processing EIS evaluates four alternatives: small tank precipitation; ion exchange; solvent extraction; and direct disposal in grout. The proposed MOX facility cumulative impact analysis includes maximum impacts of the solvent extraction process as selected in the DOE ROD for this project (66 FR 201, p. 52752, October 17, 2001).

• *Environmental Assessment for the Construction and Operation of the Highly Enriched Uranium Blend-Down Facilities at the Savannah River Site,* DOE/EA-1233 (DOE 2000e). DOE plans to construct and operate a low-enriched uranium (LEU) loading station and modifications to the existing HEU blend-down facilities. The process will convert off-specification HEU (60% uranium-235) to less than 20% uranium-235 for use as commercial fuel. The environmental assessment (EA) for this facility indicated that impacts would be either negligible or unmeasurable. A Finding of No Significant Impact was issued on November 3, 2000.

• *Draft Supplemental Programmatic Environmental Impact Statement on Stockpile Stewardship and Management for a Modern Pit Facility,* DOE/EIS-236-S2 (DOE 2003b). A modern pit facility (MPF) has been proposed by DOE's National Nuclear Security Administration to manage and maintain the U.S. nuclear weapons stockpile. DOE has prepared a Supplement to the Programmatic Environmental Impact Statement on Stockpile Stewardship and Management for a Modern Pit Facility. This MPF EIS evaluates the environmental impacts associated with constructing a new MPF at four alternate sites, including the SRS, and across a range of pit production capabilities. The MOX facility cumulative impact analysis incorporates the impacts of the highest pit production rate (450 pits/year).

For all impact areas but employment, it was conservatively assumed that the impacts of past, present, and future activities would occur simultaneously. In reality, there would be less overlap of impacts in time (e.g., the impacts of some projects would be declining during the operational life of the facility), and cumulative impact, therefore, actually would be less than is presented here. Impacts to the MEI were also determined using a conservative approach that assumed the same MEI would be exposed to all concurrent actions (see Section 4.3.1.1.2 for the location of MEI for the proposed MOX facility). In reality, the MEIs for different activities vary and are dependent on the location of the activity (Simpkins 2000).

4.5.1.1 Cumulative Impacts of the MOX, PDCF, and WSB Facilities

Cumulative impacts of the facilities at the SRS were evaluated in detail for (1) air quality; (2) human health; (3) waste generation; (4) resource use (land, electricity, and water); and (5) employment. These impacts were evaluated on the basis of the anticipated effects of facility construction and normal operations (as presented in Section 4.3) and the potential for contributions to existing cumulative impacts on the SRS. The analysis focused primarily on normal facility operations over an assumed 10-year operating period. Construction impacts were considered in the cumulative impact analysis only with respect to the amount of land developed, because other construction impacts would be too short-lived to contribute substantially to cumulative impacts to any resources. Additionally, standard mitigation practices employed during construction (e.g., dust control measures, erosion control) would likely reduce these impacts to negligible levels.

Impacts to water quality, geologic resources, ecological resources, aesthetic and scenic resources, and cultural resources are not treated explicitly in the cumulative impact analysis because direct and indirect impacts to these resources are expected to be small (see Sections 4.3 and Appendix H). Facility operations would not contribute to the cumulative impacts of SRS activities on water quality because liquid effluents would be discharged to surface water under existing NPDES permit guidelines. No impacts are anticipated to aesthetic and scenic resources because the facilities would be visually consistent with surrounding SRS industrial facilities and would not be visible from off-

> **Topics Evaluated and Impact Criteria Used in the Cumulative Impact Analysis**
>
> - *Air quality:* % NAAQS for criteria pollutants.
>
> - *Human health:* Radiological dose to off-site MEI, off-site population, and SRS workers and resultant latent cancer fatalities.
>
> - *Waste generation:* Generation rate of various waste types relative to existing SRS capacity.
>
> - *Resource use:* Amount of land developed relative to total SRS area; amount of electricity and water used relative to existing SRS capacity.
>
> - *Employment:* Number of jobs at the SRS.

site. Impacts to geologic, ecological, and cultural resources are expected to be small and would be limited to the immediate vicinity of the facilities (which would be located on a partially developed site), thus reducing the potential for cumulative impact. Any cumulative impacts to these resources would be proportional to the cumulative impact projected for land development at the SRS.

Cumulative impacts to air quality were evaluated for five pollutants — TSP, PM_{10}, NO_2, SO_2, and CO. Normal operations of the MOX, PDCF, and WSB facilities would result in small contributions (2% or less) to cumulative concentrations of these air pollutants (see Table 4.24). For four air pollutants (annual total suspended particulates, 24-hour PM_{10}, 3-hour SO_2, and 24-hour SO_2), the cumulative total concentrations would be above 90% of the NAAQS and, therefore, approaching noncompliance. However, even without the contributions from operations of the proposed facilities, the cumulative totals for these four pollutants would be above 90% of the NAAQS. The cumulative total concentration of $PM_{2.5}$ could not be

Table 4.24. Estimated cumulative impacts to air quality from MOX, PDCF, and WSB facility operations and other activities at the SRS[a]

Source	TSP, annual	PM10		NO2, annual	SO2			CO	
		24 h	annual		3 h	24 h	annual	1 h	8 h
SRS baseline[b, c]	74.6	138	25.9	26.3	1,246	355	31.1	10,363	6,867
MOX facility, PDCF, and WSB	0.002	1.3	0.002	0.07	22	4.9	0.004	116	26
SNF management[d]	0.02	0.1	0.02	3.4	1.0	0.1	0.02	9.8	1.3
HEU disposition[e]	0.05	0.01	0.01	0.01	0.7	0.3	0.02	0.1	0.07
Tritium extraction facility[f]	0.0002	0.01	0.00009	0.006	0.09	0.001	0.00009	3.6	0.5
Plutonium residues[g]	0.0	0.0	0.0	0.04	0.0	0.0	0.0	0.0	0.0
Salt processing[h]	0.001	0.07	0.001	0.03	0.4	0.05	0.0005	18.0	2.3
Tank closure[i]	0.005	0.08	0.004	0.03	0.2	0.04	0.002	1.2	0.3
Modern Pit Facility[j]	0.18	0.33	0.07	2.4	1.9	0.83	0.17	6.8	4.7
Total concentration (µg/m³)	74.9	139.9	26.0	32.2	1,247.6	361.5	31.3	10,518.2	6,902.5
MOX, PDCF, and WSB contribution (%)	0.00	0.9	0.01	0.23	1.8	1.4	0.01	1.1	0.4
NAAQS (µg/m³)	75	150	50	100	1,300	365	80	40,000	10,000
% of standards	99.6	93.3	52.0	32.3	96.0	99.0	39.2	26.3	69.0

Pollutant concentrations (µg/m³)

[a]Maximum predicted off-site cumulative ground-level concentrations of nonradiological pollutants.

[b]SRS baseline includes the impacts of existing SRS facilities (SRS maximum) and regional emissions (background) from Table 4.8. These values are hypothetical levels that are based on maximum permitted emissions from SRS sources and do not necessarily represent actual air quality conditions.

[c]Includes Defense Waste Processing Facility operations.

[d]*Source:* DOE (2000c).

[e]*Source:* DOE (1996b).

[f]*Source:* DOE (1999c).

[g]*Source:* DOE (1998).

[h]*Source:* DOE (2001) using maximum impact alternative.

[i]*Source:* DOE (2002b).

[j]*Source:* DOE (2003b).

determined because information was not available for many of the future actions considered here. However, the facilities would contribute a very small amount of $PM_{2.5}$ (0.009% of the annual standard) and only when emergency generators were used. It should be noted that all of the air quality analyses are based on very conservative assumptions (e.g., maximum concentrations for all facilities), and it is not likely that NAAQS exceedances would occur at the SRS.

During normal operations, the contribution of the MOX, PDCF, and WSB facilities to cumulative radiological dose to the public would be small (7% or less of total dose; see Table 4.25). The cumulative dose to an MEI would increase by 1% as a result of facility operations. The estimated risk of a LCF resulting from cumulative dose to the MEI is extremely small (4×10^{-7}). The estimated number of LCFs resulting from cumulative collective dose to the off-site population is 0.02. These very small numbers mean that statistically, radiological doses from plant operations would not be expected to cause any latent cancer fatalities in the off-site population.

Cumulative collective dose to workers at SRS would increase approximately 9% as a result of MOX, PDCF, and WSB facility operations. The number of expected LCFs among workers resulting from cumulative dose (that resulting from dose contributions from the SRS baseline, the proposed action, and other reasonably foreseeable future actions) is 1.7. For most types of waste, facility operations would contribute relatively small volumes to the cumulative waste generation volumes at the SRS (see Table 4.26), and existing waste treatment facilities at the SRS have sufficient capacity to treat this cumulative total (see Section 4.3.4.2). The largest proportionate increase would be in the amount of nonhazardous solid waste (approximately 19% increase).

The cumulative impacts of the facilities to land development, electricity usage, and groundwater usage at the SRS would be quite small and well within existing SRS capacity (see Table 4.27). Construction of the facilities would result in a slight increase (1.7%) in the amount of developed land at the SRS, but the cumulative amount of developed land on the SRS would remain quite small (3.9% of the total site). Facility operations would use 186,000 MWh/yr of electricity (3.6% of SRS capacity). Cumulative electricity demand resulting from facility operations and all existing and planned actions would be only 28% of SRS capacity. Facility operations would use 76 million L/yr (20.1 million gal/yr) of groundwater (0.02% of SRS capacity). Cumulative groundwater demand would be only 4.8% of SRS capacity.

Determination of the cumulative impacts on the SRS workforce is complicated by the fact that employment is not expected to be constant during the life of the facility and other existing and planned actions at the SRS discussed in the beginning of Section 4.5.1. The analysis presented here considered the time lines of workforce projections for the SRS baseline and reasonably foreseeable future actions and the year in which the workforce would be highest. The results of these conservative analyses are presented in Table 4.27. Overall, employment at the SRS has decreased from 22,070 in September 1993 to 14,193 in September 2000. Projections indicate that site employment will continue to decline to approximately 10,000 by

Table 4.25. Estimated annual cumulative radiological dose and latent cancer fatalities resulting from MOX, PDCF, and WSB facility operations and other activities at the SRS

Source	Dose to maximally exposed individual[a]				Collective dose to off-site population				Collective dose to workers	
	Air pathway (rem)	Liquid pathway (rem)	Total dose (rem)	Latent cancer fatalities[b]	Air pathway (person-rem)	Liquid pathway (person-rem)	Total dose (person-rem)	Latent cancer fatalities[b]	Total dose (person-rem)	Latent cancer fatalities[b]
SRS baseline[c]	4.0×10^{-5}	1.4×10^{-4}	1.8×10^{-4}	1.1×10^{-7}	2.3	3.9	6.2	3.7×10^{-3}	163	0.1
MOX, PDCF, and WSB	4.0×10^{-6}	—d	4.0×10^{-6}	2.4×10^{-9}	1.6	—d	1.6	9.4×10^{-4}	262	1.6×10^{-1}
SNF management[e]	1.5×10^{-5}	5.7×10^{-5}	7.2×10^{-5}	4.3×10^{-8}	0.6	0.2	0.8	4.5×10^{-4}	55	3.3×10^{-2}
HEU disposition[f]	2.5×10^{-6}	—d	2.5×10^{-6}	1.5×10^{-9}	0.2	—d	0.2	9.6×10^{-5}	11.3	6.8×10^{-3}
Tritium extraction facility[g]	2.0×10^{-5}	—d	2.0×10^{-5}	1.2×10^{-8}	0.8	—d	0.8	4.6×10^{-4}	4.0	2.4×10^{-3}
Plutonium residue management[h]	5.7×10^{-7}	—d	5.7×10^{-7}	3.4×10^{-10}	0.006	—d	0.006	3.7×10^{-5}	7.6	4.6×10^{-3}
Defense waste processing facility[i]	1.0×10^{-6}	—d	1.0×10^{-6}	6.0×10^{-10}	0.07	—d	0.07	4.2×10^{-5}	118	7.1×10^{-2}
Salt processing[j]	3.1×10^{-4}	—d	3.1×10^{-4}	3×10^{-9}	18.1	—d	18.1	1.1×10^{-2}	29	1.7×10^{-2}
DOE complex miscellaneous components[k]	4.4×10^{-6}	4.2×10^{-8}	4.4×10^{-6}	2.7×10^{-11}	0.007	2.4×10^{-4}	0.007	4.3×10^{-6}	2	1.2×10^{-3}
Tank closure[l]	2.5×10^{-8}	—d	2.5×10^{-8}	1.5×10^{-11}	0.0014	—d	0.0014	8.4×10^{-7}	1,600	1.0
Modern Pit Facility[m]	8.0×10^{-9}	—d	8.0×10^{-9}	4.8×10^{-12}	1.3×10^{-6}	—d	1.3×10^{-6}	7.8×10^{-10}	560	3.4×10^{-1}
Vogtle Nuclear Power Plant[n]	5.4×10^{-7}	5.4×10^{-5}	5.5×10^{-5}	3.3×10^{-8}	0.04	0.003	0.05	2.7×10^{-5}	—d	—d
Total	4.0×10^{-4}	2.5×10^{-4}	6.5×10^{-4}	3.9×10^{-7}	23.6	4.1	27.7	0.02	2,812	1.7
MOX, PDCF, and WSB contribution to total (%)	1.0	0.00	0.62	0.62	6.7	0.00	5.7	5.7	9.3	9.3

See next page for footnotes.

Table 4.25. Continued

[a]The MEIs for different facilities for the same pathway and the MEIs for different pathways for the same facility are likely to be different individuals. Therefore, simple addition of doses for all MEIs to estimate the total MEI dose is not accurate, but it is shown here to be conservative, (i.e., to present impacts that are overestimates of what would actually happen).

[b]Latent cancer fatalities are calculated by multiplying dose by the FGR-13 health risk conversion factor of 6×10^{-4} fatal cancer per person-rem (Eckerman 1999).

[c]SRS baseline includes the impacts of existing facilities and the residual impacts of past activities. Values are from Arnett and Mamatey (2001).

[d]Less than minimum reportable levels.

[e]*Source:* DOE (2000c); SNF = spent nuclear fuel.

[f]*Source:* DOE (1996b).

[g]*Source:* DOE (1999c).

[h]*Source:* DOE (1998).

[i]*Source:* DOE (1994).

[j]*Source:* DOE (2001).

[k]*Source:* DCS (2002a).

[l]*Source:* DOE (2002b).

[m]*Source:* DOE (2003b).

[n]*Source:* NRC (1996).

Table 4.26. Estimated cumulative waste generation at the SRS resulting from operation of the MOX, PDCF, and WSB facilities and other activities at the SRS

Source	Total waste generation over 30-year period (m³)			Annual waste generation	
	Low-level waste	Hazardous-mixed waste	Transuranic waste	Nonhazardous solid waste (m³)	Nonhazardous liquid waste (L)
SRS baseline[a,b]	120,000	3,900	6,000	6,670	4.2×10^8
MOX, PDCF, and WSB facilities[c]	28,838	120	4,431	4,140	6.0×10^7
Salt processing[d]	920	56	0	–[e]	Negligible
Environmental restoration and D&D activities[d]	62,000	6,200	0	NA[f]	NA
Modern Pit Facility[g]	150,900	290	33,900	6,900	8.2×10^7
Other future actions[h,i]	21,750	4,013	10,100	4,105	2.2×10^7
Total volume	384,408	14,580	54,521	21,815	5.8×10^8
MOX, PDCF, and WSB contribution to total (%)	7.5	0.8	8.1	19.0	10.4
SRS treatment capacity	534,900	534,900	–[j]	–[k]	1.0×10^9
Total volume as % of SRS capacity	71.9	2.7	–[j]	–[k]	58.0

[a]SRS baseline includes the impacts of existing facilities and the residual impacts of past activities.

[b]High-level, low-level, hazardous-mixed, and transuranic waste volumes from DOE (2001); nonhazardous solid and liquid waste volumes from DCS (2002a).

[c]Total waste generation for MOX, PDCF, and WSB operations over a 10-year period, the operational period of these facilities.

[d]*Source:* DOE (2001).

[e]Value presented in DOE (2001) as 61 metric tons/yr.

Footnotes continued on next page.

Table 4.26. Continued

[f]NA = not available.

[g]Source: DOE (2003b).

[h]30-year waste generation volumes include life-cycle waste associated with DWPF operations (DOE 1994), HLW tank closure (DOE 2002b), SNF management (DOE 2000c), Tritium Extraction Facility (DOE 1999c), plutonium residues (DOE 1998), HEU disposition (DOE 1996b), commercial light water reactor waste, and weapons components that could be processed at the SRS. Values presented were derived from values provided in DOE (2001), but were adjusted to remove the contribution from SPD facilities (included in salt processing values) and sodium-bonded SNF management, which no longer involves SRS operations.

[i]Nonhazardous waste volumes include waste generated by activities associated with HEU disposition (DOE 1996b), Tritium Extraction Facility (DOE 1999c), DWPF operations (DOE 1994), and HLW tank closure (DOE 2002b).

[j]Transuranic waste is transported off-site for disposal at the WIPP facility.

[k]Nonhazardous solid waste is recycled or disposed of at on-site and off-site facilities.

Table 4.27. Estimated cumulative impacts to resource use and employment from MOX, PDCF, and WSB facility operations and other activities at the SRS

Source	Land area		Electricity		Groundwater		Employment	
	Developed area (acres)	% Total SRS area	Average annual usage (MWh/yr)	% Total SRS capacity	Average annual usage (L/yr)	% Total SRS capacity	Number of workers	% SRS total
SRS baseline[a]	7,241[b]	3.7	411,000	9.3	1.7×10^{10}	4.7	13,227	78.2
MOX, PDCF, and WSB facilities	123	0.06	186,000	4.2	7.6×10^{7}	0.02	490	2.9
SNF management[b]	0	0.00	15,800	0.4	2.1×10^{8}	0.06	520	3.1
HEU disposition[c]	0	0.00	5,000	0.1	1.9×10^{7}	0.005	125	0.7
Tritium extraction facility[d]	3	0.002	20,600	0.5	NA[e]	NA	400	2.4
Plutonium residue management[f]	0	0.00	1,329	0.03	1.6×10^{7}	0.005	NA	NA
Defense waste processing facility[g]	105	0.05	32,000	0.7	NA	NA	60	0.4
Salt processing[h]	0	0.00	24,000	0.6	1.2×10^{7}	0.003	220	1.3
Tank closure[i]	0	0.00	0	0.0	8.7×10^{6}	0.002	85	0.5
Modern pit facility[j]	171	0.09	545,600	12.4	5.0×10^{8}	0.1	1,797	10.6
Total	7,643	3.9	1,241,329	28.2	1.8×10^{10}	4.8	16,924	100.0

[a]SRS baseline includes the impacts of existing facilities and the residual impacts of past activities.

[b]*Source:* DOE (2000c); SNF = spent nuclear fuel.

[c]*Source:* DOE (1996b).

[d]*Source:* DOE (1999c).

[e]NA = not available.

[f]*Source:* DOE (1998).

[g]*Source:* DOE (1994).

[h]*Source:* DOE (2001).

[i]*Source:* DOE (2000a).

[j]*Source:* (DOE 2003b).

2010 (DOE 1999c). Facility construction would result in a peak workforce of 1,000 in 2005. Facility operations would support 490 workers annually (3.2% of the total projected for the SRS).

4.5.1.2 Cumulative Impacts of the No-Action Alternative

The no-action alternative would be a decision by the NRC not to approve the proposed MOX facility. Because all the surplus plutonium would remain at the DOE sites, the facilities planned for processing this surplus plutonium at the SRS — the proposed MOX facility, PDCF, and the WSB — would not be constructed. Since none of the surplus plutonium from other DOE sites would be stored at the SRS, none of the projected impacts of these facilities (as presented in Section 4.5.1.1) would occur.

4.5.2 Cumulative Impacts of Transportation

Cumulative impacts of transportation were estimated by adding the contributions from four sources:

- Historical shipments of spent nuclear fuel and radioactive waste;

- Reasonably foreseeable future actions involving the transportation of radioactive materials;

- Spent fuel shipments to a geological repository at Yucca Mountain, Nevada;

- General transportation of radioactive materials not related to any particular action; and

- Transportation of surplus plutonium and depleted uranium to the SRS, fresh MOX fuel from the SRS to a surrogate Midwest nuclear power plant, and TRU waste to the WIPP.

Estimates of contributions from the first four sources to the collective occupational dose and dose to the general population were summarized in the EIS for a geological repository at Yucca Mountain (DOE 2002d). These estimates are presented in Table 4.28. The future SNF shipments listed in Table 4.28 include potential spent MOX fuel shipments to the repository.

The shipment risks from spent MOX fuel are similar to those for typical SNF. Therefore, these risks are expected regardless of the fuel type, normal LEU or MOX, that will be used in existing nuclear power plants in the future. The estimated dose resulting from the proposed action is similar to that resulting from historical shipments of spent nuclear fuel and radioactive waste, 100 times smaller than that resulting from reasonably foreseeable future actions and 1,000 times less than general transportation. The contribution to cumulative occupational and general population dose associated with the proposed action is expected to be insignificant.

**Table 4.28. Estimated cumulative transportation impacts
of facility operations and shipment of radioactive
materials from other sources (1943 to 2048)**

Category	Collective occupational dose [person-Sv (person-rem)]	Latent cancer fatalities	Collective dose to the general population [person-Sv (person-rem)]	Latent cancer fatalities
Historical shipments[a]	3.3 (330)	0.2	2.3 (230)	0.1
Reasonably foreseeable future actions[a]	197 (19,670)	12	498 (49,770)	30
Spent fuel shipments to geologic repository[a]	88 (8,800)	5	16 (1,600)	1
General transportation (1943 to 2048)[a]	3,300 (330,000)	198	2,900 (290,000)	174
MOX shipments[b]	2.1-5.3 (210-530)	0.1-0.3	3.3-5.6 (330-560)	0.2-0.4
Total	3,600 (360,000)	200	3,400 (340,000)	200

[a]*Source:* DOE (2002d).

[b]Doses represent total for all shipments associated with the MOX program. (See Table 4.20 [total campaign].)

4.6 Cost-Benefit Analysis

4.6.1 Introduction

This section compares the costs and benefits of the proposed action with the costs and benefits of the no-action alternative. The cost-benefit analysis sets forth the various environmental impacts (both negative and positive) of the proposed action, and the economic costs and benefits of building and operating the proposed MOX facility, the PDCF, and the WSB. Costs and benefits are assessed at both the national and regional levels. At the national level, the overall costs of proposed MOX facility construction and operation are compared with the benefits of plutonium supply reduction. The benefits to national security from plutonium supply reduction are substantial, but these benefits are not quantifiable in terms of dollars and cents.

The national benefits associated with the proposed action that are quantifiable include project expenditures during construction and operation of the proposed MOX facility, the PDCF, and the WSB. Various sectors in the national economy would provide the materials, equipment, and services needed to build and operate these facilities. However, because of the preliminary nature of the data needed to calculate impacts, no quantitative estimate of the impacts of construction and operation of the proposed MOX facility on the national economy was included in this EIS. A significant national benefit of the proposed action would be the avoided cost of continued plutonium storage. These costs are estimated to be approximately $256 million per year (2003 dollars) (NNSA 2002). Another national benefit of the proposed action would be the generation of additional supplies of electricity. However, this analysis does not assign a specific economic value to the electricity that would be generated by the irradiation of MOX fuel given the uncertainty surrounding the associated costs, in particular, the cost of power plant infrastructure upgrades.

There would also be regional costs and benefits associated with construction and operation of the proposed MOX facility. At the regional level, excluding costs and benefits that cannot be quantified, the proposed MOX facility would produce an overall net benefit of $1,940 million (see Table 4.29).

4.6.2 National Costs and Benefits

The primary national benefit of construction and operation of the proposed MOX facility would be a reduction in the supply of weapons-grade plutonium available for unauthorized use. Once the plutonium component in MOX fuel has been irradiated in commercial nuclear reactors, the isotopic composition of the plutonium would be more proliferation resistant. Moreover, since the plutonium would then be part of the resultant high-level nuclear waste, the plutonium would no longer be available for other uses. Compared with the no-action alternative — in which the weapons-grade plutonium would continue to be stored at several existing DOE locations — converting surplus plutonium into MOX fuel and irradiating it better ensures its security, since it would reduce the number of locations where the various forms of plutonium are stored (DOE 1997a). Converting surplus weapons-grade plutonium into MOX fuel is thus viewed as better ensuring that weapons-usable material would not be obtained by rogue states and terrorist groups. Implementing the proposed action would promote the above nonproliferation objectives.

A significant benefit of the MOX program would be the avoided cost of continuing to store the plutonium inventory. These costs are estimated to be approximately $256 million per year (2003 dollars) (NNSA 2002).

For the no-action alternative, although the costs and benefits of continued storage of plutonium in the present DOE locations are not re-evaluated in this analysis, these issues are discussed in the SPD EIS (DOE 1999a). Some of the impacts of the no-action alternative represent impacts of each entire DOE site, not just the impacts of continued storage. Continued storage of plutonium by the DOE at its present locations would not be expected to produce additional LCFs. Annual LCFs of approximately 0.002 in the surrounding population of the storage sites were estimated. The annual collective dose to members of the public (i.e., those living and working within 80 km [50 mi] of the SRS) produced by routine operation of the proposed MOX facility, the PDCF, and the WSB would be expected to result in an LCF rate of approximately 0.0009/yr or less. Therefore, continued storage would result in higher annual impacts.

**Table 4.29. Summary of project costs and benefits in the REA
(in millions of 2003 dollars, except where noted)**

Item	MOX facility[a]
Costs	
Internal costs	
Construction	6
Operation	3
Short-term external costs (construction)	
Housing shortages	2% of vacant rental housing units would be required
Overcrowding in local public facilities	Minimal
Inflation	Minimal
Noise and congestion	Minimal
Water and sewage systems	Minimal
Long-term external costs (operations)	
Housing values	Less than 1% of vacant owner occupied housing would be required
Cost of providing public services	Less than 1% increase in revenues would be required
Deterioration in recreational values	Minimal
Restrictions to water and land	Minimal
Aesthetic values	Minimal
Cultural and historical sites	Minimal
Total REA costs	9
Benefits	
Avoided cost of continued plutonium storage	14
Total tax revenues	110
Economic activity in the REA	
Construction	
Annual average employment	1,020 jobs
Total income	370
Total regional product	760
Operations	
Annual average employment	1,270 jobs
Total Income	640
Total regional product	1,180
Other benefits	
Enhancement of recreational values	Minimal
Increased knowledge of the environment	Minimal
Total REA benefits	1,950
Net REA benefit	+1,940

[a]Data may not add to totals because of independent rounding.

The national costs associated with the proposed action are the total life-cycle costs, which include research and development and pre-capital costs, design and construction costs, operating costs, deactivation costs, and contingency costs. Decommissioning costs are not included given the uncertainty surrounding their magnitude. The total cost of the proposed action is estimated to be $4,064 million (in 2003 dollars), with $2,238 million to cover the cost of the proposed MOX facility and $1,825 million for the PDCF and WSB (NNSA 2002). A significant item included in the estimated total cost of the proposed facilities is the credits associated with the value of the MOX and HEU fuel. These items amount to $1,002 million over the life of the project (NNSA 2002).

4.6.3 Regional Costs and Benefits

The various quantifiable costs and benefits of the proposed MOX facility in the REA are identified in Table 4.29. Costs and benefits are presented for construction and operation, including decommissioning, over a 20-year project life. On balance, the proposed MOX facility would provide a net benefit (total benefits minus total costs) to the REA. The net benefit of the proposed MOX facility would be approximately $1,940 million. Sections 4.6.3.1 and 4.6.3.2 provide a more detailed description of the costs and benefits of the proposed MOX facility.

4.6.3.1 Regional Costs

Both potential internal and external costs are included in the assessment. Potential external costs include both long-term and short-term costs. Long-term external costs can also be associated with potential accidents at the proposed facilities. The impacts of accidents associated with the proposed facilities on agriculture, water, and fisheries resources, and subsequently on the economies of communities surrounding SRS, would be small. In the case of the most serious accidents, potential damage to crops under the plume in the event of an airborne release and the subsequent damage to water resources from the associated runoff would be small because the amount of radioactive material deposited per unit area would be relatively small. Dilution of runoff would occur fairly rapidly in the affected rivers and streams and would not cause any significant risk to the economies of the communities downstream of the location of the proposed facilities. Any interdiction of crops as a result of the deposition of radioactive material would be a limited, one-time event, and if it were to occur at all, would only affect a small number of farm communities.

Although the probability of severe accidents is very low, if such accidents did occur, the people living within 80 km (50 mi) of the SRS would likely be affected. The extent to which the surrounding population would be affected would depend on the amount of material released and the direction and speed at which airborne material was dispersed by wind conditions at the time of the accident. While the overall risk to the surrounding population would be very low (since the probability of severe accidents occurring would be very low), the greatest short-term risk of exposure would be to population groups located to the west-northwest of SRS, while the greatest 1-year risk would be to the southwest of SRS from crop contamination.

Routine operation of the proposed facilities is expected to produce an annual latent cancer risk of about 1 in 250 million for the maximally exposed member of the public. The annual collective dose (associated with the facilities) to members of the public living and working within 80 km (50 mi) of SRS is expected to produce an LCF risk of approximately 0.0009 or less.

No adverse impacts from chemical exposure of workers at the proposed facilities are anticipated. Less than one fatality and approximately 410 worker injuries are expected during the 10-year operating period of the proposed facilities.

Routine proposed facilities operations are expected to produce insignificant impacts to air quality and would not exceed any ambient air quality standards for criteria pollutants at SRS. Maximum levels of $PM_{2.5}$ in the vicinity of SRS already exceed the applicable levels, and facility construction would create an additional 0.07% of the present standard; facility operations would contribute 0.009%.

Water consumption during operation of the proposed MOX facility, PDCF, and WSB would represent an increase of about 5% of the water demand for the A-Area loop in 2000 and about 2% of the excess A-Area loop capacity. Discharges to surface water from the WSB during facility operations would comply with the NPDES permit guidelines.

Waste management systems at SRS would not be adversely affected by wastes generated by the proposed facilities. Adequate storage capacity and handling procedures are in place at SRS to process hazardous wastes generated during both construction and operation. Nonhazardous liquid and solid wastes would not adversely affect the Central Sanitary Waste Treatment Facility.

Other long-term external costs would include the potential impact of the proposed MOX facility, PDCF, and WSB (proposed facilities) on deterioration in recreational values, access restrictions to water or land (including any income lost), aesthetic impacts, impacts on local cultural and historical sites, decreased housing values, and the increased cost of providing local public services.

No impacts to recreational values, local aesthetic quality, or local water or land access would be expected from the proposed facilities. The location of the proposed facilities is close to the center of the SRS, and no recreation opportunities are currently available to the public in the vicinity. The proposed facilities would not change the industrial nature of the F-Area, and since the closest viewing location is about 8 km (5 mi) to the south, no changes in aesthetic quality would be expected (see Appendixes G and H). Construction of the facilities would occur on land already owned by the federal government and would have no impact on water or land access.

Impacts to housing values resulting from facility construction and operation, or to the cost of providing local public services are unlikely because of the relatively small number of long-term new residents that would be expected to move into the REA from elsewhere. Sufficient local housing is likely to be available to absorb new residents. Only 2% of vacant rental housing would be needed for workers during construction and less than 1% of vacant owner-occupied housing would be needed during operations. Changes in local public expenditures to maintain existing levels of public services would likely be small, with five additional local public service employees likely to be required (see Appendixes G and H).

The impacts of MOX fuel transportation, including those on property values, were not considered because of uncertainty surrounding the routes that would be used and the timing of shipments.

Short-term external costs include the contribution of the proposed facilities to housing shortages; local inflation, noise, and congestion; impacts on the local water supply and sewage systems; and crowding in local public schools, hospitals, and other local public facilities.

The proposed facilities would not produce any significant costs in the REA at the SRS in the short term. Sufficient vacant rental units would be available in the REA for use by construction workers, and sufficient owner occupied units would be available to operations employees (see Section G.2.7 in Appendix G). Inflation in prices in the local area is not likely because much of the equipment, materials, and services required would be specialized, and a significant portion would be obtained from outside the REA. Material and equipment expenditures assumed to be made locally would not likely push local industries to capacity, and no labor shortages would be likely. Any construction and managerial positions not filled from within the local labor market would be taken by workers moving to the area from other labor markets in the southeastern United States (see Appendixes G and H).

Noise and congestion from construction activities for the proposed facilities would likely be minor. Additional traffic generated during construction and operation would be unlikely to cause any additional traffic congestion on the major road segments surrounding the site, given the relatively small incremental increase in traffic from the proposed action (see Appendix H). Relatively small utility requirements would mean that no impacts would be expected on the local water supply and sewage systems. Local public schools, hospitals, and other local public facilities are not expected to suffer any overcrowding because of the relatively small number of new residents expected during the construction and operation under the proposed action (see Appendix H).

Internal costs are the life-cycle costs of design, construction, and operation of the project borne by the federal government. The internal costs of the proposed action in the REA are approximated using a cost localization factor that apportions total life-cycle project costs on the basis of the ratio of REA population to total national population. Internal costs apportioned to the REA using this method are small, amounting to $9 million for the proposed action.

4.6.3.2 Regional Benefits

The potential benefits of construction and operation of the proposed facilities include economic benefits — such as employment, income, and gross regional product — and various additional potential benefits — such as enhancement of recreational values, environmental enhancement in support of the protection of wildlife and wildlife habitat, and increased knowledge of the environment.

A significant benefit of the proposed action would be the avoided costs of continued plutonium storage. At the national level, these costs are estimated to be approximately $256 million per year (NNSA 2002) and would be incurred for as long as the material continued to be stored. Application of the same localization factor used in Section 4.6.3.1 to estimate the regional portion of plutonium storage costs avoided with the construction and operation of a MOX facility indicates that $14 million would be saved over what it would cost if plutonium was stored in existing facilities for an additional 25 years.

The measurement of the local employment and income economic benefits is based on the use of regional economic multipliers. These multipliers capture the indirect (off-site) effects of on-site activities associated with construction and operation.

To estimate employment benefits, life-cycle cost estimates were used (NNSA 2002) in association with data on the relationship between direct and indirect (off-site) employment benefits associated with construction and facility operations at the SRS. Data on the relationship between direct and indirect employment for a MOX facility were taken from the SPD EIS (DOE 1999a; see Appendix F, Section 9.2 for more information on the methodology used). By using direct (on-site) facilities employment data taken from the ER (DCS 2002a) as the basis for calculation, the indirect employment impacts during the construction and operation of the proposed facilities were estimated by application of the direct-to-indirect employment multiplier for the project at the SRS from the SPD EIS. The direct impacts of no action were estimated by using the relationship between total annual cost during construction and operation and direct employment for the proposed action. Indirect impacts were then estimated by application of the direct-to-indirect employment multiplier for a proposed MOX facility at the SRS from the SPD EIS (DOE 1999a).

The impacts on regional income of construction and operation were estimated by using employment impact estimates together with average regional income multipliers for the REA taken from IMPLAN regional economic data (MIG Inc. 2001). IMPLAN input-output economic accounts show the flow of commodities to industries from producers and institutional consumers. The accounts also show consumption activities by workers, owners of capital, and imports from outside the region. The IMPLAN model contains 528 sectors representing industries in agriculture, mining, construction, manufacturing, wholesale and retail trade, utilities, finance, insurance and real estate, and consumer and business services. The model also includes information for each sector on employee compensation; proprietary and property income; personal consumption expenditures; federal, state, and local expenditures; inventory and capital formation; and imports and exports.

The benefits of the proposed facilities to the economy of the REA would be significant (see Table 4.29). In the peak year of construction, 1,820 workers would be required for the proposed action. On average, 1,020 jobs would be created for the proposed facilities during the construction period. During operations, 1,270 workers would be required in each year. The facility would also contribute significantly toward personal income within the REA. The proposed facilities would produce $370 million in income over the construction period and $640 million during operations (see Appendix H).

No taxes are paid by the federal government (income, property, or sales taxes), and contractors constructing and operating a facility on behalf of the federal government are currently exempt from local sales taxes in Georgia and South Carolina. Although local tax revenues, primarily state income and sales tax revenues, paid by federal government employees, contractors, and their employees would increase, the increase would be relatively small. During both construction and operation, the proposed facilities would produce approximately $110 million in tax revenues in the REA.

The gross regional product (GRP) provides the best measure of the overall benefits of both alternatives to the economy of the REA. The GRP is the sum of value added in the production of all goods and services in a year and measures the overall level of economic activity in the REA. The proposed facilities would produce $1,950 million in GRP in the REA over the entire life of the project.

4.7 Resource Commitment

Construction of the proposed facilities would result in some impacts that cannot be avoided. Impacts may be irreversible if the future uses of the resource are limited. This section addresses unavoidable, irreversible, and irretrievable impacts of constructing and operating the facility and the relationship between short-term uses of F-Area and the SRS for the facility and long-term productivity. A summary of unavoidable impacts is presented in Table 4.30.

4.7.1 Unavoidable Adverse Environmental Impacts

Geology and Soils. Impacts to geology and soils from construction and operation of the proposed MOX facility, PDCF, and WSB are expected to be insignificant. Restoration work, consisting of final grading and revegetation, would reclaim over half of the 41.9 ha (103.5 acres) of land in the F-Area that would be disturbed during construction. The 41.9-ha (103.5-acre) disturbed area is assumed to include 2 ha (4.9 acres) for laydown area for constructing the PDCF, and 9.7 ha (24 acres) for a laydown area for constructing the WSB.

Some land in the area would be permanently altered because of constructing buildings, roads, and parking lots. The proposed MOX facility would permanently alter 6.9 ha (17 acres) of land, the PDCF would permanently alter 1.2 ha (3 acres), and the WSB would permanently alter about 2.5 ha (6.2 acres). Because soils in the affected areas are not unique within the SRS, and the permanently altered areas represent only about 7% of the land available in F-Area (160 ha [395 acres]) and only about 0.01% of the 80,292 ha (198,400 acres) of land area at SRS (DCS 2002a), overall physical impacts on soil would be insignificant.

Table 4.30. Unavoidable impacts of constructing and operating the proposed facilities

Resource	Unavoidable impacts
Geology and soils	• Construction excavation work may result in release of contaminated materials
Surface water	• Potential impacts to surface water quality by release of nonhazardous discharge effluent, sediment, contaminated runoff, or accidental release of oil or construction equipment fuel
Ecology	• Initial loss of up to 50.0 ha (123.4 acres) of woodland and grassland habitat in F-Area. Over 30 ha (75 acres) would be landscaped following construction.
Land use	• A worst-case accident at the facility could result in minor land use impacts outside of the SRS
Cultural and paleontological resources	• Construction would directly affect two prehistoric archaeological sites that are eligible for listing on the *National Register of Historic Places*
Waste management	• Small impact to waste management system at the SRS
	• Volumes of TRU and hazardous waste produced by facilities would represent 3% of the WIPP disposal capacity and 2% of the SRS treatment and storage capacity, respectively.
	• Nonhazardous liquids produced would be about 6% of the capacity at SRS.
Human health risk	• Annual radiological impacts to SRS employees from exposure to radioactive air pollutants are expected to be small at 1×10^{-5} LCFs/yr for the MOX facility and WSB collectively and 2×10^{-5} for the PDCF. The risk from the public's exposure to radioactive air pollutants is also expected to be small, at 4×10^{-5} annual LCFs for MOX and WSB combined, and 9×10^{-4} for the PDCF facilities.
	• MOX facility workers would have an expected lifetime LCF of about 1 chance in 1,000.
	• 122 lost workday injuries annually during a 3-5-year construction period
	• 41 lost workday injuries annually during 10 or more years of operations
Socioeconomics	• Increase in employment of 0.1 of a percentage point during construction
	• In-migrating workers during construction and operations would require 2% and <1% of vacant housing in the ROI

The potential exists that accidental releases of contaminated material during construction and normal operations might adversely affect receiving soils. However, if good engineering practices were used and any accidental spills were cleaned up promptly and thoroughly, chemical impacts to soil would be insignificant.

Surface Water. Impacts to surface water are expected to be negligible. Because surface water would not be used to supply water for construction or operations, there would be no impacts to surface water levels or flows.

Surface water quality could potentially be impacted by nonhazardous discharge effluent, sediment, contaminated runoff, and accidental releases. However, good engineering practices, compliance with existing NPDES permits, and prompt, thorough cleanup of accidental releases would help to ensure that impacts to surface water quality during construction and normal operations would be insignificant.

Groundwater. The groundwater system beneath the SRS would be directly affected (i.e., used) during construction and normal operations of the proposed facilities because it is the only source of water for these activities. However, the impact to existing groundwater supplies would be small. Projected total water use for the proposed and existing facilities in the A-Area loop, which obtains water from wells in both A-Area and F-Area, represents about 3% of the existing capacity during the construction phase. There would be no releases to underlying aquifers.

No direct impacts to groundwater quality (as opposed to quantity) are expected from construction or normal operations; there would be no releases to underlying aquifers. Water use during operation of the facilities represents an increase of about 5% of the water demand for A-Area loop in 2000 and about 2% of the excess A-Area loop capacity. Groundwater quality could be impacted by discharges to an NPDES outfall and accidental releases of contaminated material. However, impacts are expected to be negligible because of good engineering practices, prompt and thorough cleanup of any spills, and adherence to NPDES permit requirements.

Air Quality. Emissions associated with the construction and normal operation of the proposed facilities would have a negligible effect on air quality. Concentrations of pollutants would remain below standard levels. For both construction and normal operations, contributions of the proposed facilities to TSP, PM_{10}, $PM_{2.5}$, CO, SO_2, NO_2, and PAH concentrations would be 5.0% or less of applicable standard levels.

Noise. Potential noise impacts from construction and operation of the proposed facilities should be negligible at all off-site locations.

Ecology. Impacts of construction on ecological resources would primarily result from the loss and alteration of up to 50.0 ha (123.4 acres) of habitat. The woodland and grassland habitats that would be impacted represent a small fraction of those types of habitats at the SRS. Overall, the adverse impacts related to construction are expected to be limited to the immediate project vicinity and should not affect the viability of any vegetation types or wildlife populations at the SRS.

Sediment and erosion control measures implemented during site preparation and construction should prevent impacts to surface waters, aquatic and wetland resources, and protected fish species. No federally listed species have been reported in the areas that will be disturbed by construction. The SRS has established habitat management areas for the federally and state-endangered red-cockaded woodpecker, but the proposed facilities would not be located within any of these areas.

No adverse impacts to ecological resources are expected from operations of the proposed facilities.

Land Use. Land use of the entire F-Area is currently classified as developed/industrial. Since the facilities would be industrial, no adverse effects to land use would result from their construction or routine operation. If an operational accident occurred, F-Area would remain in developed/industrial land use. A worst-case accident could result in minor impacts to lands outside of the SRS. Future F-Area land use is expected to remain developed/industrial.

Cultural and Paleontological Resources. Construction of the proposed facilities would directly affect two prehistoric sites that are eligible for listing on the NRHP. Data recovery plans have been implemented, excavation has been completed, and monitoring will be conducted during ground-disturbing construction activities. Five additional eligible sites are located in the vicinity of the construction area. Mitigation measures would be taken to ensure that these sites were not disturbed directly or indirectly by construction activities.

No historic structures, traditional cultural property, or fossil-bearing strata have been identified in the project area; therefore, there would be no MOX-related impacts to such resources during construction.

Routine operations are unlikely to affect archaeological resources. However, the potential exists that storm-water detention releases resulting from a heavy rainfall could cause erosion in the area of an eligible site. Periodic monitoring of this site may be required.

An operational accident might affect archaeological resources by restricting access to sites that require regular monitoring. Such an accident might also affect traditional plant resources that might be present on the SRS.

Transportation. The existing road network at the SRS can readily accommodate the additional traffic expected during construction. In addition, the increased construction traffic would have negligible impacts on noise and air emissions. For operations, the impacts of transportation of the uranium and plutonium metal feed materials to the SRS, shipping fresh MOX fuel to a surrogate nuclear power plant site, shipping TRU waste to WIPP, and shipping spent MOX fuel were considered.

For routine transportation, the expected LCFs from radiation exposure could be up to 0.3 each for the public and transportation crews. A total of up to 2 latent fatalities were estimated from vehicle emissions. Thus, up to 2 fatalities might be expected from routine transportation activities.

It is estimated that the radiological transportation risk from accidents is 0.01 LCF over the course of the entire shipping campaign. Chemical impacts from accidents would be negligible: 1.3×10^{-7} irreversible adverse effect (approximately 1×10^{-9} fatality) from depleted UF_6 is expected for the entire shipping campaign. None of the chemicals that might be released in any transportation accident are known to be carcinogens.

Total fatalities from direct physical trauma from accidents were estimated to range as high as 0.20. This value indicates that no fatalities are expected from accidents for the entire shipping campaign.

Infrastructure. Construction activities and normal operational activities are not expected to adversely impact current SRS infrastructures. Projected electrical power, water, and fuel needs are well within existing capacities. The existing infrastructure would require a coordinated upgrading to support all phases of the surplus disposition program at the SRS: the proposed MOX facility, PDCF, and the WSB.

Waste Management. The impacts of facility construction waste on existing SRS waste management capacities would be minimal. The types and volumes of wastes generated would be similar to those that would be expected during the construction of an industrial facility. These wastes would be managed in accordance with current SRS waste management practices. Hazardous waste would be shipped off-site to commercial RCRA permitted facilities. The nonhazardous liquid waste generated would represent less about 6% of the SRS capacity for treatment. Solid waste would be shipped to off-site facilities for recycling or disposal.

Wastes generated by facility operations would have a small to moderate impact on the waste management system at the SRS. Estimated volumes for TRU waste would represent about 13% of SRS storage capacity and 2.6% of the WIPP storage capacity. Estimated volumes for solid low-level waste and hazardous waste would represent about 21% and less than 2% of the SRS disposal and storage capacities, respectively. Nonhazardous liquid wastes generated by facility operations are estimated to be about 6% of the capacity of the Central Sanitary Wastewater Treatment Facility. Nonhazardous solid wastes would be shipped off-site for recycling or disposal.

Human Health Risk. Less than 1 facility annually is predicted during the construction and normal operation phases of the facility. An estimated 122 lost workday injuries would occur annually over the 5-year construction period, and 41 annually over the assumed 10 or more years of operations.

No radiological impacts or adverse health impacts from emissions of toxic air pollutants are expected during the construction phase of the proposed facilities, and no adverse impacts to SRS employees and the public from exposure to emissions of toxic air pollutants are expected during normal operations. Annual radiological impacts to SRS employees for exposure to air emissions from the MOX and WSB facilities collectively and the PDCF are expected to be very small, approximately 1×10^{-5} and 2×10^{-5} LCF/yr, respectively. Similarly, the risk to the public would be small at 3×10^{-10} and 9×10^{-10} LCF/yr.

Hydrazine is the only chemical, aside from the radionuclides, that would be used in MOX processing that is listed as a hazardous air pollutant under the Clean Air Act. During routine operations, off-gas treatment systems would be expected to keep hydrazine emissions to very low levels that would not cause adverse health impacts to the off-site public or noninvolved workers.

Socioeconomics. The potential socioeconomic impacts from constructing and operating the proposed facilities would be insignificant. The increase in the annual average employment growth rate would be less than 0.1 of a percentage point over the duration of construction; even less during the operation phase.

In-migration of 350 people during the peak construction year would have only a marginal effect on population growth requiring 2.0% of the available vacant rental housing units in the region of influence (ROI) for construction and less than 1% of the available vacant owner occupied housing units for facility operations.

There would be no significant impact on public finances or the need for additional local public service employees during construction or normal operation.

Minor impacts would occur to agriculture and commercial fishing as demand for their products increase during construction and normal operation. No significant impacts on agriculture and downstream fisheries are expected from facility operations.

Any impacts associated with the transportation of fresh MOX fuel, including impacts on property values, would be minimal.

Environmental Justice. There would be no unavoidable environmental justice impacts from routine operations.

Aesthetics. The addition of the proposed facilities would not adversely affect the overall aesthetics of the F-Area or the SRS. The size and appearance of facility structures would be similar to those of existing buildings adjacent to the F-Area and would maintain the industrial nature of the F-Area.

Cumulative Impacts. Cumulative impacts of normal operations of the proposed facilities at the SRS were evaluated for air quality, health and safety, waste generation, resource use, and employment. Cumulative impacts for water quality, geologic resources, ecological resources, aesthetic resources, and cultural and paleontological resources were not explicitly addressed because direct and indirect impacts to these resources are expected to be negligible.

Cumulative impacts to air quality from proposed facility operations are not expected to be significant. On the basis of conservative assumptions, facility operations are projected to contribute 2% or less to cumulative concentrations of criteria air pollutants.

During normal operations, the facilities' contribution to cumulative radiological doses to the off-site population would be low (5.7% of the total). A cumulative dose to a MEI would increase by 1.0%. No LCFs are expected from the cumulative dose to the MEI or to the off-site population. Transportation of radioactive materials associated with facility operations would not contribute significantly to cumulative impacts (collective occupational dose, dose to the general public, and LCFs).

For most types of waste, facility operations would contribute 10% or less of the cumulative waste volumes generated at the SRS; existing waste treatment facilities will be able to handle this cumulative total. The largest proportionate increase would be in the amount of nonhazardous solid waste (18.8% of total).

The cumulative impacts of the proposed facilities to land development, electricity usage, and groundwater usage at the SRS would be quite small and well within existing SRS capacities.

Construction activities would result in a peak workforce of 1,000 in the peak construction year, or about 6% of the cumulative SRS employees. Facility operations would support 490 workers annually (2.9% of the total projected workforce for the SRS) and result in a cumulative total of 16,924 employees at the SRS.

4.7.2 Irreversible and Irretrievable Commitments of Resources

This section addresses the major irreversible and irretrievable commitments of resources associated with the no-action alternative and proposed action as described in Chapter 2. A commitment of a resource is irreversible when its primary or secondary impacts limit the future options for a resource. An irretrievable commitment refers to the use or consumption of resources neither renewable nor recoverable for use by future generations.

The 23.6 ha (58.3 acres) within which the proposed MOX facility, PDCF, and WSB would be built and the estimated 15.5 ha (38.3 acres) needed for infrastructure upgrades (e.g., pipeline and powerline rights-of-way, storm-water basin, batch plant, and roads) would be precluded from other uses until the NRC license to operate the facility was terminated (i.e., about 20 years into the future). About 3.6 ha (8.9 acres) of mostly woodland vegetation surrounding the proposed MOX facility site border would require grading for facility construction. Existing habitats would be eliminated, and ecological succession that would typically lead to progression from grassland to woodland vegetation would not occur. Although ultimate decommissioning of the facility could result in removal of all structures and paved surfaces, it is unlikely that woodland habitat comparable in quality to that north and west of the F-Area could become reestablished in less than 50 to 70 years.

Construction and operation activities would involve use of materials that could not be recovered or recycled. Soil excavated to produce the cement used in concrete would be irretrievably lost. Concrete and steel represent the bulk of construction materials. Other major construction materials that would be irretrievably lost or difficult to recycle include aluminum, lumber, piping materials, and electric wires and cables (DCS 2002a).

Water would be used for dust suppression during construction. Except for the water chemically bound in the production of concrete, water needed for construction and operation would eventually be recycled through the atmosphere and surface waters for distribution elsewhere. Water used during operation would be treated and discharged to the environment. Water obtained from groundwater supplies would be replaced through natural recharges of local aquifers. An estimated 760 million L (201 million gal) of water would be needed during the 10-year operating life of the facilities. Construction water requirements would total about 695 million L (185 million gal). A list of resources that would be required for the proposed MOX, PDCF, and WSB facilities is provided in Table 4.31.

Construction, operation, deactivation, and decommissioning of the project site would require a commitment of financial and human resources. Commitments of machinery, construction equipment vehicles, and fossil fuels (e.g., fuel oil and diesel oil) would be needed during the life of the project. None of these resources is expected to be in short supply in the vicinity of the SRS.

No valuable mineral resources are known to be present at the project site or immediate vicinity that could by affected by facility construction and operation security requirements in the F-Area.

4.7.3 Relationship between Short-Term Uses of the Environment and Long-Term Productivity

Short-term uses of the environment for the proposed action include (1) using a 23.6-ha (58.3-acre) area in F-Area for the proposed facilities, and (2) using an additional 15.5 ha (38.3 acres) of land for infrastructure upgrades and a process pipeline right-of-way needed to transport liquid high-level alpha waste from the proposed MOX facility. These uses would allow the U.S. government to fulfill its obligations in a September 2000 agreement with the Russian government to convert surplus weapons-grade plutonium no longer needed for defense purposes into MOX fuel for irradiation in nuclear reactors.

The proposed action would result in favorable short-term effects for the local economy, specifically for the nearby communities of Aiken and North Augusta, South Carolina, and Augusta, Georgia. These communities would benefit from the increase in income generated by direct jobs and workers in support industries in the SRS vicinity.

The use of 39.1 ha (96.6 acres) of land (up to 50.0 ha [123.4 acres] would be disturbed by construction) on the SRS for the facility is consistent with the SRS Long Range Comprehensive Plan (DOE 2000b) and use of the F-Area for processing nuclear materials. The proposed project would require clearing of up to 14.8 ha (36.4 acres) of woodland. Clearing would

Table 4.31. Irreversible and irretrievable commitments of resources for the proposed MOX, PDCF, and WSB facilities

Resource	Consumption
Construction[a]	
Electricity	85,500 MWh
Fuel oil	7.624 million L (1,960,000 gal)
Water	695 million L (185 million gal)
Concrete	149,300 m^3 (195,240 yd^3)
Steel	36,367 MT (40,100 tons)
Operations[b,c]	
Electricity	1,860,000 MWh
Water	760 million L (201 million gal)
Fuel oil	5,362,600 L (1,376,000 gal)
Plutonium	34 MT (37.5 tons)
Depleted uranium	665 MT (726 tons)
Argon	3.7 m^3 (129 million ft^3)
Argon-methane	103,930 m^3 (3.67 million ft^3)
Dodecane	29,144 L (7,700 gal)
Helium	96,570 m^3 (3.41 million ft^3)
Hydrogen	105,070 m^3 (3.71 million ft^3)
Hydrogen peroxide	20,060 L (5,300 gal)
Hydrazine (35%)	15,140 L (4,000 gal)
Hydroxylamine nitrate	348,220 L (92,000 gal)
Manganese nitrate	45.4 kg (100 lb)
Nitric acid	49,205 L (13,000 gal)
Nitrogen	45,310 million m^3 (1.6 billion ft^3)
Nitrogen tetroxide	37,380 million m^3 (1.32 million ft^3)
Oxalic acid dehydrate	40,363 kg (89,000 lb)
Oxygen	20,110 m^3 (710,000 ft^3)
Porogen	2,993 kg (6,600 lb)
Silver nitrate	1,088 kg (2,400 lb)
Sodium carbonate	1,995 kg (4,400 lb)
Sodium hydroxide (10M)	189 L (50 gal)
Tri-butyl phosphate	28,009 L (7,400 gal)
Zinc stearate	2,798 kg (6,170 lb)

[a]Consumption amounts are based on a 5-year construction period.

[b]Represents total volumes for the MOX and PDCF facilities.

[c]Consumption amounts are based on facility operations for an assumed 10-year period. The data on chemicals are only for the proposed MOX facility.

eliminate wildlife habitat in these woodlands. Infrastructure upgrades for electrical supply and additional roadways built for the proposed project would have long-term benefit to F-Area for ongoing and future projects. If DOE decides to decommission the proposed facilities and remove all structures and paved surfaces, the site could be reclaimed to woodland vegetation. Reclamation would require about 50 to 70 years to establish woodlands comparable in species composition to areas that would be cleared for construction.

4.8 References for Chapter 4

Arnett, M.W., and A.R. Mamatey (eds.) 2001. *Savannah River Site Environmental Report for 2000.* WSRC-TR-2000-00328. Westinghouse Savannah River Company, Aiken, SC.

Birch, M.L. 2001. Letter with attachments from Birch (ES&H Manager, Duke Cogema Stone & Webster, Charlotte, NC) to J.B. Davis (U.S. Nuclear Regulatory Commission, Washington, DC). June 19.

Biwer, B.M., et al. 1997. *Transportation Impact Analyses in Support of the Depleted Uranium Hexafluoride Programmatic Environmental Impact Statement.* Attachment to memorandum from Biwer to H.I. Avci, Argonne National Laboratory, Argonne, IL. May 21.

BTS (Bureau of Transportation Statistics) 2002. *National Transportation Statistics; 2001.* BTS02-06. U.S. Department of Transportation, Washington, DC.

Craig, D.K. 2002. *Revision 19 of ERPGs and TEELs for Chemicals of Concern.* WSMS-SAE-02-0171. Westinghouse Safety Management Solutions, Inc., Aiken, SC. Dec.

Craig, D.K., et al. 2000. "Derivation of Temporary Emergency Exposure Limits." *Journal of Applied Toxicology* 20:11-20.

DCS (Duke Cogema Stone & Webster) 2000a. *Mixed Oxide Fuel Fabrication Facility Environmental Report.* Docket Number 070-03098. Charlotte, NC.

DCS 2000b. *MOX Fuel Fabrication Facility Site Geotechnical Report.* DCSO1-WRS-DCS-NTE-G-00005-A. Charlotte, NC.

DCS 2001. *Responses to Request for Additional Information for the Duke Cogema Stone & Webster (DCS) Mixed Oxide (MOX) Fuel Fabrication Facility (FFF) Environmental Report (ER).* Report with letter and attachments on CD-ROM. Submitted by P.S. Hastings (DCS, Charlotte, NC) to U.S. Nuclear Regulatory Commission, Washington, DC. July 12.

DCS 2002a. *Mixed Oxide Fuel Fabrication Facility Environmental Report, Revision 1 & 2.* Docket Number 070-03098. Charlotte, NC.

DCS 2002b. *Amended Construction Authorization Request for the Mixed Oxide Fuel Fabrication Facility.* Docket Number 070-03098. Charlotte, NC.

DCS 2002c. *Responses to the Request for Additional Information on the Environmental Report Revisions 1 & 2.* DCS-NRC-000116, Docket Number 070-03098. Charlotte, NC. Oct. 29.

DCS 2002d. *Corrections to Responses to the Environmental Report Revisions 1 & 2.* DCS-NRC-000118, Docket Number 070-03098. Charlotte, NC. Nov. 15.

DCS 2003a. *Mixed Oxide Fuel Fabrication Facility Environmental Report, Revision 3.* Docket Number 070-03098. Charlotte, NC. June.

DCS 2003b. *Mixed Oxide Fuel Fabrication Facility Environmental Report, Revision 4.* Docket Number 070-03098. Charlotte, NC. Aug.

DCS 2004a. *Mixed Oxide Fuel Fabrication Facility Environmental Report, Revision 5.* Docket Number 070-03098. Charlotte, NC. June 10.

DCS 2004b. *Mixed Oxide Fuel Fabrication Facility Construction Authorization Request, Revision 6/10/04.* Docket Number 070-03098. Charlotte, NC.

DCS 2004c. *Mixed Oxide Fuel Fabrication Facility Environmental Report, Revision 5, Responses to Request for Additional Information.* Docket Number 070-03098. Charlotte, NC. Sept. 17.

DOE (U.S. Department of Energy) 1994. *Final Defense Waste Processing Facility Supplemental Impact Statement.* DOE/EIS 0082-S. Savannah River Operations Office, Aiken, SC.

DOE 1995. *Savannah River Site Waste Management Final Environmental Impact Statement.* DOE/EIS-0217. Savannah River Operations Office, Aiken, SC. July.

DOE 1996a. *Storage and Disposition of Weapons-Usable Fissile Materials Final Programmatic Environmental Impact Statement.* DOE/EIS-0229. Office of Fissile Materials Disposition, Washington, DC. Dec.

DOE 1996b. *Disposition of Surplus Highly Enriched Uranium Final Environmental Impact Statement.* DOE/EIS-0240. Office of Fissile Materials Disposition, Washington, DC.

DOE 1997a. *Final Waste Management Programmatic Environmental Impact Statement for Managing Treatment, Storage, and Disposal of Radioactive and Hazardous Waste.* DOE/EIS-0200-F. Office of Environmental Management, Washington, DC. May.

DOE 1997b. *Waste Isolation Pilot Plant Disposal Phase Final Supplemental Environmental Impact Statement.* DOE/EIS-0026-S-2. Carlsbad Area Office, Carlsbad, NM. Sept.

DOE 1998. *Final Environmental Impact Statement on Management of Certain Plutonium Residues and Scrub Alloy at the Rocky Flats Environmental Technology Site.* DOE/EIS-0277F. Savannah River Operations Office, Aiken, SC.

DOE 1999a. *Surplus Plutonium Disposition Final Environmental Impact Statement.* DOE/EIS-0283. Office of Fissile Materials Disposition, Washington, DC. Nov.

DOE 1999b. *DOE Standard, Radiological Control.* DOE-STD-1098-99. Washington, DC. July.

DOE 1999c. *Final Environmental Impact Statement for the Construction and Operation of a Tritium Extraction Facility at the Savannah River Site.* DOE/EIS-0271. Savannah River Operations Office, Aiken, SC.

DOE 2000a. *Savannah River Site High-Level Waste Tank Closure Draft Environmental Impact Statement.* DOE/EIS-0303D. Savannah River Operations Office, Aiken, SC. Nov.

DOE 2000b. *Savannah River Site Long Range Comprehensive Plan.* Savannah River Site, Aiken, SC. Dec.

DOE 2000c. *Savannah River Site Spent Nuclear Fuel Management Final Environmental Impact Statement.* DOE/EIS-0279. Savannah River Operations Office, Aiken, SC. March.

DOE 2000d. *Final Environmental Impact Statement for the Treatment and Management of Sodium-Bonded Spent Nuclear Fuel.* DOE/EIS-0306. Office of the Nuclear Facilities Management, Germantown, MD. July.

DOE 2000e. *Environmental Assessment for the Construction and Operation of the Highly Enriched Uranium Blend-Down Facilities at the Savannah River Site.* DOE/EA-1322. Savannah River Operations Office, Aiken, SC. Nov.

DOE 2001. *Savannah River Site Salt Processing Alternatives Draft Supplemental Environmental Impact Statement.* DOE/EIS-0082-S2D. Savannah River Operations Office, Aiken, SC. March.

DOE 2002a. "Amended Record of Decision for the Surplus Plutonium Disposition Program." *Federal Register* 67:19432, April 19.

DOE 2002b. *Savannah River Site High-Level Waste Tank Closure Final Environmental Impact Statement.* DOE/EIS-0303. Washington, DC.

DOE 2002c. *Supplement Analysis for Storage of Surplus Plutonium Materials in the K-Area Material Storage Facility at the Savannah River Site.* Assistant Secretary for Environmental Management, Washington, DC. Feb.

DOE 2002d. *Final Environmental Impact Statement for a Geologic Repository for the Disposal of Spent Nuclear Fuel and High-Level Radioactive Waste at Yucca Mountain, Nye County, Nevada.* DOE/EIS-0250. Office of Civilian Radioactive Waste Management, Washington, DC. Feb.

DOE 2002e. "Notice of Intent to Prepare a Supplemental Programmatic Environmental Impact Statement on Stockpile Stewardship and Management for a Modern Pit Facility." *Federal Register* 67:59577, Sept. 23.

DOE 2003a. *Changes Needed to the Surplus Plutonium Disposition Program, Supplement Analysis and Amended Record of Decision.* DOE/EIS-0283-SA1. Office of Fissile Materials Disposition, Washington, DC. April.

DOE 2003b. *Draft Supplemental Programmatic Environmental Impact Statement on Stockpile Stewardship and Management for a Modern Pit Facility.* DOE/EIS-236-S2, National Nuclear Security Administration, Washington DC. May.

Eckerman, K.F., et al. 1999. *Cancer Risk Coefficients for Environmental Exposure to Radionuclides, Federal Guidance Report No. 13.* EAP 402-R-99-001. Prepared by Oak Ridge National Laboratory, Oak Ridge, TN, for U.S. Environmental Protection Agency, Office of Radiation and Indoor Air, Washington, DC. Sept.

EPA (U.S. Environmental Protection Agency) 1990. "National Oil and Hazardous Substances Pollution Contingency Plan; Final Rule (400 CFR Part 300)." *Federal Register* 55(35):6154-6176, Feb. 21.

EPA 1995. *User's Guide for the Industrial Source Complex (ISC3) Dispersion Models.* EPA-454/B-95-003a and updated data. Office of Air Quality Planning and Standards, Research Triangle Park, NC. Sept.

EPA 2003. "Air Data — Reports and Maps." Office of Air Quality Planning and Standards, Research Triangle Park, NC. June. Available at http://www.epa.gov/air/data/reports.html.

FDA (Food and Drug Administration) 1998. *Accidental Radioactive Contamination of Human Food and Animal Feeds: Recommendations for State and Local Agencies.* Center for Devices and Radiological Health, Rockville, MD. Aug. Available at www.fda.gov/cdrh.

Fledderman, P.D. 2002. *Plutonium Disposition Program (PDP) Preconstruction Environmental Monitoring Report.* ESH-EMS-2002-1141, Rev. 0. Savannah River Site, Environmental Monitoring Section. June 26.

Halverson, N.V. 1999. "Revised Cumulative Impact Data." SRT-EST-99-0328, Rev. 1. Interoffice memorandum to C. B. Shedrow, Westinghouse Savannah River Company, Aiken, SC. Aug. 17.

Hanna, S.R., et al. 1997. *HGYSYSTEM/UF6, Model Enhancements for Plume Rise and Dispersion around Buildings, Lift-Off of Buoyant Plumes, and Robustness of Numerical Solver.* K/SUB/93-XJ94/J2R1. Prepared by EARTH TECH for Lockheed Martin. Jan.

Hunter, C.H. 2001. "Clean Air Act Title V Dispersion Modeling for SRS (Revision 2)." Memo from Hunter to P.C. Carroll, Westinghouse Savannah River Company, Aiken, SC, March 15.

MIG, Inc. 2001. "IMPLAN Data Files." MIG, Inc., Stillwater, MN.

Monette, F.A., et al. 1996. *Supplemental Information Related to Risk Assessment for the Off-Site Transportation of Transuranic Waste for the U.S. Department of Energy Waste Management Programmatic Environmental Impact Statement.* ANL/EAD/TM-27. Argonne National Laboratory, Argonne, IL. Dec.

Neuhauser, K.S., and F.L. Kanipe 1992. *RADTRAN 4, Volume 3: User Guide.* SAND89-2370. Sandia National Laboratories, Albuquerque, NM. Jan.

NNSA (National Nuclear Safety Administration) 2002. *Report to Congress: Disposition of Surplus Defense Plutonium at Savannah River Site.* Office of Fissile Materials Disposition, Washington, DC. Feb. 15.

NRC (U.S. Nuclear Regulatory Commission) 1977. *Final Environmental Statement on the Transportation of Radioactive Material by Air and Other Modes.* NUREG-0170. Washington, DC.

NRC 1979. *Technology, Safety and Costs of Decommissioning a Reference Small Mixed Oxide Fuel Fabrication Plant.* NUREG/CR-0129, Vols. 1 & 2. Washington, DC.

NRC 1988. *Final Generic Environmental Impact Statement on Decommissioning of Nuclear Facilities.* NUREG-0586. Washington, DC.

NRC 1996. *Dose Commitments due to Radioactive Releases from Nuclear Power Plant Sites in 1992.* NUREG/CR 2850, Vol. 14. Washington, DC.

NRC 1997. *Environmental Assessment for Renewal of Special Nuclear Material License No. SNM-1097, General Electric Company Nuclear Production Facility, Wilmington, North Carolina.* Docket Number 70-1113. May.

NRC 1998. *Nuclear Criticality Safety Standards for Fuels and Material Facilities.* Regulatory Guide 3.71. Office of Nuclear Regulatory Research, Washington, DC. Aug.

NRC 2001. *Draft Environmental Review Guidance for Licensing Actions Associated with NMSS Programs.* NUREG-1748. Office of Nuclear Materials Safety and Safeguards, Washington, DC. Sept.

NRC 2003. *Final Environmental Review Guidance for Licensing Actions Associated with NMSS Programs.* NUREG-1748. Office of Nuclear Materials Safety and Safeguards, Washington, DC. Aug.

NSC (National Safety Council) 2001. *Injury Facts.* 2001 Edition. Itasca, IL.

Orr, M.P. 2001. "Comparison of HEPA and Deep-Bed Sand Filters for Final Air Filtration at MOX Fuel Fabrication Facility." Letter report from Orr (Advanced Technologies and Laboratories, Inc., Rockville, MD.) to W.C. Gleaves (U.S. Nuclear Regulatory Commission, Washington, DC). Nov. 15.

Policastro, A.J., et al. 1997. *Facility Accident Impact Analyses in Support of the Depleted UF6 Programmatic Environmental Impact Statement.* Environmental Assessment Division, Argonne National Laboratory, Argonne, IL. Jun.

Post, L. 1994a. *HGSYSTEM 3.0, User's Manual.* TNER.94.058. Shell Research Limited, Thorton Research Centre, Chester, United Kingdom.

Post, L. (ed.) 1994b. *HGSYSTEM 3.0, Technical Reference Manual.* TNER.94.059. Shell Research Limited, Thorton Research Centre, Chester, United Kingdom.

Reynolds, R.M. 1992. *ALOHA (Area Locations of Hazardous Atmospheres), Theoretical Description.* National Oceanic and Atmospheric Administration, Seattle, WA. Aug.

Roberson, J.H. 2002. "Declaration of Jessie Hill Roberson," in the matter of *Jim Hodges, Governor, State of South Carolina, Plaintiff, v. Spencer Abraham, Secretary, United States Department of Energy, and the United States Department of Energy, Defendants.* United States District Court, District of South Carolina, Aiken Division, Civ. 1 02 1426-22. May.

SCDHEC (South Carolina Department of Health and Environmental Control) 1995. *South Carolina Nuclear Facility Monitoring — Annual Report 1995.* Columbia, SC.

SCDHEC 2001. *Air Dispersion Modeling Summary Sheet for Westinghouse Savannah River Company* (dated April 3, 2001). E-mail from R. DuBose (Air Quality Division, Westinghouse Savannah River Company, Charlotte, NC) to A. Smith (Argonne National Laboratory, Argonne, IL). June 8.

Simpkins, A.A. 2000. *Maximally Exposed Offsite Individual Location Determination for NESHAPS Compliance.* WSRC-RP-2000-00036. Westinghouse Savannah River Company, Aiken, SC.

SRS (Savannah River Site) 2001. *SRS Emergency Response Plan.* WSRC-SCD-7, Rev. 16. Aiken, SC. July 31.

U.S. Bureau of the Census 2002. *U.S. Census American Fact Finder.* Washington, DC. Available at http://factfinder.census.gov/.

Wike, L.D. (compiler) 2000. *Site Selection for Surplus Plutonium Disposition Facilities at the Savannah River Site.* Westinghouse Savannah River Company, Aiken, SC. July.

WSRC (Westinghouse Savannah River Company) 1995. *RCRA Facility Investigation/Remedial Investigation Report, Revision 1, for the Old F-Area Seepage Basin.* WSRC-RP-94-942, Rev. 1. Prepared for U.S. Department of Energy, Savannah River Site, Aiken, SC. June.

WSRC 2000. *Savannah River Approved Site Treatment Plan, 2000 Annual Update.* WSRC-TR-94-0608, Rev. 8. Savannah River Site, Aiken, SC.

5 MITIGATION

5.1 Introduction

This chapter addresses potential means to mitigate adverse environmental impacts from the proposed action as required by Appendix A of Title 10, Part 51, of the *Code of Federal Regulations* (10 CFR Part 51). Mitigation measures for the proposed Pit Disassembly and Conversion Facility (PDCF) have been considered by the U.S. Department of Energy (DOE) in its Surplus Plutonium Disposition Environmental Impact Statement (SPD EIS) (DOE 1999) and January 11, 2000, Record of Decision (DOE 2000, 2002) and are not repeated in this document. The recent DOE supplemental analysis (DOE 2003) discusses impacts related to operation of the proposed Waste Solidification Building (WSB) but does not identify any mitigation measures for the WSB. Therefore, for completeness, the discussion of mitigation measures in this EIS includes potential measures for the WSB. A full discussion of potential mitigation measures for each resource area is provided in Section 5.2, and these measures are summarized in Table 5.1. It is important to note that while potential mitigation measures for the WSB are identified in this EIS, the NRC does not have the regulatory authority to implement mitigation measures for DOE facilities. For the purpose of reaching a final NRC staff decision on its proposed action, the NRC assumes that the DOE will not implement the mitigation measures identified herein that pertain to the proposed WSB.

Under Council of Environmental Quality (CEQ) regulation 40 CFR 1500.2(f), federal agencies shall to the fullest extent possible use all practicable means consistent with the requirements of the National Environmental Policy Act (NEPA) and other essential considerations of national policy to restore and enhance the quality of the human environment and avoid or minimize any possible adverse effects of their actions on the quality of the human environment. The CEQ regulations define mitigation to include the following: (1) avoiding the impact altogether by not taking a certain action or parts of an action; (2) minimizing impacts by limiting the degree or magnitude of the action and its implementation; (3) rectifying the impact by repairing, rehabilitating, or restoring the affected environment; (4) reducing or eliminating the impact over time by preservation and maintenance operations during the life of the action; and (5) compensating for the impact by replacing or providing substitute resources or environments. This definition has been used in defining potential mitigation measures.

The NRC staff has reviewed the mitigation measures and has concluded that no additional mitigation measures are required beyond the regulatory requirements and those measures identified by DCS.

5.2 Mitigation Measures

The NRC staff evaluated proposed mitigation measures identified by Duke Cogema Stone & Webster (DCS) (2003) and identified other potential measures that could reduce or eliminate adverse environmental impacts of the proposed mixed oxide (MOX) facility and WSB (as

Table 5.1. Summary of DCS mitigation commitments and additional measures identified by NRC staff for reducing or avoiding impacts[a]

Technical area	Mitigation	Measures proponent
Soils and Hydrology	• Control of pollutants in stormwater discharges during construction will be addressed as provided in the Storm Water Pollution Prevention Plan that Duke Cogema Stone & Webster (DCS) will file with its notice of intent to discharge stormwater during construction under the South Carolina National Pollutant Discharge Elimination System (NPDES) General Permit for stormwater discharges from construction activities (Permit No. SCR100000). Filing of a Storm Water Pollution Prevention Plan is required by Part IV, "Storm Water Pollution Prevention Plans," in Permit No. SCR100000. The South Carolina Department of Health and Environmental Control (SCDHEC) has issued the NPDES General Permit for stormwater discharges from construction activities as provided in South Carolina Regulations (SC Regulation R.61-9.122.28).	REG
	• Erosion and sediment controls will be implemented as provided in the Storm Water Pollution Prevention Plan that DCS will file with its notice of intent to discharge stormwater during construction under the South Carolina NPDES General Permit for stormwater discharges from construction activities (Permit No. SCR100000). Filing of a Storm Water Pollution Prevention Plan is required by Part IV, "Storm Water Pollution Prevention Plans," in Permit No. SCR100000. The SCDHEC has issued the NPDES General Permit for stormwater discharges from construction activities as provided in SC Regulations R.61-9.122.28.	REG
	• Creation of foundations and building of structures for the proposed mixed oxide (MOX) facility, and Waste Solidification Building (WSB) (hereafter "the facilities") will be limited to the upper soil layers, thus minimizing impacts to groundwater.	DCS
	• Good engineering practices will be used during operation and construction to minimize chemical impacts to soils.	DCS
	• Sanitary wastes generated during construction will be collected with a combination of portable toilets and semipermanent facilities connected to the Central Sanitary Waste Treatment Facility.	DCS
	• Regular monitoring of the double-walled liquid high-alpha waste pipeline will be conducted to detect leaks.	DCS

Table 5.1. Continued

Technical area	Mitigation	Measures proponent
Ecology	• The right-of-way for the 610-m (2,000-ft) pipeline to convey liquid high-alpha-activity waste and stripped uranium waste for the proposed MOX facility to the WSB will be less than 7.6 m (25 ft) wide and thus will minimize vegetation removal.	DCS
	• Before construction activities begin, the site would be surveyed for migratory bird nests.	DCS
	• Measures should be taken to protect trees on the MOX site not selected for removal and not controlled after site clearing by the U.S. Department of Agriculture (USDA) Forest Service — Savannah River. If such trees or other landscape features not controlled by the USDA Forest Service — Savannah River are accidentally scarred or damaged, they should be replaced in a manner consistent with the Savannah River Site Natural Resources Management Plan.	NRC
	• Construction crews would receive environmental briefings as appropriate to alert them to specific areas of concern (e.g., possible harassment and other adverse impacts to wildlife species during the construction period) and to explain the reasons for such concern.	NRC
	• Impacts during the clearing of vegetation should be controlled by the USDA Forest Service — Savannah River, consistent with the Savannah River Site Natural Resources Management Plan.	NRC
	• Following construction, site restoration (e.g., soil stabilization and revegetation) would be conducted in compliance with appropriate U.S. Department of Energy (DOE) policies for reclamation of construction areas.	DCS
	• Access roads should be sited on previously disturbed areas where possible to minimize sensitive vegetation removal.	NRC
Air Quality and Noise	• DCS will have a Construction Emissions Control Plan, which will implement a number of good engineering practices to reduce fugitive dust emissions consistent with the requirements in SC Regulation R.61-62.6, "Control of Fugitive Particulate Matter."	REG
	• Particulate emissions from the silo hopper and concrete mixer used in the cementation process during operation of the WSB will be required to meet the conditions specified in the SCDHEC permit.	REG

Table 5.1. Continued

Technical area	Mitigation	Measures proponent
Infrastructure	• Road upgrades for ingress and egress of the proposed MOX site will be conducted in existing traffic rights-of-way.	DCS
Land Use	• No mitigation measures are needed to reduce impacts of the proposed action on land use.	
Waste Management	• No mitigation measures are needed to reduce impacts of the proposed action on the Savannah River Site (SRS) waste management system.	
Human Health Risk	• Radiation doses to workers during construction will be kept to a minimum by using administrative limits and ALARA (as low as reasonably achievable) programs, including worker rotations.	REG
	• Exposure to hydrazine will be limited by complying with SCDHEC emission standards.	REG
	• To minimize adverse effects to facility and SRS workers from exposure to nitrogen tetroxide, DCS should comply with the requirements in the Occupational Safety and Health Administration's (OSHA's) Process Safety Management Rule (29 CFR 1910.119).	REG
	• The radiation exposure of radiographers will be monitored or badged during construction.	REG
	• The radiography contractor will follow the contractor's existing U.S. Nuclear Regulatory Commission (NRC) or agreement-state license in evaluating and monitoring radiographer exposure.	REG
	• Radiation and chemical exposures of facility workers during operations would be kept to a minimum through (1) use of engineering controls to keep airborne chemical concentrations below applicable occupational exposure limits, and (2) use of enclosed operations to the extent possible.	DCS
	• To minimize adverse effects to facility and SRS workers in the event of an accidental release of process chemicals identified as presenting moderate or high risks to workers (as identified in Section 4.3.5.3), DCS has committed in its Construction Authorization Request (CAR) to integrate any emergency preparedness plans for the proposed MOX facility with the DOE SRS Emergency Response Plan.	DCS
	• Construction workers should be protected from inadvertent radiation and chemical exposures by soil testing and analysis prior to excavation to ascertain that levels of radiation and inorganic or organic chemicals in soils would not present a health hazard during construction activities.	NRC

Table 5.1. Continued

Technical area	Mitigation	Measures proponent
Cultural and Paleontological Resources	• Periodic monitoring of nearby eligible archaeological sites shall be conducted to check for possible erosion.	DOE
	• Additional mitigation measures, such as avoidance agreements, shall be determined in consultation with the South Carolina State Historic Preservation Office (SCSHPO).	DOE
	• If inadvertent discoveries of cultural resources occur during site construction, mitigation would follow the guidelines of 36 CFR 800.11 and/or 43 CFR 10.4.	REG
Aesthetics	• No mitigation measures are necessary to reduce aesthetic impacts of the proposed action.	
Socioeconomics	• No mitigation measures are necessary to reduce impacts to socioeconomic factors.	
Environmental Justice	• DCS should work closely with SRS to implement procedures to protect low-income and minority groups in the event of an accidental chemical or radiological release from the proposed facilities that impact areas beyond the SRS boundary.	NRC
	• DCS should conduct focused public information campaigns to provide important information to low-income and minority groups/communities. Included in these campaigns would be descriptions of existing monitoring programs, and information on the nature, extent, and likelihood of any airborne release from the facility. The campaigns would also include a description of the relevant risks associated with the proposed facilities. These campaigns should include information on sheltering and other protection strategies that may be needed, including detailed descriptions of any evacuation procedures that may be required.	NRC
	• DCS should provide public information to local agencies and groups representing low-income or minority groups on existing soil or groundwater contamination monitoring programs and the nature, extent, and likelihood of surface release. Key information would include the extent of any likely damage to drinking water supplies and subsistence resources, and the relevant preventative measures that may be taken.	NRC
	• DCS should meet with local communities providing emergency response services and other emergency facilities to discuss additional measures to ensure that the low-income and minority populations in their jurisdictions are located and fully prepared in the event that sheltering or evacuation procedures are required, in addition to public information campaigns targeting low-income and minority groups. This would include the development of spatial databases providing information on the locations of low-income and minority populations, local resources available to emergency response agencies, and any evacuation routes that might be required.	NRC

Table 5.1. Continued

---

[a]The mitigation measures are commitments made by DCS that were identified in the ER (DCS 2002) and other potential measures identified by the NRC staff in preparing this EIS. Under the column "Measures proponent," "DCS" refers to the applicant, "DOE" refers to the U.S. Department of Energy, "NRC" refers to the U.S. Nuclear Regulatory Commission, and "REG" refers to a regulatory requirement or a permit/license condition that DCS would be required to implement.

indicated in Table 5.1). The applicant, DCS, has proposed design features and other activities to reduce impacts for the proposed MOX facility. In Table 5.1, the proponent for these mitigation measures is designated as "DCS." In addition, compliance with federal and state regulations, permits, and guidelines will reduce potential impacts (see Chapter 6 for a discussion of applicable environmental regulations and permits). For example, the South Carolina National Pollutant Discharge Elimination System (NPDES) general permit requires the implementation of a Storm Water Pollution and Prevention Plan that would mitigate potential impacts to surface waters from construction activities. The regulations, permits, and guidelines typically recommend best management practices. These practices (i.e., mitigation measures) would be determined during the permitting process, which would occur in the future. For that reason, general types of activities that would comprise best management practices are discussed. The proponent for these mitigation measures is designated as "REG," and for other mitigation measures proposed by the NRC staff, the proponent is designated as "NRC" in Table 5.1. Not all NRC-suggested mitigation measures are within the NRC's regulatory authority.

5.2.1 Hydrology

Surface water resources could be adversely affected by construction of the proposed MOX facility and WSB. Introducing pollutants or erosion into surface waters could impact the quality of the surface water and aquatic organisms. Several design features that would mitigate impacts to surface water were proposed by DCS and the DOE. During construction of the proposed MOX facility and WSB, no direct discharges of contaminated water into Upper Three Runs Creek, Four Mile Branch, or their tributaries, are expected to occur. Sanitary wastes would be collected with a combination of portable toilets and semipermanent facilities connected to the Savannah River Site (SRS) Central Wastewater Treatment Facility. All wastewater would be treated in the sitewide treatment system before release under existing NPDES permits, thus minimizing impacts to surface waters.

Potential impacts from stormwater discharges during construction would be mitigated by compliance with the Storm Water Pollution Prevention Plan that is required by South Carolina Department of Health and Environmental Control (SCDHEC) regulations. DCS plans to file this plan in its notice of intent to discharge storm water during construction under the South Carolina NPDES General Permit for stormwater discharges from construction activities (Permit No. SCR100000). Under the General Permit, best management practices would be followed to divert the flow of runoff water away from exposed soils, store flows, or otherwise limit runoff and the discharge of pollutants from exposed areas to the degree attainable. Such practices might include, but not necessarily be limited to, use of silt fences, earth dikes, drainage swales, sediment traps, check dams, temporary or permanent sediment basins, temporary seeding, permanent seeding, mulching, use of geotextiles, sod stabilization, vegetative buffer strips, protection of trees, and preservation of mature vegetation. Because groundwater would be used as the source of water during construction, groundwater could be adversely affected during construction of the facilities. Because the capacity of the existing wells at SRS are sufficient to meet the needs of the project, further mitigation would not significantly reduce the impacts associated with using groundwater during construction. While construction could directly impact groundwater quality if any of the buildings or structures extended below the surface of the groundwater, the design for the proposed MOX facility and WSB do not involve encroachment on groundwater. Groundwater could be indirectly impacted by infiltration of contaminated surface water or surface spills during construction. These impacts would be mitigated by following appropriate good engineering practices and following the provisions of the required Stormwater Pollution Prevention Plan as discussed above.

During normal operation of the proposed MOX facility and WSB, surface water would not be used. The primary mitigation activities for surface water quality would be ensuring that releases of effluent meet NPDES permit guidelines. Design features proposed by DCS and the DOE include this mitigation strategy. Mixed, hazardous, and radioactive wastes in liquid form would be sent off site for disposition, or sent to SRS waste management facilities, or would be treated and processed at the WSB prior to being discharged to surface waters or converted into a solid waste. See Section 4.3.4 for a further discussion of how such solid wastes would be managed. Stormwater run-off from paved areas would be collected by the stormwater system. The stormwater would be temporarily retained in a detention basin to reduce the amounts of oils and other pollutants from entering surface water. The uncontaminated HVAC condensate would also be discharged to the stormwater system in accordance with SCDHEC standard stormwater permit conditions. The detention basin would also reduce the flow into surface waters following precipitation events.

Water for normal operations would be obtained from existing SRS wells. Because the quantity of water required for operations is within the capacity of the existing wells, further mitigation would not significantly reduce the impacts of using the groundwater during operations. The design features for the project do not include direct releases to underlying aquifers. However, the quality of groundwater could be affected indirectly by receipt of contaminated surface water. As discussed above, design features have been proposed by DCS and the DOE to limit contamination of surface water. Operation of a sand filter, if used, would not directly impact groundwater because the filter would be covered to prevent infiltration and it would have a concrete wall and bottom.

Deactivation and decommissioning could also impact water resources at the site. These impacts would be mitigated by using the methods discussed above for construction.

Accidents could impact surface water and groundwater directly and indirectly. Impacts to surface water would primarily be indirect. These impacts would be produced by contaminated runoff from spill areas. DCS has committed to preparing and implementing a Spill Control and

Countermeasures Plan during operation. A similar plan would be prepared for the WSB. Mitigation would be accomplished by following best management practices in these plans that would include prompt cleanup and removal of contaminated materials. Direct impacts to groundwater could occur if there were a failure in the underground pipelines carrying liquid waste from the proposed MOX facility to the WSB. The impacts would be mitigated by regular monitoring of the system to detect leaks for the double-walled pipelines, and developing contingency plans to remediate any spills promptly and thoroughly.

Further mitigation was not identified by the NRC that would significantly reduce the impacts to surface water or groundwater.

5.2.2 Soils

Soils could be affected by construction activities, normal operations, activities associated with deactivation and decommissioning, and accidents. Several design features proposed by DCS and the DOE were considered to be mitigation measures. The locations selected for the proposed MOX facility and the WSB contain soils that are not unique to the SRS, and there are no soils classified as prime farmlands. In addition, the grading and landscape plans would be designed in part to reduce future erosion following construction activities and limit slope instability.

To a great extent, the impacts of construction on soils would be mitigated by the following SCDHEC regulations (see discussion in Chapter 6) that require installation of sediment detention basins that would catch and hold runoff water. These detention ponds would be situated in strategic locations and would be designed to control the release of storm-water runoff at a rate equal to or slightly less than that of the predevelopment conditions. In addition, following good engineering practices will be required by the Stormwater Pollution Prevention Plan that DCS will file with the State of South Carolina in its Notice of Intent to discharge stormwater during construction under the General Permit for stormwater discharges (Permit No. SCR100000). Such practices could include silt fences, sediment traps, check dams, etc., and would mitigate the consequences of construction including impacts associated with potential spills.

During normal operations, there would be no planned direct discharges of water to the soil, and stack emissions of contaminated particulates would be filtered. These mitigation measures would minimize adverse impacts to the soil.

During deactivation and decommissioning, impacts could once again occur to soils through mobilization of contaminants by water or wind. Mitigation activities for this phase of the project would be the same as those outlined for construction.

Accidents during the lifetime of the facilities could also adversely impact soils. Following the Spill Control and Countermeasures Plan as discussed in Section 5.2.1 would mitigate these potential impacts.

Further mitigation was not identified by the NRC that would significantly reduce the impacts to soils.

5.2.3 Ecology

Construction of the proposed MOX facility and WSB and associated infrastructure would disturb up to 50.0 ha (123.4 acres) of land in the F-Area of the SRS. Several design features proposed by DCS and the DOE were considered to be mitigation measures. The location of the facilities would mitigate many of the construction impacts to ecological resources. The site selected for the facilities would be largely in previously disturbed or developed locations, and there are no designated wetlands or Carolina bays within the areas to be disturbed. For example, a portion of the construction activities for the proposed MOX facility would take place in an area where spoils for previous F-Area construction has been stored, and most of the WSB would be located within "facility" land (e.g., landscaped areas). Also the new, widened, and realigned roads would be located within previously cleared rights-of-way. In addition, the facilities would not be located within either the red-cockaded woodpecker management area or its supplemental management area. Clearing of vegetation should be conducted in accordance with the Savannah River Site Natural Resources Management Plan by the U.S. Department of Agriculture (USDA) Forest Service. Complying with this plan will minimize impacts to ecological resources. Following construction, the cleared and graded areas not covered with facilities, parking lots, or roads would be landscaped. This landscaping would provide habitat for some wildlife species, mitigating the loss of habitat from constructing the facilities.

As discussed in Section 5.2.1, complying with the Storm Water Pollution Prevention Plan would mitigate impacts of ecological resources. Best management practices for soil erosion and sediment control would be used to prevent runoff and dust from entering sensitive habitats and nearby streams (e.g., unnamed tributaries to Upper Three Runs Creek), and direct construction disturbance of nearby streams would be avoided.

Potential mitigation measures to protect ecological resources were identified by the NRC. DCS should take action at the construction site to prevent the workforce from removing vegetation in excess of that needed for construction clearing. To ensure protection of vegetation during construction, DCS should designate an environmental supervisor to supervise vegetation clearance. Any accidentally scarred or damaged trees should be replaced consistent with the Savannah River Site Natural Resources Management Plan. Construction crews should also receive environmental briefings as appropriate to alter them to specific areas of concern (e.g., possible harassment and other adverse impact to wildlife species during the construction period, identification of spills and notification of supervisors) and to explain the reasons for such concerns. In addition, following construction, site restoration (e.g., soil stabilization and revegetation) should be done in compliance with appropriate DOE policies for reclamation of construction areas.

During normal operations, the major mitigation factor would be to limit releases of contaminants (chemicals and radioactive materials) to the environment. The mitigation measures discussed in Section 5.2.1 would also mitigate impacts to ecological resources.

Impacts of deactivation and decommissioning would be mitigated by using the same methods described for construction, particularly those for erosion and sediment control.

Accidents could also impact ecological resources at the proposed facilities. These impacts would be produced primarily by contaminated runoff water entering sensitive habitat. Additional impacts could occur through air emissions from an accident. Mitigation measures would include following the Spill Control and Countermeasures Plan discussed in Section 5.2.1. These mitigation measures would reduce the likelihood of bioaccumulation and biomagnification in the food chain.

The NRC staff has reviewed the mitigation measures for ecological resources and has concluded that no additional mitigation measures are required beyond the regulatory requirements and the measures identified by DCS.

5.2.4 Air Quality

During construction of the proposed MOX facility and WSB, emission of criteria pollutants (carbon monoxide, nitrogen dioxide, and sulfur dioxide [CO, NO_2, and SO_2]), total suspended particulates (TSP), and volatile organic compounds (VOCs) would require mitigation. Of these, suspended particles would be the principal concern. Suspended particles could be produced by fugitive dust from earthmoving activities, fugitive dust from the concrete batch plant, and exhaust emissions from diesel-powered construction equipment and from worker and delivery vehicles. Most of this dust would be generated within the construction site; dust created along roadways in the SRS would be naturally mitigated by dispersal. To a great extent, the impacts of construction on air quality would be mitigated by the following SCDHEC regulations (see discussion in Chapter 6). South Carolina Regulations (SC Regulations R.61-62.6, Control of Fugitive Particulate Matter) require DCS to have a Construction Emissions Control Plan. This plan would implement a number of good engineering practices to reduce fugitive dust emissions. These would include applying, as appropriate, standard dust control practices, such as watering or sweeping roads and water exposed areas. Particulate emissions from the silo hopper and concrete mixer used during the cementation process to construct the WSB would be controlled as provided in a State of South Carolina Permit to Construct the concrete batch plant. The State of South Carolina Permit to Construct would provide for controls on particulate emissions consistent with the requirements in SC Regulations R.61-62.5, Standard No. 4, "Emissions from Process Industries."

During normal operations, air quality impacts would be produced by process emissions, testing of emergency diesel generators, trucks moving materials and wastes, and employee vehicles. Several design features proposed by DCS and the DOE were considered to be mitigation measures. These impacts would be mitigated by using an air filtration system (e.g., high-efficiency air particulate [HEPA] filters or sand filter) to remove radioactive particulates prior to discharge of process exhaust air to the atmosphere and by using internal scrubbers to reduce chemical gas concentrations. Parking lots and access roads would be paved to minimize the emission of fugitive dust during normal operations.

Mitigation activities for deactivation and decommissioning would be similar to those used for construction. These strategies would be primarily aimed at reducing fugitive dust.

In the event of an accident, adverse impacts to the air would be mitigated by the air filtration systems and prompt and thorough cleanup, if necessary.

Further mitigation was not identified by the NRC that would significantly reduce the impacts to air quality.

5.2.5 Noise

Noise is unwanted sound that interferes with or interacts negatively with the human or natural environment. Construction of the proposed MOX facility and WSB could adversely affect the level of noise. These adverse impacts would be mitigated by locating the facilities away from the SRS public boundary and sensitive receptors. The siting of the facilities is considered a design feature that mitigates noise impacts. The level of noise could also be a concern for federally listed or endangered species; however, none are known to occur in F-Area. As discussed in Section H.3.1.1, noise levels could startle small mammals and frighten birds. Generally, these disturbances would be short-term and localized. Construction workers could also be adversely affected by the levels of noise. Compliance with Occupational Safety and Health Administration (OSHA) regulations to implement appropriate hearing protection measures would mitigate noise impacts to workers. These measures include the use of standard silencing packages on construction equipment, administrative controls, engineering controls, and personal hearing protection devices.

During normal operations, noise would be produced by cooling systems, vents, motors, generators, material-handling equipment, employee vehicles, and truck traffic. Impacts of these noises on the public would be mitigated by the location of the facilities (about 8.7 km [5.4 mi] from the site boundary).

Operation workers could also be exposed to noise levels higher than the acceptable limits specified by the OSHA in its noise regulation (29 CFR 1926.52). Appropriate mitigation programs would be implemented according to pertinent OSHA standards to minimize impacts on workers. These programs include the use of administrative control, engineering controls, and personal hearing protection devices.

Mitigation measures used during deactivation and decommissioning of the facilities would be similar to those employed for construction.

Further mitigation was not identified by the NRC that would significantly reduce the impacts from noise.

5.2.6 Infrastructure

Upgrades of roadways to and from the proposed MOX site would be conducted in existing traffic rights-of-way.

5.2.7 Waste Management

During construction of the proposed MOX facility and WSB, hazardous and nonhazardous wastes would be generated. Impacts of hazardous and nonhazardous wastes would be mitigated by managing them in accordance with the hazardous waste management practices in place at the SRS and following applicable state and federal regulations. These practices are discussed in Section 4.3.4. The regulations address collecting, handling, storing, sampling, treating, and disposal of the various types of waste minimize impacts to numerous resources including hydrology, soils, air quality, ecology and human health.

Impacts of wastes generated during normal operations of the facilities would be similarly mitigated by managing them in accordance with the hazardous waste management practices in place at the SRS and following applicable state and federal regulations.

During deactivation and decommissioning, impacts of generated wastes would be mitigated in the same ways as discussed above. Impacts of wastes produced by accidents would be mitigated by rapid and thorough cleanup and by following the prescribed SRS waste management practices.

Further mitigation was not identified by the NRC that would significantly reduce the waste management impacts.

5.2.8 Human Health Risk

As discussed in the previous sections, complying with various regulations will mitigate impacts to construction workers. Impacts of fugitive dust on workers would be mitigated by following the Construction Emissions Control Plan. Occupational hazards (e.g., chemical exposure, noise, physical hazards) to workers would be mitigated by following OSHA guidelines. Impacts from hazardous wastes generated during facility construction would be mitigated by appropriately packaging and shipping the material off-site for commercial recycling, treatment, or disposal. Exposure to hazardous materials such as paints and solvents would be mitigated by following good engineering practices, such as using good ventilation and cleaning up small spills promptly and thoroughly. Wastewater generated during construction would be transported to the CSWTF for treatment prior to release.

During construction of the proposed MOX facility and WSB, workers could be adversely affected by exposure to soil or groundwater previously contaminated by radioactivity or chemicals. Potential mitigation measures were identified by the NRC staff to mitigate the possibility that workers could be exposed to the previously disturbed soils that may be contaminated. As discussed in Section 4.3.1, DCS has conducted limited testing of the previously disturbed soils. Impacts from contaminated soil should be mitigated by conducting further sampling of the soil for radioactive contamination before excavation begins at the site. In addition, workers should be monitored, as appropriate, to ensure that radioactive doses are maintained at levels as low as reasonably achievable.

During normal operations of the proposed MOX facility and WSB, workers could be impacted by exposure to internal and external radiation. These impacts would be mitigated by complying with NRC regulations including instituting monitoring, enforcing administrative limits, and developing ALARA programs that would include worker rotations. DCS has incorporated several design features into the proposed MOX facility design to mitigate exposure to workers and the public. These include, but are not limited to, containment (e.g., gloveboxes), shielding, and air filtration.

During normal operations, workers at the proposed MOX facility and WSB could also be impacted by chemical exposure. Complying with OSHA guidelines and SCDHEC regulations would mitigate adverse impacts from chemicals. Health risks from occupational exposures through all pathways (i.e., inhalation, skin contact [dermal], and ingestion) would be mitigated by using enclosed operations (e.g., gloveboxes) as much as possible. In addition, workplace exposure to such chemicals as hydrazine, that are used in the plutonium polishing process to separate plutonium from the solvent, would be monitored to ensure that airborne concentrations within the facility were kept below the occupational exposure limit. Off-gas treatment systems would be used to limit hydrazine emissions to very low levels that would mitigate adverse human health impacts.

During the fuel fabrication process at the proposed MOX facility, purified plutonium dioxide would be mixed with depleted uranium dioxide. Impacts from this process would be mitigated by performing the mixing in closed containers located in gloveboxes that would confine contamination to inaccessible areas. Air exhaust from the gloveboxes would be passed through HEPA filters to collect particulate emissions.

During normal operations, occupational hazards to workers at the proposed MOX facility and WSB would be mitigated by following OSHA guidelines.

DCS has committed to establishing a protocol with the DOE to integrate DCS's emergency plans with the existing SRS emergency preparedness program. The consequences of accidents (fire, explosion, load handling, and criticality) on human health would be mitigated by following SRS emergency procedures. For fires, key features would include fire barriers, minimizing combustibles and ignition sources, installing ventilation systems with fire dampers and HEPA filters, using nitrogen blanket systems, providing only qualified canisters and containers, incorporating fire suppression and detection systems, developing and following appropriate emergency procedures, providing worker training, and equipping and training local fire brigades. For explosions, the following mitigation devices would be available: scavenging air systems, hydrogen monitoring systems, temperature control systems, chemical addition and concentration control systems, sampling systems, process shutdown controls, operator training, and operations and maintenance procedures. Key mitigation features for load handling include load path restrictions, crane-operating procedures, maintenance procedures, operator training, qualified canisters, reliable load-handling equipment, and ventilation systems with HEPA filters. Key mitigation features for criticality accidents include geometry, mass, and moderation controls.

Mitigation activities for the deactivation and decommissioning of the facilities would be essentially the same as those discussed for construction.

The NRC staff has reviewed the mitigation measures for human health impacts and has concluded that no additional mitigation measures are required beyond the regulatory requirements and the measures identified by DCS.

5.2.9 Cultural, Historical, and Paleontological Resources

Construction of the proposed MOX facility and WSB would directly impact two prehistoric archaeological sites that are eligible for listing on *National Register of Historical Places.* There are no known fossil-bearing strata within the area of the project, and although there are about 400 historic sites or sites with historic components, none of them are located within the location of the proposed facilities.

Impacts of construction to two prehistoric archaeological sites were mitigated in part through data recovery as described in a data recovery plan that was submitted and approved by the South Carolina State Historic Preservation Office (SCSHPO) (Long 2002). When construction activities begin, the removal of fill on the site areas will be monitored by staff members of the SRARP (Gould 2002).

Five additional eligible sites are located in the vicinity of the planned construction, but no direct impacts to these sites are expected. However, indirect impacts could still affect these sites. Possible mitigation activities for these indirect impacts include awareness training for workers so that they would not inadvertently disturb the sites, possible restrictions on where heavy machinery is allowed, and periodic monitoring by staff members of the SRARP to check for possible surface erosion or evidence of other impacts from an increase in F-Area activities (e.g., unauthorized pedestrian or vehicle activity at the archaeological sites). The need for an avoidance agreement for one site or additional mitigation activities for potential erosion at several of the sites should be determined in consultation with the SCSHPO.

Inadvertent discoveries of cultural resources could also occur during site construction. Mitigation of any adverse impacts to these sites would follow the guidelines of 36 CFR 800.11 (historic properties) and/or 43 CFR 10.4 (Native American human remains, funerary objects, objects of cultural patrimony, and objects that are sacred).

During normal operations, archaeological resources are unlikely to be affected. Therefore, no mitigation activities would be required.

Potential impacts of deactivation and decommissioning eligible archaeological sites or historic structures would have to be evaluated at the time of decommissioning. Mitigation measures would be determined in consultation with the SCSHPO.

Further mitigation was not identified by the NRC that would significantly reduce the impacts to cultural, historical, and paleontological resources.

5.2.10 Aesthetics

Construction, operation, deactivation, and decommissioning of the structures associated with the proposed MOX facility and WSB would have a minimal effect on the scenic character of the surrounding area and would be consistent with the VRM Class IV designation of the area. The buildings would be low-rise structures with heights of less than 30 m (100 ft). This height would be similar to that of other buildings in the area. The tallest new structure would be a stack that is 37 m (120 ft) above the existing grade. Impacts of these buildings on visual resources would be mitigated by the presence of trees and rolling terrain that would effectively screen them from view, and the distance of the facility from the nearest publicly accessible viewpoints located on State Highway 125 and SRS Road 1, both approximately 6 km (4 mi) away.

Further mitigation was not identified by the NRC that would significantly reduce the impacts.

5.2.11 Socioeconomics

Construction of the proposed MOX facility and WSB would have a minor beneficial socioeconomic impact on the region. Therefore, further mitigation would not significantly reduce the impacts. Although the region should benefit from the construction, the peak demand for workers could adversely affect other construction activities in the area. These impacts would be mitigated by the short duration of the peak demand for workers (a few months). In addition, given that a majority of workers would be hired from the existing regional labor pool, impacts from worker relocation to area businesses, public services, and facilities would be mitigated.

Transportation impacts during construction would be primarily associated with construction labor. To minimize conflicts with other SRS activities, the work schedule would be coordinated and staggered with other SRS activities to minimize the number of vehicles entering and exiting the SRS during peak commuting periods.

Normal operations of the facilities would require approximately 480 new permanent positions and an additional 780 indirect jobs. Given the population and its rate of growth, no significant socioeconomic impacts are expected, and further mitigation would not significantly reduce the impacts.

The impacts of deactivation and decommissioning of the facility would be similar to those for construction, and mitigation activities would be similar to those previously discussed. No mitigation of socioeconomic impacts would be required for accidents, unless residents were evacuated and prevented from quickly returning to their homes. Such impacts would be mitigated, to the extent possible, by rapid cleanup of the accident.

5.2.12 Environmental Justice

As discussed in Section 4.3.7, impacts to the environmental justice community would not be high and adverse from construction and normal operations associated with the proposed action. Mitigation measures discussed above in Section 5.2.8 would mitigate impacts to the general public including the environmental justice community. Therefore, further mitigation would not significantly reduce impacts specific to the environmental justice community.

Section 4.3.7 discusses possible impacts to the environmental justice community from accidents. In developing mitigation measures for these potential impacts, the NRC considered that accident impacts are different from impacts from construction or normal operations. That is construction and normal operations impacts would occur, if the facilities were authorized to be constructed, but the likelihood of accident impacts is less certain. In addition, mitigation of accident impacts for the general public would also mitigate potential impacts to the environmental justice community. Considering these factors, the NRC identified the following potential mitigation measures specifically to address disproportionate impacts to the environmental justice community from potential accidents.

Various procedures might be used to reduce the potential impacts to low-income and minority groups in the event of an accidental chemical or radiological release from the facilities. As discussed in Sections 4.3.5 and 4.3.7, the potential impacts associated with accidents would be lower if the population exposed to population exposed to a contaminate plume did not ingest crops that could be contaminated. In addition, seeking shelter indoors would reduce the inhalation and direct exposure associated with contaminate plumes. Because the mitigation activities for part of the environmental justice community involve knowing what to do in case of an accident, the NRC believes that education and public outreach are potential methods to mitigate these potential impacts. The potential mitigation activities include development and implementation of the following:

- Focused public information campaigns to provide technical and environmental health information directly to low-income and minority groups, or to local agencies and representative groups; and

- Additional programs directed at local communities providing emergency response services and other emergency facilities to incorporate additional measures to protect low-income and minority populations.

Included in the public information campaigns would be descriptions of existing air and groundwater monitoring programs; the nature, extent, and likelihood of any future airborne or groundwater release from the facilities; and the likely characteristics of environmental and health impacts. Key information would include the extent of any likely damage to drinking water supplies and subsistence resources and the relevant preventive measures that may be taken.

The additional programs under the second group of measures would ensure that the low-income and minority population in local government jurisdictions are located and fully prepared in the event that sheltering or other protection strategies may be required and would ensure that detailed descriptions of evacuation routes that may be used have been developed and distributed. In addition to public information campaigns targeting low-income and minority groups, these programs would include the development of spatial database programs for use by local emergency response planners. These databases would provide information on the locations of low-income and minority populations and the locations of relevant local resources available to emergency response agencies, and would have detailed descriptions of evacuation routes that might be required.

The NRC staff has reviewed the mitigation measures for environmental justice and has concluded that no additional mitigation measures are required beyond the regulatory requirements and the measures identified by DCS.

5.3 References for Chapter 5

DCS (Duke Cogema Stone & Webster) 2002. *Mixed Oxide Fuel Fabrication Facility Environmental Report, Revision 1 & 2.* Docket Number 070-03098. Charlotte, NC.

DCS 2003. *Mixed Oxide Fuel Fabrication Facility Environmental Report, Revision 4.* Docket Number 070-03098. Charlotte, NC. Aug.

DOE (U.S. Department of Energy) 1999. *Surplus Plutonium Disposition Final Environmental Impact Statement.* DOE/EIS-0283. Office of Fissile Materials Disposition, Washington, DC. Nov.

DOE 2000. "Record of Decision for the Surplus Plutonium Disposition Final Environmental Impact Statement." *Federal Register* 65:1608, Jan. 11.

DOE 2002. "Amended Record of Decision for the Surplus Plutonium Disposition Program." *Federal Register* 67:19432, April 19.

DOE 2003. *Changes Needed to the Surplus Plutonium Disposition Program Supplement Analyses and Record of Decision.* DOE/EIS-0283-SA1. Office of Fissile Materials Disposition, Washington, DC, April.

Gould, A.B. 2002. Letter from Gould (Director, Environmental Quality Division, DOE, Savannah River Operations Office, Aiken, SC) to C.C. Long (South Carolina State Historic Preservation Office, Columbia, SC). October 24.

Long, C.C. 2002. Letter from Long (South Carolina State Historic Preservation Office, Columbia, SC) to A.B. Gould (Director, Environmental Quality Division, DOE, Savannah River Operations Office, Aiken, SC). October 24.

6 ENVIRONMENTAL REGULATIONS AND PERMITS

The proposed project would be subject to many federal, state, local, and other legal requirements, and a variety of permits, licenses, and approvals would have to be obtained. Many of these requirements are identified and their status summarized in Table 6.1. For items that are the responsibility of the facility owner or operator, Table 6.1 presents requirement status on the basis of information obtained from the environmental report (ER) (DCS 2002a; 2003a,b; 2004). No independent evaluation was made of the status of consents not discussed in the ER that are the responsibility of the facility owner or operator. For items that are the responsibility of the U.S. Nuclear Regulatory Commission (NRC), references are made to other sections of this environmental impact statement (EIS) that discuss their status.

Because of the early stage of project design, the information in Table 6.1 should not be considered comprehensive or binding. It may later be determined that the facility is subject to additional requirements that are not listed in Table 6.1 or qualifies for exemptions or exclusions from some requirements that are listed.

For ease of reference, the information in Table 6.1 has been divided into the following categories:

- Civilian Use of Nuclear Material,

- Air Quality Protection and Noise Control,

- Protection of Water Resources,

- Waste Management and Pollution Prevention,

- Biotic Resources,

- Cultural Resources,

- Transportation, and

- Other.

Table 6.1. Applicable environmental regulations and consents or activities

Responsible agency	Authority	Requirement	Status
Civilian Use of Nuclear Material			
NRC	Atomic Energy Act of 1954, as amended (AEA) (42 U.S.C. 2011 et seq.); 10 CFR Part 40	*Part 40 License* to receive, possess, use, and transfer depleted uranium	DCS has satisfied this requirement by specifying depleted uranium activities in the Construction Authorization Request for its Part 70 License (DCS 2001, Sections 1.2.2 and 1.2.3, and 2002b).
NRC	AEA; 10 CFR Part 70	*Part 70 License* to receive, possess, use, and transfer plutonium	DCS has applied for this consent by filing a Construction Authorization Request and an Environmental Report with the NRC (DCS 2002a; 2003a,b).
South Carolina Department of Health and Environmental Control (SCDHEC)	AEA; South Carolina (SC) Regulations R.61-63	*Radioactive Materials License* to receive, use, possess, transfer, and dispose of radioactive material, including depleted uranium	DCS has satisfied this requirement by applying for a Part 70 License from the NRC.
Air Quality Protection and Noise Control			
SCDHEC	Clean Air Act (CAA) Section 165 (42 U.S.C. 7475); SC Regulations R.61-62.5 Standard No. 7	*Prevention of Significant Deterioration (PSD) Permit* to construct and operate a new major stationary source of air pollution in an area that complies with National Ambient Air Quality Standards for carbon monoxide, lead, nitrogen dioxide, ozone, sulfur oxides, particulate matter with aerodynamic diameter less than or equal to 10 μm (PM_{10}), and $PM_{2.5}$	DCS has determined that gaseous emissions from the facility would not be enough to trigger the requirement for a PSD Permit (DCS 2002a, Section 7.2.1.1). Section 4.3.2.2 discusses impacts of facility operations on air quality.

Table 6.1. Continued

Responsible agency	Authority	Requirement	Status
SCDHEC	CAA, Title V, Sections 501 - 507 (42 U.S.C. 7661 - 7661f); SC Regulations R.61-62.70	*Title V Operating Permit* for a new or existing stationary source that is a major source; a source subject to National Emission Standards for Hazardous Air Pollutants (NESHAPs); a source subject to New Source Performance Standards (NSPS); or an affected source under the Acid Rain Program	DCS has determined that the quantity of criteria and hazardous air pollutants (other than radionuclides) expected to be emitted during facility operation would not be enough to trigger the requirement for a Title V Operating Permit (DCS 2002a, Section 7.2.1.1). Even so, DCS has initiated consultation with the SCDHEC and plans to submit any permit forms necessary to augment the existing Title V Operating Permit held by the DOE SRS (DCS 2002a, Section 7.2.1.1).
SCDHEC	CAA, Section 112 (42 U.S.C. 7412); 40 CFR Part 61; SC Regulations R.61-62.63	*Approval for Construction* of a new source or modification that is subject to NESHAPs	DCS has determined that the proposed facility would be subject to NESHAPs requirements in 40 CFR Part 61, Subpart H, which govern radionuclide emissions from all DOE-owned or DOE-operated facilities, whether or not they are licensed by the NRC. However, EPA Region IV and SCDHEC approved an alternate calculation methodology that exempted the facility from preparing an application for NESHAPs construction approval (DCS 2002a, Section 7.2.1.1). Section 4.3.2.2 discusses impacts on air quality during routine operation.
SCDHEC	CAA, Section 111 (42 U.S.C. 7411); 40 CFR Part 60; SC Regulations R.61-62.60	*Demonstration of Compliance* with applicable NSPS	DCS has determined that the facility would not trigger the requirement to comply with any NSPS (DCS 2002a, Section 7.2.1.1).

Table 6.1. Continued

Responsible agency	Authority	Requirement	Status
SCDHEC	CAA, Section 112(r) (42 U.S.C. 7412); 40 CFR Part 68, Subpart G; SC Regulations R.61-62.68	*Risk Management Plan* for any stationary source that has more than a threshold quantity of a regulated substance in a process	DCS has determined that a Risk Management Plan is not required because the projected quantities of regulated substances at the facility would not be greater than threshold levels (DCS 2002a, Section 7.1.2).
SCDHEC	SC Pollution Control Act (SC Code of Laws, 1976, as amended, Title 48, Chapter 1); SC Regulations R.61-62.1, Section II.A	*State Construction Permit* to construct, alter, or add to a source of air contaminants within South Carolina, if the emission limits imposed would be more restrictive than those imposed by other federal or state air permitting requirements	DCS plans to develop a Construction Emissions Control Plan and to submit standard permit application forms required by the SCDHEC in order to evaluate the applicability of all state air permitting requirements (DCS 2002a, Section 7.2.1.1).
NRC	CAA, Section 176 (42 U.S.C. 7506); 40 CFR Part 93, Subpart B	*Determination of Conformity* with applicable air quality implementation plans	No air quality implementation plans apply to the area where the facility is located.

Protection of Water Resources

SCDHEC	Clean Water Act of 1977 (CWA) (33 U.S.C. 1251 et seq.); SC Regulations R.61-9	*National Pollutant Discharge Elimination System (NPDES) Permit for Storm Water Discharges during Construction* for discharges of storm water from any land disturbance activity affecting an area greater than 5 acres	DCS has determined that the facility construction activities would be covered by the South Carolina NPDES General Permit for storm-water discharges from construction activities within the state (Permit No. SCR100000), provided that a notice of intent, supported by a Storm Water Management Pollution Prevention Plan is filed before construction activities are initiated (DCS 2002a, Section 7.2.1.2). DCS plans to submit the notice of intent and required plans at the appropriate time.

Table 6.1. Continued

Responsible agency	Authority	Requirement	Status
SCDHEC	CWA (33 U.S.C. 1251 et seq.); SC Regulations R.61-9	*NPDES Permit for Storm Water Discharges from Industrial Activity Areas* for discharges of storm water from any facility or activity classified as "associated with industrial activity"	DCS has determined that the South Carolina NPDES General Permit for storm-water discharges from industrial activities within the state (Permit No. SCR000000) would cover runoff exposed to pollutants in an industrial activity area at the facility after construction is complete, provided that a notice of intent, supported by a Storm Water Management Pollution Prevention Plan, is filed (DCS 2002a, Section 7.2.1.2). DCS plans to submit the notice of intent and required plan at the appropriate time.
SCDHEC	CWA (33 U.S.C. 1251 et seq.); SC Regulations R.61-9	*NPDES Permit for Wastewater Discharges* for discharges to surface waters of wastewater from industrial facilities	DCS has determined that the facility would not discharge process wastewater. Accordingly, DCS has consulted with the SCDHEC regarding the need for an NPDES permit and plans, as appropriate, to file a notice of intent to discharge non-process wastewater covered by the South Carolina NPDES general permit for utility water discharges (Permit No. SCG 250000) (DCS 2004, Section 7.2.1.2).
SCDHEC	SC Pollution Control Act (SC Code of Laws, 1976, as amended, Title 48, Chapter 1); SC Regulations R.61-67	*State Construction Permit* to construct, alter, or add to wastewater treatment facilities within South Carolina	DCS has initiated consultation with the SCDHEC and at the appropriate time, plans to obtain a permit to construct the tie-in between the existing SRS Central Sanitary Waste Treatment Facility and the sanitary wastewater system from the facility (DCS 2004, Section 7.2.1.2).

Table 6.1. Continued

Responsible agency	Authority	Requirement	Status
SCDHEC	SC Safe Drinking Water Act (SC Code of Laws, 1976, as amended, Title 44, Chapter 55); SC Regulations R.61-58	*Public Water System Construction Permit* for construction, modification, or expansion of any public water system	DCS has initiated consultation with the SRS Environmental Protection Department, which is responsible for compliance with SCDHEC requirements applicable to the existing drinking water systems at the SRS. DCS plans to obtain the necessary permit before construction begins on a tie-in between the existing SRS drinking water system and the facility drinking water system (DCS 2002a, Section 7.2.1.3).
SCDHEC	SC Safe Drinking Water Act (SC Code of Laws, 1976, as amended, Title 44, Chapter 55); SC Regulations R.61-58	*Public Water System Operating Approval* for placing a new, modified, or expanded public water system into service	DCS has initiated consultation with the SRS Environmental Protection Department, which is responsible for compliance with SCDHEC requirements applicable to the existing drinking water systems at the SRS. DCS plans to obtain the necessary operating approval before beginning operation of the tie-in between the existing SRS drinking water system and the facility drinking water system (DCS 2002a, Section 7.2.1.3).
U.S. Environmental Protection Agency (EPA)	CWA (33 U.S.C. 1251 et seq.); 40 CFR Part 112	*Spill Prevention Control and Countermeasures (SPCC) Plan* for any facility that could discharge oil in harmful quantities into navigable waters	DCS plans to prepare the required SPCC Plan (DCS 2002a, Section 7.2.1.2).
SCDHEC	CWA (33 U.S.C. 1251 et seq.); SC Regulations R.61-101	*State Water Quality Certification* certifying that the applicable state water quality standards will not be violated as a result of discharges to navigable waters by an activity authorized by a federal license	The SCDHEC has notified DCS that a State Water Quality Certification in accordance with SC regulation R.61-101 is not required (SCDHEC 2003).

Table 6.1. Continued

Responsible agency	Authority	Requirement	Status
NRC; U.S. Army Corps of Engineers	CWA (33 U.S.C. 1251 et seq.); Executive Order 11988 (42 FR 26951; May 24, 1977) as amended by Executive Order 12148 (44 FR 43239; July 20, 1979)	*Floodplain Assessment* to evaluate the effects of issuing a Part 70 License on any floodplain	DCS has completed a floodplain assessment and incorporated its results into the design of the facility (DCS 2002a, Section 7.1.3 and Table 7-1). Section 3.3.1 discusses the results of the floodplain assessment.
U.S. Department of the Interior (National Park Service); NRC	Wild and Scenic Rivers Act, as amended (16 U.S.C. 1271 et seq.)	*Wild and Scenic Rivers Assessment* to ensure that issuing a Part 70 License will not result in activities that would adversely affect the values for which a river is being studied or has been designated as a wild and scenic river	DCS has determined that no river that is being studied or has been designated as a national wild and scenic river occurs within the SRS (DCS 2002a, Section 4.4.2.1).
U.S. Army Corps of Engineers	CWA (33 U.S.C. 1251 et seq.)	*Section 404 Permit* to discharge dredged or fill material into waters of the United States, including wetlands	DCS has determined that no wetlands are present on the facility site and that no other discharge of dredged or fill material into water of the United States would occur at the facility site (DCS 2002a, Section 4.6.2.2). Therefore, DCS has concluded that no Section 404 permit is required from the U.S. Army Corps of Engineers (DCS 2002a, Section 7.1.3 and Table 7-1).

Table 6.1. Continued

Responsible agency	Authority	Requirement	Status
Waste Management and Pollution Prevention			
EPA; SCDHEC	Resource Conservation and Recovery Act (RCRA), as amended by the Hazardous and Solid Waste Amendments of 1984 (HSWA) (42 U.S.C. 6901 et seq.), Subtitle C; SC Regulations R.61-79.262	*EPA Identification Number* to identify a hazardous waste generator	DCS has determined that the facility would generate small quantities of hazardous wastes. Therefore, DCS plans to file a notice of hazardous waste activity with EPA and obtain an EPA identification number when hazardous waste activities commence at the site (DCS 2002a, Section 7.2.1.4). Hazardous waste generated during facility operations is discussed in Section 4.3.2.4.
SCDHEC	RCRA, as amended by HSWA (42 U.S.C. 6901 et seq.), Subtitle C; SC Regulations R.61-79.270	*Hazardous Waste Facility Permit* for a facility that will store hazardous wastes beyond the allowed accumulation periods, treat hazardous wastes, or dispose of hazardous wastes	DCS has determined that the facility will not store hazardous waste beyond the allowed accumulation time. Also, DCS does not plan to treat or dispose of hazardous waste at the facility. Therefore, DCS has concluded that the facility would not require a hazardous waste facility permit (DCS 2002a, Section 7.2.1.4).
SCDHEC	RCRA, as amended by HSWA (42 U.S.C. 6901 et seq.), Subtitle I; SC Regulations R.61-92	*Underground Storage Tank Installation and Operation Permits* to install and operate an underground storage tank that will contain regulated substances, including petroleum products and other substances defined in Section 101(14) of the Comprehensive Environmental Response Compensation and Liability Act (CERCLA)	DCS has initiated consultation with the SCDHEC regarding underground storage tanks for managing regulated substances at the facility and plans to obtain the necessary permits at the appropriate time (DCS 2002a, Section 7.2.1.4).

Table 6.1. Continued

Responsible agency	Authority	Requirement	Status
Biotic Resources			
NRC; U.S. Fish and Wildlife Service; South Carolina Department of Natural Resources; Georgia Department of Natural Resources	Endangered Species Act of 1973, as amended (16 U.S.C. 1531 et seq.); Migratory Bird Treat Act of 1918 (MBTA), as amended (16 U.S.C. 703-712); Nongame and Endangered Species Conservation Act (SC Code of Laws, 1976, as amended, Title 50, Chapter 15); Endangered Wildlife Act of 1973 (Georgia Laws 1973, p. 932, et seq.); Wildflower Preservation Act of 1973 (Georgia Laws 1973, p. 333, et seq.)	*Consultation* between the NRC, the U.S. Fish and Wildlife Service, and affected states to ensure that activities resulting from issuance of a Part 70 License (1) are not likely to jeopardize the continued existence of any species listed at the federal or state level as endangered or threatened, or result in destruction of critical habitat of such species and (2) will include appropriate precautions to mitigate adverse effects on birds protected by the MBTA	DCS has obtained declarations from the U.S. Fish and Wildlife Service and the South Carolina Department of Natural Resources indicating that facility construction and operation would have no effect on threatened and endangered species under their jurisdictions (DCS 2002a, Sections 7.1.6 and 7.2.3).

Table 6.1. Continued

Responsible agency	Authority	Requirement	Status
Cultural Resources			
NRC; Advisory Council on Historic Preservation; South Carolina State Historic Preservation Officer	National Historic Preservation Act of 1966, as amended (16 U.S.C. 470 et seq.); Archaeological and Historical Preservation Act of 1974 (16 U.S.C. 469-469c-2); Antiquities Act of 1906 (16 U.S.C. 431 et seq.); Archaeological Resources Protection Act of 1979, as amended (16 U.S.C. 470aa-mm)	*Archaeological and Historical Resources Consultation* between the NRC and the State Historic Preservation Officer or Tribal Historic Preservation Officer before allowing federally licensed activities to proceed in an area where archaeological or historic resources might be located	DCS has determined that, while there are no historic sites located within the facility site, there are two prehistoric archaeological sites that are eligible for listing on the *National Register of Historical Places* (DCS 2002a, Section 4.8.2). Mitigation of these sites was completed during August 2002 (DCS 2002a, Table 7-1). Sections 3.7 and 4.3.7.8 describe the required consultations.
NRC	American Indian Religious Freedom Act of 1978 (42 U.S.C. 1996); Native American Graves Protection and Repatriation Act of 1990 (25 U.S.C. 3001, et seq.)	*Native American Resources Consultation* between the NRC and Native Americans to ensure that activities resulting from issuance of a Part 70 License have been designed to protect access to, physical integrity of, and confidentiality of Native American sites	DCS reports that consultation has been initiated with appropriate Native American groups to identify concerns about construction activities associated with a facility such as the MOX facility at the SRS (DCS 2002a, Section 4.8.4). Sections 3.7.3 and 4.2.6.3 discuss the status of this consultation.

Table 6.1. Continued

Responsible agency	Authority	Requirement	Status
Transportation			
U.S. Department of Transportation (DOT); NRC	Hazardous Materials Transportation Act, as amended by the Hazardous Materials Transportation Uniform Safety Act of 1990 and other acts (49 U.S.C. 1501, et seq.); Atomic Energy Act of 1954, as amended (42 U.S.C. 2011, et seq.); 49 CFR 172, 173, 174, 177, and 397; 10 CFR 71	*Packaging, Labeling, and Routing Requirements for Radioactive Materials*	At the appropriate time, DCS will comply with DOT and NRC requirements for packaging, labeling, and routing of radioactive materials. DCS has identified no specific permits, licenses, or approvals that will be required for transportation of materials to or from the facility.
Other			
NRC; U.S. Natural Resource Conservation Service	Farmland Protection Policy Act (7 U.S.C. 4201 et seq.); 7 CFR Part 658	*Prime Farmland Assessment* to consider alternatives to address the adverse effects on prime farmland of activities resulting from issuance of a Part 70 license	DCS has determined that none of the land on the facility site has been identified as prime farmland because the land is not available for agricultural production (DCS 2002a, Section 7.1.7 and Table 7-1).
NRC	National Environmental Policy Act of 1969, as amended (NEPA) (42 U.S.C. 4321 et seq.); 40 CFR 1500 - 1508; 10 CFR Part 51	*Environmental Impact Statement (EIS)* to evaluate the potential environmental impacts of a proposed major federal action that may significantly affect the quality of the human environment, and to consider alternatives to the proposed action	This EIS meets the requirements of the NEPA.

Table 6.1. Continued

Responsible agency	Authority	Requirement	Status
OSHA; South Carolina Department of Labor, Licensing, and Regulation	Occupational Safety and Health Act, as amended (29 U.S.C. 651, et seq.); 29 CFR 1910.119; SC Regulations, Chapter 71, Article 1, Subarticle 6, "South Carolina Occupational Safety and Health Standards for General Industry and Public Sector Marine Terminals"	*Process Hazard Analysis* to identify, evaluate, and control the hazards of a process involving a flammable liquid or gas, hydrocarbon fuel, or highly hazardous chemical at or above the specified threshold quantity	Before operating the proposed facility, DCS would be required to perform a process hazard analysis for nitrogen tetroxide, which would be present at the proposed MOX facility in a quantity greater than the specified threshold quantity.

6.1 References for Chapter 6

DCS (Duke Cogema Stone & Webster) 2001. *Construction Authorization Request for the Mixed Oxide Fuel Fabrication Facility.* Docket Number 070-03098. Charlotte, NC.

DCS 2002a. *Mixed Oxide Fuel Fabrication Facility Environmental Report, Revision 1 & 2.* Docket Number 070-03098. Charlotte, NC.

DCS 2002b. *Amended Construction Authorization Request for the Mixed Oxide Fuel Fabrication Facility.* Docket Number 070-03098. Charlotte, NC.

DCS 2003a. *Mixed Oxide Fuel Fabrication Facility Environmental Report, Revision 3.* Docket number 070-03098. Charlotte, NC. June.

DCS 2003b. *Mixed Oxide Fuel Fabrication Facility Environmental Report, Revision 4.* Docket Number 070-03098. Charlotte, NC. Aug.

DCS 2004. *Mixed Oxide Fuel Fabrication Facility Environmental Report, Revision 5.* Docket Number 070-03098. Charlotte, NC. June 10.

SCDHEC (South Carolina Department of Health and Environmental Control) 2003. "Duke Cogema Stone and Webster (DCS) Mixed Oxide Fuel Fabrication Facility 401 Water Quality Certification." Letter from Q. Epps (Section Manager, Water Quality Certification Standards, Navigable Waters, and Wetlands Programs, SCDHEC, Columbia, SC) to M.L. Birch (Manager, Environment, Safety and Health, DCS, Charlotte, NC) Mar. 3.

7 GLOSSARY

7Q10 flow: The 7-day low flow, 10-year recurrence flow for a river. This flow is the lowest recorded over any 7 consecutive days within any 10-year period.

absorbed dose (*dose*[1]): The amount of energy deposited in any material by ionizing radiation. The unit of absorbed dose, the rad, is a measure of energy absorbed per gram of material.

accident: An unplanned sequence of events resulting in undesirable consequences, such as the release of radioactive or hazardous material to the environment.

accident risk: Risk based on both the severity of an accident (consequence) and the probability that the accident will occur. High-consequence accidents that are unlikely to occur (low probability) may pose a low overall risk. For purposes of comparison, accident risk is typically calculated by multiplying the accident consequence (for example, dose or expected fatalities) by the probability of the accident's occurring.

accident severity categories: A method of characterizing all the possible types of accident scenarios that might occur according to their likely outcome and the probability of occurrence. The *Nuclear Regulatory Commission* method, which is used in this environmental impact statement, divides the spectrum of accidents into eight categories. Category I accidents are the least severe but the most frequent; Category VIII accidents are very severe but very infrequent.

accident source term: The amount of radioactive or hazardous material released to the environment following an accident.

acute: Resulting in immediate impacts.

acute health endpoint: A human health impact involving immediate injury or fatality.

administrative outfall: An authorized liquid waste *outfall* that discharges no pollutants.

Advisory Council on Historic Preservation: Under the National Historic Preservation Act of 1966, the Council reviews federal undertakings that may affect historic structures, sites, or archeological artifacts. Second contact in sequential review that begins with the State Historic Preservation Officer.

An independent federal agency that serves as the chief policy advisor to the President and Congress on matters concerning historic preservation. Included on the 20 member Council are the heads of several federal agencies, including the Secretaries of the Interior and Agriculture.

---

[1] Italicized words and phrases are entries in this glossary.

aerosol: Particles of solid or liquid matter that can remain suspended in air from a few minutes to many months, depending on the particle size and weight.

aerosolize: The process of converting a solid or a liquid into an airborne suspension of fine particles (an *aerosol*).

affected environment: For an environmental impact statement (EIS), a description of the existing environment covering information necessary to assess or understand the impacts. It must contain enough detail to support the impact analyses and must highlight environmentally sensitive resources (for example, floodplains, wetlands, threatened and endangered species, archeological resources).

aggregate: The sum total.

air pollutant: Any substance in air which could, if in high enough concentration, harm humans, other animals, vegetation, or material. Pollutants may include almost any natural or artificial composition of matter capable of being airborne.

air quality: A measure of the quantity of pollutants, measured individually, in the air. These levels are often compared to regulatory standards.

Air Quality Control Region (AQCR): An interstate or intrastate area designated by the *U.S. Environmental Protection Agency* for the attainment and maintenance of *National Ambient Air Quality Standards*.

air quality standards: The legally prescribed level of constituents in the outside air that cannot be exceeded during a specific time in a specified area.

air toxics (hazardous air pollutants): Substances that have adverse impacts on human health when present in the *ambient air*.

ALARA (as low as reasonably achievable): An approach to keep radiation exposures (both to the workforce and the public) and releases of radioactive material to the environment at levels that are as low as social, technical, economic, practical, and public policy considerations allow. ALARA is not a dose limit; it is a practice whose objective is the attainment of dose levels as far below applicable limits as possible.

algorithm: A formula or set of steps used to solve a problem.

ALOHA model: A computer model used to assess the impacts of potential chemical releases.

alpha particle (α): A positively charged particle made up of two protons and two neutrons that is emitted in the radioactive decay of certain atoms. An alpha particle is identical to the nucleus of the helium atom. It is easily stopped by a sheet of paper. Since they cannot penetrate human skin, alpha particles are not considered an external exposure hazard. Alpha particles within the body can cause harm, however.

ambient: Undisturbed, natural conditions, such as ambient temperature; surrounding conditions.

ambient air: The surrounding atmosphere, usually the outside air, as it exists around people, plants and structures. It is not the air in immediate proximity to emissions sources.

Ambient Air Quality Standards: Regulations prescribing the levels of airborne pollutants that may not be exceeded during a specified time in a defined area.

American Indian Religious Freedom Act: States that the policy of the United States is to protect and preserve for American Indians their inherent rights of freedom to believe, express, and exercise the traditional religions of the American Indian, Eskimo, Aleut, and Native Hawaiians. These rights include, but are not limited to, access to sites, use and possession of sacred objects, and the freedom to worship through ceremony and traditional rites.

anthropogenic: Produced by human activities.

aqueous process: An operation involving chemicals dissolved in water.

aquifer: A geologic formation that can yield significant quantities of groundwater to wells and springs.

aquitard: A geologic unit that is not permeable enough to transmit significant quantities of water. Aquitards transmit water at a very slow rate to or from an adjacent aquifer.

Archaeological and Historic Preservation Act: A federal law directed at the preservation of historic and archaeological data that would otherwise be lost as a result of federal construction. It authorized the U.S. Department of the Interior to undertake recovery, protection, and preservation of archaeological and historic data.

Archaeological Resources Protection Act of 1979: A federal act protecting cultural resources on federally owned lands. This act requires a permit for archaeological excavations or the removal of any archaeological resources on public or Native American lands.

archaeological site: Any location where humans have altered the terrain or discarded *artifacts* during prehistoric or historic times.

artifact: An object produced or shaped by human beings and of archaeological or historical interest.

as low as reasonably achievable: See *ALARA*.

atom: The smallest unit of an element that is capable of entering into a chemical reaction and displays the properties of the element.

Atomic Energy Act of 1954: A federal law that created the Atomic Energy Commission, which later split into the *Nuclear Regulatory Commission* and the Energy and Research and Development Administration (ERDA). ERDA became part of the *Department of Energy* in 1977. This act encouraged the development and use of nuclear energy and research for the general welfare and the security of the United States. This act authorized the Nuclear Regulatory Commission (NRC) to regulate and license fuel fabrication facilities that seek to receive, possess, use, or transfer special nuclear material.

atomic number: The number of positively charged protons in the nucleus of an atom and the number of electrons on an electrically neutral atom.

attainment area: An area considered to have air quality as good as or better than the National Ambient Air Quality Standards for a given pollutant. An area may be in attainment for one pollutant and nonattaining for others.

attenuate: To lessen the magnitude or severity of an impact or effect.

background radiation: Radiation that is part of our natural world. It can originate from naturally occurring radioactive materials within the Earth and from outer space (cosmic sources). Background radiation also includes global fallout as it exists in the environment from the testing of nuclear explosive devices. Background radiation varies considerably with location.

becquerel (Bq): A unit used to measure radioactivity. One Becquerel is that quantity of a radioactive material that will have one transformation in one second. There are 3.7×10^{10} Bq in one *curie* (Ci).

beta particle (β): Beta particles are electrons except they are not bound to an atom. They cannot travel far from their radioactive source (about one half inch in human tissue and a few yards in air).

beyond design basis accident: An accident generally with more severe impacts to on-site personnel and the public than a *design basis accident*. This accident is used for estimating the impacts of a facility or process.

bioaccumulation: The net accumulation of a chemical by an organism as a result of uptake from all routes of exposure.

biomagnification: The tendency of some chemicals to accumulate to higher concentrations at higher levels in the food chain through dietary accumulation.

biota: The plant and animal life of a region.

blackwater stream: A freshwater stream that has a dark color because of organic debris and tannin-containing compounds.

borrow material: Material such as soil or sand that is removed from one location and used as fill material in another location.

borrow pits: An excavated area from which earthy material has been removed, typically for construction purposes.

bound: To estimate or describe a lower or upper limit on a potential environmental or health consequence when uncertainty exists.

bounding: In the case of accident analysis, that which represents the maximum reasonably foreseeable event or impact.

breach: A general term referring to a hole in a cylinder or container. A breach may be caused by corrosion or by mechanical forces.

bryozoa: Bryozoa are microscopic aquatic animals that live in large colonies of interconnected individuals. Bryozoa are abundant in modern marine environments and are also an important part of the fossil record. They are commonly referred to as sea mats, moss animals, or lace corals.

calcareous sand: Sand containing calcium carbonate, calcium, or limestone; it is usually white or tan.

cancer: A group of diseases characterized by uncontrolled cellular growth. Increased incidence of cancer can be caused by exposure to radiation and some chemicals.

candidate species: Species for which substantial information is available to support proposing that they be added to the federal threatened and endangered species list.

CANDU (Canadian deuterium-uranium reactor): A heavy-water reactor that uses natural uranium as a fuel and heavy water as a *moderator* and a coolant.

canister: A container (generally stainless steel) into which immobilized radioactive waste is placed and sealed.

canopy: The upper forest layer of leaves consisting of the tops of individual trees whose branches sometimes cross each other.

canyon building: A term for a chemical separations plant, inspired by the building's long, high, narrow structure. Chemical separation is a process for extracting uranium and plutonium from dissolved spent nuclear fuel and irradiated targets.

capable fault: A *fault* is described as capable if it has had movement at or near the ground surface at least once within the past 35,000 years, or recurrent movement within the past 500,000 years.

capping: The process of installing a layer of clay or other impermeable material over the top of a closed landfill to prevent entry of rainwater and to minimize the escape of chemicals into the surrounding soil.

carbonate: Rocks and associated minerals that contain carbonate ion, as in calcium carbonate.

carbon monoxide (CO): A colorless, odorless gas that is toxic if breathed in high concentrations over an extended period. Carbon monoxide is a *criteria air pollutant*. One source of carbon monoxide is engine exhaust.

carcinogen: A substance that is capable of producing or inducing cancer.

cargo-related impacts: Transportation risks associated with the nature of the cargo itself.

Carolina bays: Closed, elliptical-shaped depressions capable of holding water. They are a type of *wetland*.

cask (for radioactive materials): A heavily shielded container that meets all applicable regulatory requirements for shipping *spent nuclear fuel* or *high-level waste*.

Category I Resources: Resources (for example, waters) defined by the U.S. Department of the Interior as unique and irreplaceable on a national or eco-regional basis.

Cenozoic: A geologic era dating from approximately 65 million years ago to the present. It is known as the age of mammals.

census blocks: Census blocks are defined by the U.S. Bureau of Census and are the smallest geographic unit for which the Census Bureau tabulates data. Blocks contain data from the 2000 Census of Population, including total population, population by race and ethnicity, age, marital status, population density and the number and composition of households, and information on housing unit types. Many blocks correspond to individual city blocks bounded by streets, but blocks – especially in rural areas – may include many square miles and may have some boundaries that are not streets. The Census Bureau established blocks covering the entire nation for the first time in 1990. Over 8 million blocks are identified for Census 2000.

census block groups: Census block groups are geographic entities consisting of groups of individual census blocks. Census blocks are grouped together so that they contain between 250 and 550 housing units.

census tract: An area usually containing between 2,500 and 8,000 persons that is used for organizing and monitoring census data. The geographic dimensions of census tracts vary widely, depending on population density. Census tracts do not cross county borders.

clay: A rock or mineral fragment of any composition that is smaller than very fine silt grains, having a diameter of less than 0.00016 in. (1/256 mm).

Class II water source: Current and potential drinking water, as classified by the EPA.

Clean Air Act: A federal law that mandates and provides for the enforcement of air pollution control standards from various sources. Its purpose is to protect the health and welfare of the public by controlling air pollution.

closed canopy: A forest *canopy* that is dense enough that the tree crowns fill or nearly fill the canopy layer so that light cannot reach the forest floor directly.

cloudshine: The exposure pathway of direct external exposure from radioactive material suspended in air.

Code of Federal Regulations (CFR): A publication in codified form of all federal regulations in force.

collective dose: The sum of individual doses received by all those exposed to a specified source of radiation in a given period of time. (Also referred to as population dose.)

collective population risk: A measure of possible loss or injury in a group of people that takes into account the probability that the hazard will cause harm and the consequences of that event. The collective population risk does not express the risk to specific individual members of the population.

committed effective dose equivalent (CEDE): The sum of the committed dose equivalents to various tissues of the body, each multiplied by its weighting factor. It does not include contributions from external doses. Committed effective dose equivalent is expressed in units of rem and provides an estimate of the lifetime radiation dose to an individual from radioactive material taken into the body through either inhalation or ingestion.

Comprehensive Environmental Response, Compensation, and Liability Act (CERCLA) of 1980 (Superfund): An act providing the regulatory framework for the *remediation* of past contamination from hazardous waste. If a site meets the act's requirements for designation, it is ranked along with other Superfund sites on the National Priorities List. This ranking is the U.S. Environmental Protection Agency's way of determining the priority of sites for cleanup.

conservative estimates: Conservative estimates lean on the side of pessimism and toward maximizing estimates of negative impacts.

consortium: A group (of companies) formed to undertake an enterprise beyond the resources of any one member.

contact-handled transuranic waste: Transuranic waste with a surface radiation dose rate not greater than 200 millirems per hour. It can be safely handled without any shielding other than that provided by the waste container itself.

conversion: An operation for changing material from one form, use, or purpose to another.

cooling water: Water circulated through a *nuclear reactor* or processing plant to remove heat.

cosmic radiation: Streams of highly penetrating, charged particles composed of protons, *alpha particles*, and a few heavier nuclei that bombard the earth from outer space. Cosmic radiation is part of the natural background radiation.

cost-benefit analysis: A formal quantitative procedure comparing costs and benefits of a proposed project or act under a set of preestablished rules.

Council on Environmental Quality: The President's Council on Environmental Quality (CEQ) was established by the enactment of *National Environmental Policy Act* (NEPA). The CEQ is responsible for developing regulations to be followed by all federal agencies in developing and implementing their own specific NEPA implementation policies and procedures.

criteria pollutants: Common air pollutants for which *National Ambient Air Quality Standards* have been established by the U.S. Environmental Protection Agency (EPA) under Title I of the *Clean Air Act*. Criteria *pollutants* include *sulfur dioxide, nitrogen oxides, carbon monoxide, ozone, particulate matter* (PM_{10} and $PM_{2.5}$), and *lead*. Standards for these pollutants were developed on the basis of scientific knowledge about their health effects.

critical habitat: Specific areas within the geographical range of an *endangered species* that is formally designated by the U.S. Fish and Wildlife Service under the Endangered Species Act as essential for conservation of the species.

criticality: A state in which a self-sustaining nuclear chain reaction is achieved.

cultural resources: *Archaeological sites,* architectural structures or features, traditional-use areas, and Native American scared sites or special use areas.

cumulative impacts: Potential impacts when the proposed action is added to other past, present, and reasonable foreseeable future actions. Cumulative impacts can result from individually minor but collectively significant actions taking place over a period of time.

curie (Ci): The unit used to describe the intensity of radioactivity in a sample of material. A curie is equal to 37 billion disintegrations per second, which is approximately the activity of one gram of radium. It is also a quantity of any nuclide or mixture of nuclides having one curie of radioactivity.

D&D (deactivation and decommissioning): The removal of the facility safely from service and reduction of residual radioactivity to a level that permits release of the property to a specified end state.

deactivation: The process of removing a facility from operation and placing it in a safe and stable condition. Deactivation involves removal hazardous and radioactive materials.

decibel (dB): A standard unit for measuring sound-pressure levels based on a reference sound pressure of 0.0002 dyne per square centimeter. This is the smallest sound a human can hear. In general, a sound doubles in loudness with every increase of slightly more than 3 decibels.

decibel, A-weighted (dBA): A measurement of sound approximating the sensitivity of the human ear and used to characterize the intensity or loudness of sound.

decommissioning: The process of decontaminating and dismantling a facility following deactivation and returning the site to an end state that meets the prescribed regulatory criteria.

deep dose equivalent (DDE): The dose equivalent derived from external radiation at a depth of 1 cm in tissue.

deionized water: Water from which both negative and positive ions have been removed by an ion exchange process.

Department of Energy (DOE): A federal agency whose mission is to achieve efficiency in energy use, diversity in energy sources, a more productive and competitive economy, improved environmental quality, and a secure national defense. It was created in 1977.

depleted uranium: Uranium whose content of the isotope uranium-235 is less than 0.7%, which is the uranium-235 content of naturally occurring uranium.

depleted uranium hexafluoride (UF$_6$): A compound of *uranium* and fluorine from which most of the uranium-235 isotope has been removed.

dermal absorption: Entry of a substance into the body through the skin.

design basis accident: For nuclear facilities, an assumed abnormal event used to establish the performance requirements of structures, systems, and components that are necessary to keep the facility in a safe shutdown condition indefinitely, or to prevent or mitigate the consequences of such an event, so as to ensure that the public and operating staff are not exposed to radiation in excess of appropriate guideline values.

detention ponds: Engineered depressions in the land that contain storm-water runoff until it can slowly seep back into the ground or evaporate.

direct impact: An effect that results solely from the construction or operation of a proposed action without intermediate steps or processes. Examples include habitat destruction, soil disturbance, air emissions, and water use.

direct jobs: The number of workers required at a site to implement an alternative.

disposition: A process of use or disposal of materials that results in the remaining material being converted to a form that is substantially and inherently more *proliferation* resistant than the original form.

disproportionately high and adverse environmental impact: An adverse environmental impact determined to be unacceptable or above generally accepted norms. A disproportionately high impact refers to an environmental hazard with a risk or rate of exposure for a low-income or minority population that exceeds the risk or rate of exposure for the general population.

disproportionately high and adverse human health effect: Any effect on human health from exposure to environmental hazards that exceeds generally accepted levels of risk and affects low-income and minority populations at a rate that appreciably exceeds the rate for the general population.

dissolution: The chemical dispersal (dissolving) of a solid throughout a liquid medium.

dose (radiation dose): In a general sense, dose is a measure of the amount of energy from *ionizing radiation* deposited in a material. Dose is affected by the type of radiation, the amount of radiation, and the physical properties of the material itself. Radiation dose to humans is measured in units of *sieverts* (Sv) or *rem* (1 Sv = 100 rem).

drainage basin: An aboveground area of the Earth's surface that supplies the water to a particular stream.

ecology: The study of the interrelationships of organisms and their environment.

ecosystem: A group of organisms and their physical environment.

effective dose equivalent: The sum of the products of the dose equivalent to various organs or tissues and the weighting factors applicable to each of the body organs or tissues that are irradiated. This sum is a risk-equivalent value that can be used to estimate the risk of health effects to the exposed individual. The effective dose equivalent includes the *dose* from *radiation* sources internal and/or external to the body and is expressed in units of *rem* or *sievert*.

effluent: A gas or fluid discharged into the environment, treated or untreated. Most frequently, the term applies to wastes discharged to *surface waters*.

emissions: Substances that are discharged into the air.

endangered species: Any species (plant or animal) that is in danger of extinction throughout all or a significant part of its range. Requirements for declaring a species endangered are found in the *Endangered Species Act*.

Endangered Species Act of 1973: An act requiring federal agencies, with the consultation and assistance of the Secretaries of the Interior and Commerce, to ensure that their actions will not likely jeopardize the continued existence of any endangered or threatened species or adversely affect the habitat of such species.

environmental impact statement (EIS): A document required of federal agencies by the *National Environmental Policy Act* for major proposals or legislation that will or could significantly affect the environment. It describes the positive and negative effects of the proposed and alternative actions.

environmental justice: The fair treatment of people of all races, cultures, incomes, and educational levels with respect to the development, implementation, and enforcement of environmental laws, regulations, and policies. Fair treatment implies that no population of people should be forced to bear a disproportionate share of the negative environmental impacts of pollution or environmental hazards due to a lack of political or economic strength.

Environmental Protection Agency (EPA): A federal agency that is responsible for setting, or working with state and local governments, to set standards that help control and prevent pollution and minimize the potential health effects in areas of solid and hazardous waste, pesticides, water, air, drinking water, and toxic and radioactive substances. It was created in 1970.

Eocene: A geologic epoch early in the Cenozoic era, dating from approximately 56 to 34 million years ago.

epicenter: The point on the Earth's surface directly above the focus of an earthquake.

equivalent dose: The equivalent dose is a measure of the effect that radiation has on humans. It takes into account the type of radiation and the *absorbed dose*. Not all types of radiation produce the same effects. For example, when considering beta, x-ray, and gamma ray radiation, the equivalent dose (in rem) is equal to the absorbed dose (in rads). For alpha radiation, the equivalent dose is assumed to be 20 times the absorbed dose.

erosion: The removal and transport of materials by wind, ice, or water on the Earth's surface.

exposure: Contact of an organism with a chemical, radiological, or physical agent.

exposure pathways: A route or sequence of processes by which a radioactive or hazardous material may move through the environment to humans or other organisms. Each exposure pathway includes a source or release from a source, an exposure point, and an exposure route.

external exposure: Exposure to radiation or hazardous substance that originates from sources outside of the body.

facility: Any building, structure, system, process, equipment, or activity that fulfills a specific purpose on a site.

facility workers: Persons working at the *Mixed Oxide* Fuel Fabrication Facility who are directly involved with the handling of radioactive or hazardous materials.

fault (geologic): A fracture in rock along which movement of one side relative to the other has occurred.

fauna: Animals, especially those of a specific region, considered as a group.

Federal Facilities Compliance Act of 1992: A federal law that amended the *Resource Conservation Recovery Act* with the objectives of bringing all federal facilities into compliance with applicable federal and state hazardous waste laws, waiving federal sovereign immunity under those laws, and allowing the imposition of fines and penalties. The law requires the U.S. Department of Energy to submit an inventory of all its mixed waste and to develop a treatment plan for mixed waste.

FIREPLUME: A computer code used to evaluate atmospheric dispersion of contaminants in an airborne release plume.

fissile nuclear material: Nuclear materials that are fissionable by slow (thermal) neutrons. Fissile materials include uranium-233, uranium-235, and plutonium-239.

fission: The splitting of a heavy atomic nucleus into at least two nuclei of lighter elements, accompanied by the release of energy and generally one or more neutrons. Fission can occur spontaneously or be induced by neutron bombardment.

floodplain: The lowlands adjoining inland and coastal waters and relatively flat areas, including, at a minimum, that area inundated by a 1% or greater-chance flood in any given year. The level area adjoining a river or stream that is sometimes covered by flood water. The base floodplain is defined as the 100-year (1%) floodplain.

flora: Plants, especially those of a specific region, considered as a group.

fly-ash: Small solid ash particles from the noncombustible portion of fuel that are small enough to escape with the exhaust gases.

forb: An herb other than grass.

fossil: An impression or trace of an animal or plant of past geologic ages that has been preserved in the Earth's crust.

fossil fuel: Natural gas, petroleum, coal, and any form of solid, liquid, or gaseous fuel derived from such materials for the purpose of creating useful heat.

Fujita Scale: The official classification system for tornado damage. The scale ranges from F0 (gale tornado, minor damage, winds up to 72 mph) to F6 (inconceivable tornado, winds 319-379 mph). F2 on the Fujita scale indicates a significant tornado causing significant damage.

fugitive dust: The dust released into the air from activities associated with construction, manufacturing, or vehicles operating on open fields or dirt roads. It is a subset of *fugitive emissions*.

fugitive emissions: Emissions that are not caught by a capture system. They are often caused by equipment leaks, evaporative processes, and windblown disturbances.

full-time equivalent (FTE): Equivalent to a full-time worker. For example, two people, each working half time, constitute one FTE.

gamma radiation (γ): High-energy, short-wavelength electromagnetic radiation emitted from a radioactive nucleus during decay. Gamma radiation frequently accompanies alpha and beta emissions and always accompanies fission. Gamma rays are very penetrating and are best stopped or shielded by dense materials such as lead or uranium. Gamma rays are similar to X-rays but are more energetic.

Gaussian model: An air dispersion model based on the assumption that the time-averaged concentration of a substance emitted from a point source has a Gaussian distribution about the mean centerline. A Gaussian distribution is represented by a symmetrical bell-shaped curve.

glauconitic sand: Sand that contains the mineral glauconite, which consists of a dull green earthy iron potassium silicate.

GENII: A computer software code used to evaluate dose from the migration of radionuclides introduced into the environment that may eventually affect humans through *ingestion*, *inhalation*, or direct radiation.

geologic repository: An underground facility intended for the disposal of nuclear waste. The waste is isolated by placing it in mined cavities in a continuous, stable geologic formation at depths typically greater than 300 m (984 ft).

geology: The science that deals with the study of the materials, processes, environments, and history of the Earth, including the rocks and their formation and structure.

glovebox: An airtight box used to work with hazardous material. It is vented to a closed filtering system, and has gloves attached inside to protect the worker.

gravitational acceleration (g): An acceleration equal to the Earth's gravitational acceleration at sea level (32 feet /second/second).

gross alpha: The total (or gross) radioactivity in a sample due to emission amount of *alpha particles*. It includes both naturally occurring and man-made radiation.

groundshine: Radiation from ground-deposited *radionuclides*.

groundwater: The supply of water found beneath the Earth's surface, usually in *aquifers*, which may supply wells and springs. Generally, all water contained in the ground.

grout: A cementing or sealing mixture of cement and water to which sand, sawdust, or other additives (sometimes waste) may be added. In terms of waste management practices, grouting is used to reduce the mobility of a waste material. In-situ grout is used to stabilize contaminated soil without having to remove it.

habitat: Area where a plant or animal lives.

half-life (radiological): The time in which half the atoms of a radioactive substance decay to another nuclear form. It varies for different radioisotopes from millionths of a second to billions of years.

hazard Index (HI): A measure of the noncancer risk involved in human exposure to a chemical substance. It is the sum of the *hazard quotients* for all chemicals to which an individual is exposed. A Hazard Index value of 1.0 or less means that no adverse human health effects (noncancer) are expected to occur.

hazard quotient (HQ): A comparison of the estimated intake level or dose of a chemical in air, water, or soil with its reference dose; expressed as a ratio.

hazardous waste: According to the *Resource Conservation and Recovery Act*, a waste that because of its characteristics may (1) cause or significantly contribute to an increase in mortality or an increase in serious irreversible illness, or (2) pose a substantial hazard to human health or the environment when improperly treated, stored, transported, disposed of, or otherwise managed. Hazardous wastes possess at least one of the following characteristics: ignitability, corrosivity, reactivity, or toxicity. Hazardous waste is nonradioactive.

headwaters: The source of a flowing body of water.

health risk conversion factors: Estimates of the expected number of health effects cause by exposure to a given amount of radiation. Health risk conversion factors are multiplied by the estimated radiation dose received by a given population in order to estimate the number of health effects expected to occur as a result of an exposure.

heavy combination trucks: Rigs composed of a separable tractor unit containing the engine and one to three freight trailers connected to each other and the tractor. They are typically used for shipping radioactive wastes.

herpetofauna: Reptiles and amphibians.

HGSYSTEM: A computer code used to assess hazardous chemical impacts.

high-efficiency particulate air (HEPA) filters: A filter designed to remove 99.97% of particles as small as 0.3 micrometers in diameter from a flowing air stream.

high-level (radioactive) waste (HLW): The highly radioactive waste material that results from the reprocessing of spent nuclear fuel, including liquid waste produced directly in reprocessing and any solid waste derived from the liquid. High-level waste contains a combination of *transuranic waste* and *fission* products in concentrations requiring permanent isolation. High-level waste may include other highly radioactive material that the U.S. Nuclear Regulatory Commission, consistent with existing law, determines by rule requires permanent isolation.

highly enriched uranium: Uranium enriched in the isotope uranium-235 to 20% or above, which thus becomes suitable for nuclear weapons use.

HIGHWAY: A transportation routing model.

historic structures: A standing structure that has historic significance.

human health risk: The likelihood that a given exposure or series of exposures will damage the health of individuals.

hydrazine: A highly reactive and corrosive chemical that is a *carcinogen* and a reproductive hazard. It is the only chemical that would be used in the MOX process that is listed as a hazardous air pollutant under the *Clean Air Act.*

hydrogen fluoride: A colorless, toxic, fuming, corrosive liquid or gas. It is produced when uranium hexafluoride (UF_6) comes in contact with water, such as humidity in the air. It is often a by-product when UF_6 is converted to another chemical form.

hydrology: The study of water, including groundwater, surface water, and rainfall.

ICRP (International Commission on Radiological Protection): An international body tasked with providing an overview of radiation standards and regulations and information to help standardize these regulations.

immobilization: A process used to stabilize waste, thus inhibiting its release into the environment.

impoundment: A natural or artificial body of water confined by a dam, dike, floodgate, or other barrier.

in attainment: In compliance with air quality standards. Areas that are in attainment have air quality that is as good as or better than specified in the *National Ambient Air Quality Standards* for a given pollutant. An area may be in attainment for one pollutant and nonattaining for others.

incremental impact: The impact due to an emission source (or group of sources) in isolation, without including background levels.

indirect impact: An effect that is related to, but removed from a proposed action by an intermediate step or process. An example would be surface-water quality changes resulting from soil erosion at construction sites.

indirect jobs: Jobs generated or lost in related industries within a *regional economic area* as a result of a change in direct employment.

infrastructure: The basic facilities, services, and utilities needed for the functions of an industrial facility or site. Transportation and electrical systems are part of the infrastructure.

ingestion: To take in by mouth. Material that is ingested enters the digestive system.

inhalation: To take in by breathing. Material that is inhaled enters the lungs.

in-migration: People moving into an area, in this case, the region of influence.

in situ: In its natural position or place.

internal exposure: The radiation dose to internal organs and tissues of the body from the ingestion or inhalation of radioactive contaminants in air, water, food, or soil.

invertebrates: Animals without a backbone (insects, for example).

ion: An atom that has too many or too few electrons, causing it to have an electrical charge, and therefore to be chemically active.

ion exchange: A process that removes specific chemicals and radionuclides from a liquid stream (usually water) for the purposes of purification or decontamination. In this process, salts present as charged ions in water are attached to active groups on and in an ion exchange resin, and other ions are discharged into water allowing separation of the two groups of ions.

ionizing radiation: Radiation that has enough energy to remove electrons from atoms, causing them to become charged or ionized.

irradiate: Expose to some form of radiation, usually a nuclear reactor. Irradiated reactor components and fuel are subjected to neutron radiation and become radioactive themselves or produce *isotopes*.

ISCST3: Version 3 of the Short-Term Industrial Source Complex model. It was used to estimate potential air quality impacts from MOX facility construction and operation activities.

isotope: An atom of an element with a specific atomic number and atomic mass. Isotopes of the same element have the same number of protons (atomic number) but different numbers of neutrons (atomic mass). For example, uranium-235 is an isotope of uranium with 93 protons and 143 neutrons; uranium-238 is an isotope of uranium with 92 protons and 146 neutrons.

kaolinitic clay: A fine, usually white *clay* that contains the mineral kaolinite, a hydrous silicate of aluminum.

L_{dn}: A 24-hour average sound level that gives additional weight to noise that occurs during the night (10:00 p.m. to 7:00 a.m.).

L_{eq}: For sounds that vary with time, L_{eq} is the steady sound level that would contain the same total sound energy as the time-varying sound over a given period.

$L_{eq}(24)$: L_{eq} averaged over 24 hours.

Land Disposal Restrictions: Part of the Hazardous and Solid Waste Amendments to RCRA. They restrict land disposal of certain hazardous wastes; these wastes may be land disposed only if they meet specified treatment standards.

land use: A characterization of land surface in terms of its potential utility for various activities.

latent: Occurring some time (usually several years) after exposure.

latent cancer fatalities (LCFs): Deaths resulting from cancer that has become active after a latent period following exposure to a cancer-causing agent. Latent cancer fatalities are similar to naturally occurring cancer and may be expressed at any time after the initial exposure.

latent cancers: Cancers that occur after a latency period of about 10 or more years from the time of exposure.

latency period: The average period of time between exposure to an agent and the onset of a health effect.

latent fatalities (latent mortality): Fatalities that result from acute or chronic environmental exposures to hazardous substances or radiation but that do not occur immediately after exposure.

lead: A gray-white metal that is listed as a criteria air pollutant. Health effects from exposure to lead include brain and kidney damage and learning disabilities.

linear/no threshold hypothesis: A hypothesis that implies, in part, that even small doses of radiation cause some risk of inducing cancer, and doubling of the radiation dose would mean doubling of the expected number of cancers.

listed species: Species that are considered threatened or endangered.

loam: A soil consisting of an easily crumbled mixture of *clay*, *silt*, and sand.

low-enriched uranium (LEU): Uranium enriched in the isotope uranium-235, greater than 0.7% but less than 20% of the total mass. Naturally occurring uranium contains about 0.7% uranium-235, almost all the rest is uranium-238.

low-level (radioactive) waste: Waste that contains radioactivity and is not classified as *high-level waste*, *transuranic waste*, or *spent nuclear fuel*.

low-specific-activity (LSA) drum: A container, such as a 55-gallon drum, that is used to package low-*specific-activity* material. The depleted uranium considered in this EIS is low-specific-activity material.

macroinvertebrates: Small animals, such as larval aquatic insects, that are visible to the naked eye and have no vertebral column.

magnitude: A measure of the total energy released by an earthquake. It is commonly measured in numerical units on the *Richter scale*. Each unit is different from an adjacent unit by a factor of 30.

marsh: An area of low-lying wetlands dominated by grasslike plants.

maximally exposed individual: A hypothetical person who — because of proximity, activities, or living habits — could receive the highest possible dose of radiation or of a hazardous chemical from a given event or process.

meteorology: The science dealing with the atmosphere and its phenomena, especially as relating to weather.

metric ton: A unit of mass equal to approximately 1.1 short (U.S.) tons, or 2,200 pounds.

millirem (mrem): A unit of radiation exposure equal to one-thousandths of a *rem*.

Miocene: A geologic epoch of the *Cenozoic era* dating from approximately 24 to 5 million years ago.

mitigation: A series of actions implemented to ensure that projected impacts will result in no net loss of habitat value or wildlife populations. The purpose of mitigative actions is to avoid, minimize, rectify, or compensate for any adverse environmental impact.

mixed low-level (radioactive) waste: Low-level waste that also contains hazardous chemical components regulated under the *Resource Conservation Recover Act*.

mixed oxide: For the purposes of this EIS, a physical blend of uranium oxide and plutonium oxide.

mixed transuranic waste: Transuranic waste that also contains hazardous chemical components regulated under the *Resource Conservation and Recovery Act*.

mixed waste: Waste that contains both hazardous and radioactive components.

model: A conceptual, mathematical, or physical system obeying certain specified conditions, whose behavior is used to understand the physical system it is attempting to mimic. Models are often used to predict the behavior or outcome of future events.

moderator: A material (usually water, heavy water, or graphite) used in some nuclear reactors to slow down high-velocity neutrons, thereby increasing the likelihood of *fission*. Moderation controls are a factor in mitigating *criticality* accidents.

Modified Mercalli Intensity Scale: A measure of the perceived intensity of earthquake ground shaking, originally developed in Italy nearly a century ago. It includes 12 degrees of shaking from I (not felt by people) to XII (nearly total damage).

molar concentration: The amount of a substance dissolved per unit volume of solution.

National Ambient Air Quality Standards (NAAQS): Air quality standards established by the *Clean Air Act*, as amended. The primary NAAQS are intended to protect the public health with an adequate margin of safety; and the secondary NAAQS are intended to protect the public welfare from any known or anticipated adverse effects of a pollutant.

National Emission Standards for Hazardous Air Pollutants (NESHAPs): A set of national emission standards for listed hazardous pollutants emitted from specific classes or categories of new and existing sources. These standards were implemented in the *Clean Air Act* Amendments of 1977.

National Environmental Policy Act (NEPA) of 1969: A federal law constituting the basic national charter for protection of the environment. The act calls for the preparation of an environmental impact statement (EIS) for every major federal action that may significantly affect the quality of the human or natural environment. The main purpose is to ensure that environmental information is provided to decision makers so that their actions are based on an understanding of the potential environmental and socioeconomic consequences of a proposed action and the reasonable alternatives.

National Historic Preservation Act: A federal law providing that property resources with significant national historic value be placed on the *National Register of Historic Places*. It does not require permits; rather, it mandates consultation with the proper agencies whenever it is determined that a proposed action might impact a historic property.

National Pollutant Discharge Elimination System (NPDES): A federal permitting system controlling the discharge of effluents to surface waters of the United States and regulated through the *Clean Water Act*, as amended.

National Register of Historic Places (NRHP): A list of districts, sites, buildings, structures, and objects of prehistoric or historic local, state, or national significance. The list is maintained by the Secretary of the Interior.

nitrogen oxides (NO$_x$): The oxides of nitrogen, primarily nitrogen oxide (NO) and nitrogen dioxide (NO$_2$), that are produced in the combustion of fossil fuels. Nitrogen dioxide emissions constitute an air pollution problem, because they contribute to acid deposition and the formation of atmospheric ozone. Nitrogen oxides are *criteria air pollutants*.

noise: Any sound that is undesirable because it interferes with speech and hearing, is intense enough to damage hearing, or is otherwise annoying (unwanted sound).

Noise Control Act of 1972: A federal law directing all federal agencies to carry out programs in a manner that furthers the national policy of promoting an environment free from noise that jeopardizes health or welfare.

nonattainment area: The U.S. Environmental Protection Agency's designation for an air quality control region (or portion thereof) in which ambient air concentrations of one or more criteria pollutants exceed *National Ambient Air Quality Standards*.

normal operations: Conditions during which facilities and processes operate as expected or designed. In general, normal operations include the occurrence of some infrequent events that, although not considered routine, are not classified as accidents.

Notice of Intent: A notice that an environmental impact statement will be prepared and considered. It describes the proposed action and provides information on issues and potential impacts and invites comments and suggestions on the scope of the environmental impact statement.

nuclear power plant: A facility that converts nuclear energy into electric power. Heat produced in a *nuclear reactor* is used to make steam, which drives a turbine connected to an electric generator.

nuclear reactor: A machine in which a fission chain reaction is maintained for the purpose of irradiating materials or producing heat for the generation of electricity.

Nuclear Regulatory Commission (NRC): The NRC is an independent regulatory agency created out of the Atomic Energy Commission in 1975 to regulate civilian uses of nuclear material. It is responsible for ensuring that activities associated with the operation of nuclear power and fuel cycle plants and the use of radioactive materials in medical, industrial, and research applications are carried out with adequate protection of public health and safety, the environment, and national security.

Nuclear Waste Policy Act of 1982: The act that authorized federal agencies to develop a geologic repository for the permanent storage of *spent nuclear fuel* and *high-level radioactive waste*.

off-link population: Persons living or working within 0.8 km (0.5 mi) of each side of a transportation route.

Oligocene: A geologic epoch of the *Cenozoic era* dating from approximately 34 to 24 million years ago.

on-link population: Persons sharing a transportation route.

order of magnitude: A range of numbers extending from some value to 10 times that value. If, for example, a number is two orders of magnitude greater than another, it is 100 times greater.

organic compounds: A large group of chemical compounds containing mainly carbon, hydrogen, nitrogen, and oxygen. All living organisms are made up of organic compounds.

outfall: The discharge point of a drain, sewer, or pipe into a body of water.

oxide: A compound formed when an element (for example, plutonium) is bonded to oxygen.

ozone: A strong-smelling, reactive toxic chemical gas consisting of three oxygen atoms chemically attached to each other. It is the product of the photochemical process involving the sun's energy and ozone precursors, such as hydrocarbons and oxides of nitrogen. In the stratosphere, ozone protects the Earth from the sun's ultraviolet rays, but in lower levels of the atmosphere, ozone is considered an air pollutant and can cause irritation of the eyes and respiratory tract. Ozone is one of the criteria air pollutants specified under Title I of the *Clean Air Act* and is a major constituent of smog.

PM_{10}: Particulate matter with a diameter less than 10 μm (0.0004 in.). Particles less than this diameter are small enough to be breathed and could be deposited in the lungs. PM_{10} is one of the six criteria air pollutants specified under Title I of the *Clean Air Act.*

$PM_{2.5}$: Particulate matter with a diameter less than 2.5 μm (0.0001 in.). A standard for this material as a *criteria pollutant* has been defined but not yet implemented.

Paleocene: The earliest epoch in the *Cenozoic era,* dating from approximately 65 to 56 million years ago.

paleontology: The study of plant and animal life that existed in former geologic times, particularly through the analysis of *fossils.*

Paleozoic: The longest era of geologic time, dating from approximately 544 to 248 million years ago. Seed-bearing plants, amphibians, and reptiles first appeared in the Paleozoic era.

parameters: Data or values that are input to computer codes or equations. They are quantifiable or measurable characteristics like wind speed, temperature, pH, vehicular speed, duration of exposure, etc.

particulate matter (PM): Fine liquid or solid particles such as dust, smoke, mist, fumes, or smog, found in air or emissions. The size of the *particulates* is measured in micrometers (µ); a micrometer is 1 millionth of a meter (0.000039 in.). Particle size is important because the *Environmental Protection Agency* has set standards for *PM$_{10}$* and *PM$_{2.5}$* designed to protect human health and welfare. Particulate matter is a *criteria pollutant*.

particulates: Solid particles and liquid droplets small enough to become airborne.

Pascal (Pa): A unit of measurement for pressure in the International System of Units (SI). 1 pascal = 0.0001450 pounds per square inch.

Pasquill atmosphere stability class: A classification scheme that describes the degree of atmospheric turbulence. Categories range from extremely unstable (A) to extremely stable (F). Unstable conditions promote the rapid dispersion of atmospheric contaminants and result in lower contaminant air concentrations compared with stable conditions.

permitted outfalls: *Outfalls* that are regulated by permits.

person-rem: A unit used to measure the radiation exposure to an entire group and to compare the effects of different amounts of radiation on groups of people; it is the product of the average dose equivalent (in *rem*) to a given organ or tissue multiplied by the number of persons in the population of interest.

person-sievert: A unit of radiation exposure. One person-*sievert* is equivalent to 100 *person-rem*.

person-year: The sum of the number of years each person in a study population is at risk; a metric used to aggregate the total population at risk, assuming that 10 people at risk for 1 year is equivalent to 1 person at risk for 10 years.

physiographic province: A region in which the landforms are similar in geologic structure and differ significantly from the landform patterns in adjacent regions.

physiographic regions: Geographic regions based on geologic setting.

pit: The core element of a nuclear weapon's fission component.

plasma arc cutting: Plasma arc cutting uses a high-velocity jet of electrically charged gas to cut metal at temperatures up to 50,000°F.

plume: The elongated pattern of contaminated air or water originating at a point source such as a smoke stack or a hazardous waste disposal area.

plutonium: A heavy, radioactive, metallic element with the atomic number 94. It is produced artificially in a reactor by the bombardment of uranium with neutrons and is used in the production of nuclear weapons. Weapons-usable plutonium consists mainly of plutonium-239.

point source: A source of effluents that is small enough in dimensions that it can be treated as if it were a point. A point source can be either a continuous source or a source that emits effluents only in puffs for a short time.

pollutant: Any material entering the environment that has undesired effects.

pollution: The addition of an undesirable agent to the environment in excess of the rate at which natural processes can degrade, assimilate, or disperse it.

pollution prevention: The use of any process, practice, or product that reduces or eliminates the generation and release of pollutants, hazardous substances, contaminants, and wastes, including those that protect natural resources through conservation or more efficient utilization.

polycyclic aromatic hydrocarbons (PAHs): Organic compounds that include only carbon and hydrogen with a fused ring structure containing at least two benzene (six-sided) rings. Some PAHs are potent human carcinogens. The combustion of organic substances is a common source of atmospheric PAHs.

Prevention of Significant Deterioration (PSD): A program used in development of permits for new or modified industrial facilities in an area that is already in attainment. The intent is to prevent an attainment area from becoming a *non-attainment area*. Allowable increases are lowest in Class I areas (national parks and wilderness areas); the rest of the country is subject to PSD II increments.

Price Anderson Act: First enacted into law in 1957, it limits the liability of the nuclear power industry in the event of an accident.

primary contact recreations: Activities such as swimming and diving where there is direct contact with the water.

prime farmland: Land with the best combination of physical and chemical characteristics for economically producing high yields of food, feed, forage, fiber, and oilseed crops with minimum inputs of fuel, fertilizer, pesticides, and labor. Prime farmland includes cropland, pastureland, rangeland, and forestland.

probable maximum flood: Flood levels predicted for hydrological conditions that maximize the flow of surface waters.

proliferation: The spread of nuclear, biological, and chemical capabilities and the weapons (e.g., missiles) capable of delivering them.

proprietary income: Income from self-employment.

protected species: Species that are protected by federal legislation, such as the Endangered Species Act or the Migratory Bird Treaty Act.

radiation: Energy radiated in the form of waves or particles through matter and space. Radiation comes from radioactive material or from equipment such as X-ray machines. Radiation may be either *ionizing radiation* or non-ionizing radiation.

radiation dose: See *dose*.

radioactive waste: Materials that are radioactive or are contaminated with radioactive materials and for which use, reuse, or recovery are impractical.

radioactivity: The spontaneous decay or disintegration of unstable atomic nuclei, accompanied by the emission of *radiation*. Eventually the unstable nuclei reach a stable state.

radionuclide: An atom that exhibits radioactive properties. Standard practice for naming a radionuclide is to use the name or atomic symbol of the element, followed by its atomic weight. (For example, cobalt-60 [Co-60], a radionuclide of cobalt with an atomic weight of 60.) Radionuclides can be man-made or naturally occurring, can have a long life, and can have potentially mutagenic or carcinogenic effects on the human body.

RADTRAN 4: A computer code that calculates population risks associated with the transport of radioactive materials by truck, rail, air, ship, or barge.

raffinate: The decontaminated salt solution produced by removal of radionuclides from a high-level waste solution.

raptors: Birds of prey (for example, hawks, owls, eagles).

reference dose: The chemical intake level below which noncancer adverse effects are very unlikely. It is measured in units of milligrams per kilogram of body weight per day (mg/kg/d).

regional economic area (REA): A geographic area consisting of an economic node and the surrounding, economically related counties, including the places of work and residences of the labor force. The REA for this EIS is made up of the 15 counties surrounding the Savannah River Site.

region of influence (ROI): The physical area that bounds the environmental, sociological, economic, or cultural features of interest for the purpose of analysis. A site-specific geographic area that includes the counties where approximately 90% of the site's current employees reside. The ROI for this EIS consists of Columbia and Richmond Counties in Georgia and Aiken and Barnwell Counties in South Carolina.

release fraction: The portion, or fraction, of a material that could be released or spilled to the environment during an accident.

rem (roentgen equivalent man): A unit used to derive a quantity called absorbed dose. The dosage of an ionizing radiation that will cause the same biological effect as one *roentgen* of X-ray or gamma-ray exposure; 100 rem is equivalent to one *sievert*.

remediation: Action taken to permanently remedy a release, or threatened release, of a hazardous or radioactive substance to the environment, instead of or in addition to removal.

Resource Conservation and Recovery Act (RCRA): A federal law that provides for a "cradle-to-grave" regulatory program for hazardous waste, including a system for managing hazardous waste from its generation to its ultimate disposal.

Resource Management Class: Four classifications of use to describe different degrees of modification of the landscape. Class I are areas where the natural landscape is preserved, including national wilderness area and wild sections of national wild and scenic rivers; Class II are areas with very limited land development activity, resulting in visual contrasts that are seen but do not attract attention; Class III are areas in which development may attract attention, but the natural landscape still dominates; Class IV are areas in which development activities lead to major modification of the existing character of the landscape.

respirable: Able to be inhaled into the lungs.

Richter Scale: A logarithmic scale used to express the total amount of energy released by an earthquake. The scale has 10 divisions, from 1 (not felt by humans) to 10 (nearly total destruction).

risk: The likelihood of suffering a detrimental effect as a result of exposure to a hazard. In accident analysis, a quantitative or qualitative expression of possible loss that takes into account both the probability that an event will cause harm and the consequences of that event.

Record of Decision (ROD): A document separate from but associated with an environmental impact statement that publicly and officially discloses the responsible agency's decision on the EIS alternative to be implemented.

roentgen: A unit of exposure to ionizing X- or gamma radiation equal to or producing one electrostatic unit of charge per cubic centimeter of air. It is approximately equal to one rad.

runoff: The portion of rainfall, melted snow, or irrigation water that flows across the ground surface and eventually enters streams.

Safe Drinking Water Act: A federal law protecting the quality of public water supplies, water supply and distribution systems, and all sources of drinking water.

Safety Evaluation Report (SER): The SER is an NRC document, associated with a proposed action, that focuses on health and safety issues and compliance with NRC regulations. There are two SERs associated with the MOX facility: one for the construction authorization and another for the operating license application.

sanitary waste: Nonhazardous, nonradioactive liquid and solid waste generated by normal housekeeping activities.

saltstone: A cement-like solid waste form that is a blend of cement, *fly-ash*, and *slag* used to immobilize low-radioactivity salt solutions.

savanna: A grassland with widely scattered trees and shrubs.

scoping: The process of inviting public comment on what should be considered prior to preparation of an environmental impact statement. Scoping assists the preparers of an EIS in defining the proposed action, identifying alternatives, and developing preliminary issues to be addressed in an EIS.

scrub-shrub: Woody vegetation that is less than 20 feet tall, including true shrubs, young trees, and trees or shrubs that are small or stunted because of environmental conditions.

secondary contact recreations: Activities having some direct contact with water, but where swallowing of the water is not likely to occur. An example is fishing.

sedges: Perennial nonwoody plants common to most fresh water *wetlands*. They resemble grasses.

sediment: Eroded soil particles that are deposited downhill or downstream by surface runoff.

seismic: Pertaining to any earth vibration, especially that of an earthquake.

seismic zone: An area defined by the Uniform Building Code (1991) on the basis of its susceptibility to damage as the result of earthquakes. The United States is divided into six zones: Zone 0, no damage; Zone 1, minor damage; Zone 2A (Eastern United States), moderate damage; Zone 2B (Western United States), slightly more damage that 2A; Zone 3, major damage; and Zone 4, areas within Zone 3 nearer certain major fault systems.

seismology: The study of earthquakes.

shielding: Any material that is placed between a source of radiation and people, equipment, or other objects in order to absorb the radiation and reduce radiation exposure.

sievert (Sv): A unit of radiation dose used to express a quantity called *equivalent dose*. This relates the absorbed dose in human tissue to the effective biological damage of the radiation by taking into account the kind of radiation received, the total amount absorbed by the body, and the tissues involved. Not all radiation has the same biological effect, even for the same amount of absorbed dose. One sievert is equivalent to 100 *rem*.

silt: A sedimentary material consisting of fine mineral particles intermediate in size between sand and clay.

siltation: The process by which a river, lake, or other water body becomes clogged with sediment. The process of covering or obstructing with silt.

sinter: To form a homogenous mass by heating without melting.

slag: A glass-like material left as a residue by the smelting of metallic ore.

slope factor: An upper bound estimate of a chemical's probability of causing cancer, based on extent of intake and given in units of inverse intake (1/mg/Kg-d).

source term: The estimated quantities of radionuclides or chemical pollutants released to the environment from a source or group of sources.

special nuclear material: As defined in Section 11 of the *Atomic Energy Act*, " (1) plutonium, uranium enriched in the isotope 233 or in the isotope 235, and any other material which the NRC determines to be special nuclear material, or (2) any material artificially enriched by any of the foregoing."

species of concern: A native species that is not listed as endangered or threatened but that has experienced a long-term decline in population or is vulnerable to a significant decline due to low numbers, restricted distribution, dependence on limited habitat resources, or sensitivity to environmental disturbance.

specific activity: The radioactivity of the radionuclide per unit mass of the nuclide. The specific activity of a material in which the radionuclide is essentially uniformly distributed is the radioactivity per unit mass of the material.

specific conductance: Specific conductance is the electrical conductivity of water normalized to a temperature of 25°C. It is a good measure of the concentration of total dissolved solids and salinity in water.

spent (nuclear) fuel: Fuel that has been withdrawn from a nuclear reactor following irradiation and whose constituents have not been separated. Spent fuel has been burned (irradiated) in a reactor to the extent that it no longer makes an efficient contribution to a nuclear chain reaction. This fuel is more radioactive than it was before *irradiation*.

SRS employees: Persons working at the Savannah River Site but not directly involved with the handling of radioactive or hazardous materials at the MOX facility.

stability class: Stability class describes the potential of atmospheric conditions to disperse pollutants. A relatively stable atmosphere contains very little turbulence so that pollutant concentrations remain high. Unstable atmospheric conditions promote vertical mixing and, thus, lower pollutant concentrations. The original Pasquill Stability Classifications consisted of six classes; A, the most unstable, through F, the most stable.

State Historic Preservation Officer (SHPO): The state officer charged with the identification and protection of prehistoric and historic resources in accordance with the *National Historic Preservation Act*.

subsidence: The process of sinking or settling of a land surface due to natural or artificial causes.

sulfur dioxide (SO₂): A compound of sulfur produced by the burning of sulfur-containing compounds and considered to be a major air pollutant. Sulfur dioxide is a *criteria pollutant.*

surface water: Water on the Earth's surface that is directly exposed to the atmosphere, as distinguished from water in the ground (*groundwater*).

temporary emergency exposure limits (TEELs): The TEEL-1 concentration for a chemical is the maximum concentration in air below which it is believed nearly all individuals could be exposed without experiencing other than mild transient adverse health effects or perceiving a clearly defined objectionable odor. The TEEL-2 value is the maximum concentration in air below which it is believed nearly all individuals could be exposed for up to one hour without experiencing or developing irreversible or other serious health effects or symptoms that could impair their abilities to take protective action. The TEEL-3 value is the maximum concentration in air below which it is believed nearly all individuals could be exposed for up to one hour without experiencing or developing life-threatening health effects.

terrestrial: Pertaining to plants or animals living on land rather than in the water.

threatened species: Any species that is likely to become an endangered species within the foreseeable future throughout all or a significant portion of its range. Requirements for declaring a species threatened are contained in the *Endangered Species Act.*

threshold non-linear relationship: In a threshold nonlinear relationship, some low level of exposure to a harmful substance can be tolerated without causing a health effect. (See also linear/no threshold hypothesis.)

throughput: A general term that refers to the amount of material handled or processed by a facility in a specified time period.

topography: The shape of the earth's surface. The relative position and elevations of natural and man-made features of an area.

total effective dose equivalent (TEDE): The sum of the effective dose equivalent (EDE) from exposure to external radiation and the 50-year committed effective dose equivalent (CEDE) from exposure to internal radiation.

total suspended particulates (TSP): Particles of solid or liquid matter — such as soot, dust, aerosols, fumes, and mist — up to approximately 30 μm in size, that can be suspended in the air. National, South Carolina, and Georgia Ambient Air Quality Standards all set the annual primary (health-based) TSP level at 75 μg/m³ .

toxicity: The ability of a substance to cause damage to cells or tissues of living organisms when the substance is inhaled, ingested, or absorbed by the skin.

Toxic Substances Control Act (TSCA): A federal law authorizing the U.S. Environmental Protection Agency to secure information on all new and existing chemical substances and to control any of these substances determined to cause unreasonable risk to public health or the environment. This law requires that the health and environmental effects of all new chemicals be reviewed by the EPA before such chemicals are manufactured for commercial purposes.

traditional cultural properties: Places and resources important to traditional American cultures, which include, but are not restricted to, Native American cultures.

TRAGIS (Transportation Routing Analysis Geographic Information System): A GIS-based transportation and analysis computer model for rail, highway, and waterway transportation modes.

transport index: The radiation dose rate at 1 meter (approximately 3 feet) from the lateral sides of a vehicle transporting radioactive material.

transuranic: Of, relating to, or being any element whose atomic number is higher than that of uranium (that is, 92). All transuranic elements are radioactive.

transuranic (TRU) waste: Radioactive waste that contains more than 100 nanocuries per gram of alpha-emitting isotopes with atomic numbers greater than 92 and half-lives greater than 20 years. Such wastes result primarily from fuel reprocessing and from the fabrication of plutonium weapons and plutonium-bearing reactor fuel.

Triassic: The first period of the Mesozoic era, dating from approximately 246 to 213 million years ago.

trichloroethylene (TCE): An organic solvent and degreaser.

tritium: A radioactive isotope of the element hydrogen, having two neutrons and one proton. It can be taken into the body easily because it is chemically identical to natural hydrogen. Tritium decays by beta emission with a half-life of about 12.5 years.

Type A package: A type of packaging for radioactive materials. The package must withstand the conditions of normal transportation without loss or dispersal of the radioactive contents. It does not usually require special handling or transportation equipment.

Type B package: A more durable type of packaging for radioactive materials than Type A. In addition to meeting all the Type A standards, Type B packaging must also provide a high degree of assurance that the package integrity will be maintained, even during severe accidents, with essentially no loss of the radioactive contents.

unscarified seed: Seed that has not had the hard outer coat scuffed or otherwise treated to improve germination.

Upper Cretaceous: A geologic time period from about 90 to 66 million years ago. The entire Cretaceous period dates from approximately 144 million to 66 million years ago; it is known as the age of dinosaurs.

uranium: A heavy, silvery-white metallic element (atomic number 92) with many radioactive isotopes. One isotope, uranium-235, is most commonly used as a fuel for nuclear fission. Another, uranium-238, is transformed into fissionable plutonium-239 following its capture of a neutron in a nuclear reactor.

uranium dioxide (UO_2): A black crystalline powder that is widely used in the manufacture of fuel pellets for nuclear reactors.

valence: The number of electrons with which a given atom generally bonds, or the number of bonds an atom forms.

vehicle-related impacts: Transportation risks (physical trauma or emissions) that are related to the transportation vehicle itself, not the cargo it is carrying.

viewshed: The extent of the area that may be viewed from a particular location. Viewsheds are generally bounded by topographic features such as hills or mountains.

Visual Resource Management (VRM): A process devised by the Bureau of Land Management to assess the aesthetic quality of a landscape and to design proposed activities in a way that would minimize their visual impact on that landscape. The process consists of a rating of site visual quality followed by a measurement of the degree of contrast between the proposed development activities and the existing landscape.

vitrification: A process by which glass is used to encapsulate or immobilize radioactive wastes.

volatile organic compounds (VOCs): A broad range of *organic compounds*, that readily evaporate and vaporize at normal temperatures and pressures. Examples include certain solvents, paint thinners, degreasers (benzene), chloroform, and methyl alcohol. VOCs can react with other substances, principally nitrogen oxides, to form ozone. The reactions are energized by sunlight.

Waste Isolation Pilot Plant (WIPP): A national disposal site for transuranic and mixed transuranic waste, located in southeastern New Mexico.

waste management: The planning, coordination, and direction of functions related to generation, handling, treatment, storage, transportation, and disposal of waste. It also includes associated pollution prevention and surveillance and maintenance activities.

waste minimization: An action that economically avoids or reduces the generation of waste by source reduction and recycling; or reduces the toxicity of hazardous waste, improving energy usage.

waste stream: A waste or group of wastes with similar physical form, radiological properties, EPA waste codes, or associated *Land Disposal Restriction* treatment standards. A waste stream may result from one or more processes or operations. Also, a waste or group of wastes from a process or a facility with similar physical, chemical, or radiological properties.

wastewater: Water originating from human sanitary water use (domestic wastewater) and from a variety of industrial processes (industrial wastewater).

watershed area: All land and water within the confines of a *drainage basin.*

weapons-grade: Plutonium or *highly enriched uranium*, in metallic form, that was manufactured for weapons application. Weapons-grade plutonium contains less that 7% plutonium 240.

wetland: Land areas exhibiting hydric (moist) soil conditions, saturated or inundated soil during some portion of the year, and plant species tolerant of such conditions. Wetlands include swamps, marshes, and bogs.

Wild and Scenic Rivers Act: The federal law that established the National Wild and Scenic Rivers System. It was designed to preserve and protect the free-flowing condition of selected rivers having outstanding natural, cultural, or recreational features. For federally owned land within the boundaries of rivers in the system, certain activities that would have a direct and adverse effect on the river values may be controlled.

wind rose: A circular diagram showing, for a specific location, the percentage of time the wind blows from each compass direction over a specified period of record. A wind rose for use in assessing consequences of airborne releases also shows the frequency of different wind speeds for each compass direction.

8 LIST OF PREPARERS

J. Adducci — Maps and Figures; B.A. History and Geography, University of Illinois at Chicago; 7 years experience in geographic information system (GIS) applications.

T. Allison — Environmental Justice, Socioeconomics, Aesthetics, Cost-Benefit Analysis; M.S. Mineral and Energy Resource Economics, M.A. Geography, West Virginia University; B.A. Economics and Geography, Portsmouth Polytechnic (Great Britain); 15 years experience in environmental assessment.

G. Anast — Scoping Summary Report, Glossary; B.A. Biology and Mathematics, North Central College; 12 years experience in environmental assessment.

H. Avci — Project Manager; Description of MOX Facility; Ph.D., Nuclear Engineering, M.S. Nuclear Engineering, B.S. Nuclear Engineering, University of Wisconsin; 15 years experience in environmental assessment.

B. Biwer — Human Health Assessment, Radiological Risk, Transportation Risk Analysis; Ph.D. Chemistry, M.S. Chemistry, Princeton University, B.A. Chemistry, St. Anselm College; 12 years experience in environmental assessment.

M.D. Blevins — NRC Project Manager; M.S. Environmental Systems Engineering, Clemson University; B.S. Chemistry, West Virginia University; 9 years experience in environmental assessment.

D. Brown — NRC Technical Integration, Radiological Risk Assessment; M.S. Health Physics, Clemson University, B.S. Physics, Muhlenberg College; 7 years experience in environmental assessment.

Y.-S. Chang — Meteorology, Emissions, Air Quality, Noise; Ph.D. Chemical Engineering, M.S. Chemical Engineering, University of Iowa, B.S. Chemical Engineering, Seoul National University; 15 years experience in environmental assessment.

S.Y. Chen — Decommissioning; Ph.D. Nuclear Engineering, M.S. Nuclear Engineering, University of Illinois, B.S. Nuclear Engineering, Tsing Hua University; 20 years experience in environmental assessment.

B.J. Davis — NRC Project Manager; M.S. Materials Engineering, University of Maryland, B.S. Materials Engineering, Virginia Tech; 2 years experience in environmental assessment.

J.D. DePue — Technical Editing, Document Coordination; M.S. Biology, B.A. Political Science, Texas A&I University; 29 years experience in technical and environmental assessment document editing.

LIST OF PREPARERS

APPENDIX A:

PROTECTED SPECIES

APPENDIX A:

PROTECTED SPECIES

Sixty-one threatened, endangered, and other special status species listed by the federal government or the State of South Carolina may be found in the vicinity of the Savannah River Site (SRS). Protected species listed by the state for Aiken and Barnwell Counties (within which most of the SRS is located) and by Georgia for the reach of the Savannah River bordering the SRS and for Burke County across the river from the SRS are listed in Table A.1. Table A.1 also lists the status and habitat preferences for the protected species. Species from Allendale County, South Carolina, and Screven County, Georgia, are not considered because of the distance of these counties from the F-Area. No designated critical habitat for threatened or endangered species exists on the SRS (DOE 1996).

The SRS has established a proactive threatened and endangered species program that includes habitat restoration. In particular, special efforts have been enacted since 1986 to reestablish and expand the population of the federally and state-endangered red-cockaded woodpecker (*Picoides borealis*) at the SRS. The SRS has been divided into three natural resource habitat management areas: (1) a 34,858-ha (86,069-acre) red-cockaded woodpecker habitat management area, (2) a 19,508-ha (48,167-acre) supplemental red-cockaded woodpecker habitat management area, and (3) other-use areas totaling 25,965 ha (64,111 acres) (DOE 2000). Within the red-cockaded woodpecker habitat management area, harvest rotation for loblolly and longleaf pine is set at 100 and 200 years, respectively. These long rotation periods are designed to increase the number of potential cavity nesting trees. Rotation for pines within the supplemental red-cockaded woodpecker habitat management and other-use areas is set at 50 years to encourage woodpecker recovery within the designated red-cockaded woodpecker habitat management area. The bottomland hardwood, upland hardwood, and mixed pine/hardwood timber management areas that do not provide red-cockaded woodpecker habitat are managed on 100-year rotations (DOE 2000). No red-cockaded woodpecker management is practiced within the other-use area (Edwards et al. 1999).

A combination of methods has been used to improve the red-cockaded woodpecker population at the SRS. These methods have included removing southern flying squirrels from red-cockaded woodpecker nesting cavities, excavating new nesting cavities, thinning hardwood midstory trees, and augmenting the number of female red-cockaded woodpeckers at the SRS. The excavation of cavities has allowed nesting use in younger tree stands several decades before the birds would be able to do this on their own (Allen 1990a,b). The annual conversion of slash and loblolly pine areas to longleaf pine also provides a long-term benefit to red-cockaded woodpeckers and other wildlife species associated with the longleaf pine savanna ecosystem (DOE 2000).

The endangered status of the red-cockaded woodpecker is primarily related to the loss of mature pine forests in the southeastern states from logging and fire suppression; only about 1%

of the species' historical habitat remains (WSRC 1994; FWS 2001a). They prefer longleaf and loblolly pines that are more than 70 years old, often selecting those trees with red-heart disease, which softens the core of the tree. They forage in pine trees over 30 years old (WSRC 1994; USAF 1996). The woodpeckers also prefer areas with minimal midstory trees, so as to lessen potential competition (e.g., from other woodpecker species) and predation (e.g., black rat snakes) (FWS 2001a). Other species either compete for or use abandoned red-cockaded woodpecker cavity holes, including southern flying squirrels, chickadees, bluebirds, titmice, herpetofauna (amphibians and reptiles) and insects (particularly bees and wasps) (FWS 2001b).

The red-cockaded woodpecker is a social species, living in a family group that inhabits a collection of cavity trees called a cluster. Each bird in the group maintains its own cavity tree, but only one pair in the group actually nests. A cluster may include from 1 up to 20 or more cavity trees on 1.2 to 24.3 ha (3 to 60 acres), averaging about 4.0 ha (10 acres). Territory size is related to both habitat suitability and population density. The typical territory for a family group ranges from about 50.6 to 81.0 ha (125 to 200 acres), but reported extremes are as low as 24.3 ha (60 acres) and as high as 243 ha (600 acres) (FWS 2001a,b).

The SRS contains two subpopulations of the red-cockaded woodpecker. Currently 26 active clusters with almost 150 individual birds occur on the SRS. In 1985, only four birds were reported from the SRS (DOE 2000). The closest nesting area to the proposed facility site is about 5 km (3.1 mi) away (DOE 1999). The proposed area for the facility does not occur within either the red-cockaded woodpecker habitat management area or the supplemental management area. However, all areas containing pines, including those at the proposed site, provide suitable forage areas for this species.

Table A.1. Rare, threatened, and endangered species from Aiken and Barnwell Counties, South Carolina, and Burke County, Georgia

Species common name (scientific name)	Status, federal/state[a,b]	County locations[c]	Habitat
Plants			
Aethusa-like trepocarpus (*Trepocarpus aethusae*)	–/SC	A	Bottomland hardwoods
American eelgrass (*Vallisneria americana*)	–/SC	Ba	Ponds and streams, mostly in the sandhills
American nailwort (*Paronychia americana*)	–/SC	A, Ba	Sandhills, dry pinelands
Awnpetal meadowbeauty (*Rhexia aristosa*)	–/SC	Ba	Wet depressions, Carolina bays, savannas, pinelands
Bearded milkvetch (*Astragalus villosus*)	–/SC	A, Ba	Pinelands, disturbed sites

Table A.1. Continued

Species common name (scientific name)	Status, federal/state[a,b]	County locations[c]	Habitat
Biennial beeblossom (*Gaura biennis*)	–/SC	A, Ba	Streambanks, meadows, roadsides
Bog spicebush (*Lindera subcoriacea*)	–/RC	A, Ba	Evergreen-shrub bogs, acidic swamp forests, and seepage bogs
Boykin's lobelia (*Lobelia boykinii*)	–/SC	Ba	Cypress ponds, wet pinelands, Carolina bays
Canada moonseed (*Menispermum canadense*)	–/SC	Ba	Moist woods and thickets
Candby's cowbane (*Oxypolis canbyi*)	E/E, E	Ba, Bu	Peaty muck of shallow cypress ponds, wet pine savannas, and adjacent sloughs and drainage ditches
Candy's bulrush (*Scirpus etuberculatus*)	–/SC	A	Swamps and quiet or flowing shallow water
Carolina birds-in-a-nest (*Macbridea caroliniana*)	–/SC	A, Ba	Freshwater margins
Carolina bugbane (*Trautvetteria caroliniensis*)	–/SC	Ba	Woods, especially in damp or wet soils
Carolina larkspur (*Delphinium carolinianum*)	–/SC	A	Dry woods, prairies, and sandhills
Carolina wild petunia (*Ruellia caroliniensis* spp. *ciliosa*)	–/SC	A	Moist or dry woods
Collins' sedge (*Carex collinsii*)	–/SC	A	Bogs, especially white cedar swamps
Creeping St. johnswort (*Hypericum adpressum*)	–/RC	Ba	Marshes, shores, and wet meadows
Cypressknee sedge (*Carex decomposita*)	–/SC	Ba	Wooded swamps
Drowned hornedrush (*Rhynchospora inundata*)	–/SC	A, Ba	Inundated pond margins and wet peat
Durand's white oak (*Quercus sinuata*)	–/SC	Ba	Wooded slopes, edges of streams
Dwarf burhead (*Echinodorus parvulus*)	–/SC	A, Ba	Carolina bays
Eared goldenrod (*Solidago auriculata*)	–/SC	A	Fields, roadsides, open woods

Table A.1. Continued

Species common name (scientific name)	Status, federal/state[a,b]	County locations[c]	Habitat
Eastern leatherwood (*Dirca palustris*)	–/SC	A	Rich, moist woods
Eastern wahoo (*Euonymus atropurpurea*)	–/SC	A	Woodlands and thickets, usually on moist, rich soils
Elliott's croton (*Croton elliottii*)	–/SC	A, Ba	Carolina bays
Faded trillium (*Trillium discolor*)	–/SC	A	Moist woods
False rue anemone (*Enemion biternatum*)	–/RC	A	Moist woods
Flax leaf false-foxglove (*Agalinis linifolia*)	–/RC	A	Wet, sandy soils
Florida bladderwort (*Utricularia floridana*)	–/SC	Ba	Shallow ponds, often within Carolina bays
Georgia beargrass (*Nolina georgiana*)	–/SC	A, Ba	Sandhills
Georgia plume (*Elliottia racemosa*)	–/T	Bu	Sand ridges, dry oak ridges, evergreen hammocks, sandstone outcrops
Green fringed orchid (*Platanthera lacera*)	–/SC	A, Ba	Carolina bays, bottomland hardwoods
Ground juniper (*Juniperus communis*)	–/SC	A	Dry, rocky, or otherwise poor soils
Hooded pitcher plant (*Sarracenia minor*)	–/U	Bu	Acidic soils of open bog, wet savannas, pond margins, low areas in pine flatwoods, sphagnum seeps of bottomland forests, sloughs and ditches
Lance-leaf wild-indigo (*Baptisia lanceolata*)	–/SC	Ba	Pine forests, open woods
Least trillium (*Trillium pusillum* var *pusillum*)	–/NC	A	Alluvial or low woods, savannas
Leechbrush (*Nestronia umbellula*)	–/SC, T	A, Ba, Bu	Dry, open, upland forests of mixed hardwood and pines
Long sedge (*Carex folliculata*)	–/SC	A	Wet or swampy woods

Table A.1. Continued

Species common name (scientific name)	Status, federal/state[a,b]	County locations[c]	Habitat
Loose watermilfoil (*Myriophyllum laxum*)	–/RC	A, Ba	Sinkhole ponds and other shallow, freshwater ponds; and sandy, clear streams draining spring-fed swamps
Lowland brittle fern (*Cystopteris protrusa*)	–/SC	A	Moist woods
Muhlenberg maidencane (*Amphicarpum muehlenbergianum*)	–/SC	Ba	Pastures, pinelands, moist margins of woods, disturbed sites
Narrow-leaved trillium (*Trillium lancifolium*)	–/NC	A	Moist woods
Nutmeg hickory (*Carya myristiciformis*)	–/RC	Ba	Bottomland hardwoods
Pickering's morning-glory (*Stylisma pickeringii* var *pickeringii*)	–/SC	A	Scrub habitats with scant litter accumulation, sparse ground cover, and little canopy cover (scrubby oaks and pines)
Piedmont azalea (*Rhododendron flammeum*)	–/SC	A, Ba	Upland hardwood bluffs
Piedmont bladderwort (*Utricularia olivacea*)	–/SC	Ba	Shallow, acidic ponds
Piedmont cucumber tree (*Magnolia cordata*)	–/SC	A	Rich woods
Piedmont mock bishopweed (*Ptilimnium nodosum*)	E/E	A, Ba	Wet savannas and peaty fringes of pineland pools and cypress ponds
Piedmont three-awned grass (*Aristida condensata*)	–/SC	A	Sand pine scrub, sandhills, disturbed sites
Pine-leaved golden aster (*Pityopsis pinifolia*)	–/SC	A	Barrens, sandy soils
Pink ladyslipper (*Cypripedium acaule*)	–/U	Bu	Acid soils of pine woodlands, upland hardwoods with pines
Pyramid magnolia (*Magnolia pyramidata*)	–/RC	A	Low, moist situations
Red standing-cypress (*Ipomopsis rubra*)	–/SC	A, Ba	Pastures, roadsides
Relict trillium (*Trillium reliquum*)	E/E	A	Rich moist woods on bluffs and ravine slopes

Table A.1. Continued

Species common name (scientific name)	Status, federal/state[a,b]	County locations[c]	Habitat
Reticulated nutrush (*Scleria reticularis*)	–/SC	Ba	Damp, sandy soils and pine barrens
Robbins' spikerush (*Eleocharis robbinsii*)	–/SC	A, Ba	Mud or shallow water
Rose coreopsis (*Coreopsis rosea*)	–/RC	A	Wet, often sandy or acid soils, or in shallow water
Sandhill rosemary (*Ceratiola ericoides*)	–/T	Bu	Very dry, openly vegetated, scrub-oak sandhills
Sandhills milkvetch (*Astragalus michauxii*)	–/SC	Ba	Sandhills, open sandy woods
Sarvis holly (*Ilex amelanchier*)	–/SC	A	Woody streambanks in sandhills, wet depressions, Carolina bays
Scarlet beebalm (*Monarda didyma*)	–/SC	Ba	Moist woods and thickets
Shoals spiderlily (*Hymenocallis coronaria*)	–/NC	A	Major streams and rivers in rocky shoals and in cracks of exposed bedrock
Shortleaf sneezeweed (*Helenium brevifolium*)	–/RC	Ba	Swampy or boggy places and moist pine woods
Shortleaf yelloweyed grass (*Xyris brevifolia*)	–/SC	A	Pine flatwoods, pond margins
Silky camellia (*Stewartia malacodendron*)	–/R	Bu	Rich, wooded bluffs and ravine slopes, transitional areas between sandhills and creek swamps
Slender arrowhead (*Sagittaria isoetiformis*)	–/SC	A, Ba	Carolina bays
Small-flowered buckeye (*Aesculus parviflora*)	–/RC	A	Upland hardwood bluffs
Small-flowered silverbell-tree (*Halesia parviflora*)	–/SC	A, Ba	Dry, sandy, upland sites
Smooth coneflower (*Echinacea laevigata*)	E/E	A, Ba	Meadows and open woodlands on basic or near neutral soils
Southeastern sneezeweed (*Helenium pinnatifidium*)	–/SC	Ba	Wet pinelands
Spatulate seedbox (*Ludwigia spathulata*)	–/SC	A, Ba	Wet depressions, pond margins, Carolina bays
Stalkless yellowcress (*Rorippa sessiliflora*)	–/SC	A	Bottomland hardwoods

Table A.1. Continued

Species common name (scientific name)	Status, federal/state[a,b]	County locations[c]	Habitat
Striped garlic (*Allium cuthbertii*)	–/SC	A, Ba	Sandhills, marshes
Sweet pitcher plant (*Sarracenia rubra*)	–/SC, E	A, Bu	Acidic soils in open bogs, sandhill seeps, wet savannas, low areas in pine flatwoods, along sloughs and ditches
Three-angle spikerush (*Eleocharis tricostata*)	–/SC	Ba	Pine barren ponds
Tracy beakrush (*Rhynchospora tracyi*)	–/SC	Ba	Carolina bays
Upland swamp-privet (*Forestiera ligustrina*)	–/SC	A	Sandy or rocky soils
Water toothleaf (*Stillingia aquatica*)	–/SC	Ba	Grass-sedge wet depressions, bogs
White wicky (*Kalmia cuneata*)	–/NC	A	Borders of Carolina bays and bogs; between sandhills and upland swamps
Winter grape fern (*Botrychium lunarioides*)	–/SC	A	Open fields, meadows, sandy or gravelly streambanks
Yellow pipewort (*Syngonanthus flavidulus*)	–/RC	A	Wet pinelands, pond margins
Invertebrates			
Arogos skipper (*Atrytone arogos*)	–/SC	A	Open fields, meadows, prairies
Barrel floater (*Anodonta couperiana*)	–/SC	Ba	Streams, rivers
Carolina slabshell (*Elliptio congaraea*)	T/E	Ba	Streams, rivers
Eastern creekshell (*Villosa delumbis*)	–/SC	Ba	Streams, rivers
Eastern floater (*Pyganodon cataracta*)	–/SC	Ba	Streams, rivers
Paper pondshell (*Utterbackia imbecillis*)	–/SC	Ba	Streams, rivers
Rayed pink fatmucket (*Lampsilis splendida*)	–/SC	Ba	Streams, rivers

Table A.1. Continued

Species common name (scientific name)	Status, federal/state[a,b]	County locations[c]	Habitat
Southern rainbow (*Villosa vibex*)	–/SC	Ba	Streams, rivers
Yellow lampmussel (*Lampsilis cariosa*)	–/SC	Ba	Streams, rivers

Fish

Robust redhorse (*Moxostoma robustum*)	–/E	Bu	Mainstream river habitats (e.g., Augusta Shoals of Savannah River)
Shortnose sturgeon (*Acipenser brevirostrum*)	E/E, E	A, Bu	Spawns in large coastal rivers; remainder of year spent in lower reaches or river estuary

Amphibians and Reptiles

American alligator (*Alligator mississippiensis*)	T(S/A)/–	A, Ba	Savannah River Swamp, Par Pond, Beaver Dam Creek, and other streams
Bird-voiced treefrog (*Hyla avivoca*)	–/SC	A, Ba	Wooded swamps along creeks and larger waterways
Black swamp snake (*Seminatrix pygaea*)	–/SC	A	Cypress ponds
Eastern coral snake (*Micrurus fulvius*)	–/SC	A	Well-drained pine woods; open, dry, or sandy areas; pond and lake borders; and hammocks
Eastern tiger salamander (*Ambystoma tigrinum tigrinum*)	–/SC	A	Savannah River Swamp and Carolina bays
Florida green water snake (*Nerodia floridana*)	–/SC	A	Swamps, marshes, and quiet bodies of water
Gopher frog (*Rana capito*)	–/SC	A, Ba	Gopher tortoise burrows during daylight hours
Gopher tortoise (*Gopherus polyphemus*)	–/E, T	A, Bu	Sandy soil and abundant herbaceous vegetation (e.g., longleaf pine savannas); often forced to inhabit roadsides and old fields
Pine (or gopher) snake (*Pituophis melanoleucus*)	–/SC	A	Flat, sandy pine barrens, sandhills, and dry mountain ridges, usually in or near pine woods

Table A.1. Continued

Species common name (scientific name)	Status, federal/state[a,b]	County locations[c]	Habitat
Southern hognose snake (*Heterodon simus*)	–/SC	A	Sandy woods, fields, and groves, dry river floodplains, and hardwood hammocks
Spotted turtle (*Clemmys guttata*)	–/SC, U	A, Ba, Bu	Heavily vegetated, shallow wetlands with standing or slowly flowing water
Birds			
Bachman's sparrow (*Aimophila aestivalis*)	–/R	Bu	Mature open pine woods, regenerating clearcuts, old pastures with dense ground cover of grasses and forbs, palmetto scrub
Bald eagle (*Haliaeetus leucocephalus*)	T/E	A	Active nests in Pen Branch area and area south of Par Pond
Little blue heron (*Egretta caerulea*)	–/SC	A	Freshwater ponds, lakes, and marshes; coastal saltwater wetlands
Red-cockaded woodpecker (*Picoides borealis*)	E/E, E	A, Ba, Bu	Nests in mature pine forests (particularly longleaf); forages in pine forests
Wood stork (*Mycteria americana*)	E/E, E	Ba, Bu	Variety of freshwater and estuarine wetlands for breeding, feeding, and nesting; nests in trees in standing water or on islands
Mammals			
Black bear (*Ursus americanus*)	–/SC	A	Forests and swamps
Eastern fox squirrel (*Sciurus niger*)	–/SC	A, Ba	Pine forests with interspersed clearings
Eastern woodrat (*Neotoma floridana*)	–/SC	A, Ba	Hummocks, swamps, and cabbage palmetto
Hoary bat (*Lasiurus cinereus*)	–/SC	A	Wooded areas
Rafinesque's big-eared bat (*Corynorhinus rafinesquii*)	–/E	A	Roosts in or near mature forests with water nearby; forage among canopies of large trees

Table A.1. Continued

Species common name (scientific name)	Status, federal/state[a,b]	County locations[c]	Habitat
Spotted skunk (*Spilogale putorius*)	–/SC	A	Brushy or sparsely wooded areas, along streams, among boulders, prairies
Star-nosed mole (*Condylura cristata*)	–/SC	A, Ba	Low, wet ground near lakes and streams

[a]E = endangered; T = threatened; T(S/A) = threatened (similarity of appearance); NC = of concern, national (unofficial, plants only); R = rare; RC = of concern, regional (unofficial, plants only); SC = species of concern; U = unusual; – = not listed.

[b]For species listed from both South Carolina and Georgia counties, the status for South Carolina is provided first.

[c]A = Aiken County, South Carolina; Ba = Barnwell County, South Carolina; Bu = Burke County, Georgia.

Sources: Burt and Grossenheider (1976); Conant (1958); DCS (2002); DOE (1991); Fernald (1989); Flora of North America Editorial Committee (1997); Gleason and Cronquist (1991); Harrar and Harrar (1962); Knox and Sharitz (1990); National Geographic Society (1999); Ozier et al. (1999); Patrick et al. (1995); Petrides (1988); SCDNR (2001a,b); USDA (2001); Workman and McLeod (1990); Wunderlin (1982).

References for Appendix A

Allen, D.H. 1990a. *Establishment of a Viable Population of Red-Cockaded Woodpeckers at the Savannah River Site, Annual Report FY 1990.* U.S. Department of Energy, Savannah River Site, Aiken, SC, and USDA Forest Service, Southeastern Forest Experiment Station, Clemson, SC. Aug. 31.

Allen, D.H. 1990b. Letter from Allen (USDA Forest Service, Southeastern Forest Experiment Station, Clemson, SC) to P. Jackson (DOE Savannah River Operations Office, Environmental Division, Aiken, SC). Aug. 30.

Burt, W.H., and R.P. Grossenheider 1976. *A Field Guide to the Mammals.* 3rd Edition. Houghton Mifflin Co., Boston, MA.

Conant, R. 1958. *A Field Guide to Reptiles and Amphibians of the United States and Canada East of the 100th Meridian.* Houghton Mifflin Co., Boston, MA.

DCS (Duke Cogema Stone & Webster) 2002. *Mixed Oxide Fuel Fabrication Facility Environmental Report, Revision 1 & 2.* Docket No. 070-03098. Charlotte, NC.

DOE (U.S. Department of Energy) 1991. *Draft Environmental Impact Statement for the Siting, Construction, and Operation of New Production Reactor Capacity.* DOE/EIS-0144D. Office of New Production Reactors, Washington, DC. April.

DOE 1996. *Storage and Disposition of Weapons-Usable Fissile Materials Final Programmatic Environmental Impact Statement.* DOE/EIS-0229. Office of Fissile Materials Disposition, Washington, DC. Dec.

DOE 1999. *Surplus Plutonium Disposition Final Environmental Impact Statement.* DOE/EIS-0283. Office of Fissile Materials Disposition, Washington DC. Nov.

DOE 2000. *Savannah River Site Long Range Comprehensive Plan.* Savannah River Site, Aiken, SC. Dec.

Edwards, J.W., et al. 1999. *Savannah River Site Red-Cockaded Woodpecker Management Plan.* U.S. Department of Agriculture Forest Service, Savannah River Natural Resource Management and Research Institute, Aiken, SC. Feb.

Fernald, M.L. 1989. *Gray's Manual of Botany.* Dioscorides Press, Portland, OR.

Flora of North America Editorial Committee 1997. *Flora of North America North of Mexico.* Volume 3. Oxford University Press, New York, NY.

FWS (U.S. Fish and Wildlife Service) 2001a. *The Red-Cockaded Woodpecker at Carolina Sandhills NWR.* U.S. Department of the Interior. Available at http://carolinasandhills.fws.gov/rcw.html.

FWS 2001b. *Red-Cockaded Woodpecker.* U.S. Department of the Interior. Available at http://rcwrecovery.fws.gov/rcw.htm.

Gleason, H.A., and A. Cronquist 1991. *Manual of Vascular Plants of Northeastern United States and Adjacent Canada.* The New York Botanical Garden, Bronx, NY.

Harrar, E.S., and J.G. Harrar 1962. *Guide to Southern Trees.* Dover Publications, Inc., New York, NY.

Knox, J.N., and R.R. Sharitz 1990. *Endangered, Threatened, and Rare Vascular Flora of the Savannah River Site.* Savannah River Site, National Environmental Research Park Program, Aiken, SC.

National Geographic Society 1999. *Field Guide to the Birds of North America.* 3rd Edition. Washington, DC.

Ozier, J.C., et al. (proj. coords.) 1999. *Protected Animals of Georgia.* Georgia Department of Natural Resources, Wildlife Resources Division, Nongame Wildlife-Natural Heritage Section, Social Circle, GA.

Patrick, T.S., et al. 1995. *Protected Plants of Georgia.* Georgia Department of Natural Resources, Wildlife Resources Division and Georgia Natural Heritage Program, Social Circle, GA.

Petrides, G.A. 1988. *A Field Guide to Eastern Trees Eastern United States and Canada.* Houghton Mifflin Company, Boston, MA.

SCDNR (South Carolina Department of Natural Resources) 2001a. *South Carolina Rare, Threatened & Endangered Species Inventory. Species Found in Aiken County.* South Carolina Department of Natural Resources, Wildlife and Freshwater Fisheries Division, Columbia, SC. Available at http://www.dnr.state.sc.us/heritage/owa/county_species. list?pcounty=aiken.

SCDNR 2001b. *South Carolina Rare, Threatened & Endangered Species Inventory. Species Found in Barnwell County.* South Carolina Department of Natural Resources, Wildlife and Freshwater Fisheries Division, Columbia, SC. Available at http://www.dnr.state.sc.us/ heritage/owa/county_species.list?pcounty=barnwell.

USAF (U.S. Air Force) 1996. *The Red-Cockaded Woodpecker.* Air Force Materiel Command, Air Force Development Test Center, Office of Public Affairs, Eglin AFB, FL. Available at http://www.eglin.af.mil/46tw/46xp/46xpe/fact/images/wpecker.pdf.

USDA (U.S. Department of Agriculture) 2001. *The PLANTS Database, Version 3.1.* Natural Resources Conservation Service, National Plant Data Center, Baton Rouge, LA. Available at http://plants.usda.gov.

Workman, S.W., and K.W. McLeod 1990. *Vegetation of the Savannah River: Major Community Types.* SRO-NERP-19. Savannah River Site, National Environmental Research Park Program, Aiken, SC. April.

WSRC (Westinghouse Savannah River Company) 1994. "SRS Ecology Environmental Information Database." CD database prepared for the U.S. Department of Energy by Westinghouse Savannah River Company under Contract No. DE-AC09-89SR18035. May 5.

Wunderlin, R.P. 1982. *Guide to the Vascular Plants of Central Florida.* University Press of Florida, Tampa, FL.

APPENDIX B:

LETTERS OF CONSULTATION[1]

---

[1] The following letters have been reproduced from the best available copies.

UNITED STATES
NUCLEAR REGULATORY COMMISSION
WASHINGTON, D.C. 20555-0001

September 24, 2001

Mr. John Ross, Chief
United Keetowah Band
  of Cherokee Indians
PO Box 746
Tahlequah, OK 74465-0746

Dear Mr. Ross:

The U.S. Nuclear Regulatory Commission (NRC) is evaluating the potential impacts associated with the construction, operation, and deactivation of a proposed Mixed Oxide (MOX) Fuel Fabrication Facility (Facility) to be constructed at the Department of Energy's (DOE) Savannah River Site (SRS) in South Carolina. The NRC has regulatory responsibility for approving and licensing the construction and operation of the proposed MOX facility.

On March 7, 2001, pursuant to the National Environmental Policy Act, the NRC issued a Notice of Intent to prepare an Environmental Impact Statement (EIS) for its action (66 FR 13794). The EIS will address the potential environmental effects of manufacturing MOX fuel from surplus weapons plutonium. Public scoping meetings for the MOX Facility EIS were held on April 17 and 18, 2001, in North Augusta, SC and Savannah, GA, respectively, and on May 8, 2001, in Charlotte, NC.

Two maps are enclosed. The first map shows the general location of the proposed MOX Facility and the locations of two potential recipients of the MOX fuel assemblies, the Catawba Nuclear Station and the William B. McGuire Nuclear Station. Transportation corridors between SRS and the two reactor stations have not yet been identified. The second map shows the proposed location of the MOX Facility at SRS.

Argonne National Laboratory (ANL) is assisting the NRC in preparing the MOX Facility EIS and will be evaluating potential impacts to cultural resources as part of their analysis. An archaeologist from ANL is in the process of researching available documents from SRS on archaeological surveys, historic building inventories, and resources of interest to Native Americans.

The proposed project area has been surveyed for archaeological sites. These surveys have identified five archaeological sites (38AK155, 38AK330, 38AK548, 38AK546/547, and 38AK757) in the project vicinity. Only Site 38AK546/547 will likely be affected by the construction of the MOX Facility. This site has been determined to be eligible for listing on the National Register of Historic Places (NRHP), and a data recovery plan is being implemented to mitigate impacts to this site. The possibility of indirect effects to other nearby sites from activities associated with the construction or operation of the facility will be analyzed during preparation of the EIS. Sites 38AK155 and 38AK757 have also been determined eligible for the NRHP and can be avoided during construction and operation of the MOX Facility to mitigate the potential for adverse impacts to these sites. The remaining two sites were determined not

J. Ross   2

eligible. No traditional cultural properties have been identified in the project area to date. Through previous consultations with Native Americans, some interest has been expressed in traditional plant resources that could exist at SRS. However, none of these plant resources is currently known to exist at the MOX Facility construction site.

The NRC is also initiating consultations regarding the proposed project with points of contact (Tribal Historic Preservation Officers or designated representatives) from the following Native American Tribes, Bands, and Nations, as well as with the South Carolina Department of Archives and History.

> Catawba Indian Nation
> Ma Chis Lower Alabama Creek Indian Tribe
> Pee Dee Indian Association
> Muscogee Creek Nation
> National Council of the Muscogee Creek
> Yuchi Tribal Organization, Inc.

We would appreciate receiving information on concerns or issues you may have regarding the proposed project. We are especially interested in your assistance in identifying properties of known religious or cultural significance that may be affected by the construction and operation of the proposed facility. Sensitive information will remain confidential as stipulated under 36 CFR Part 800.11. Please submit comments to me within 30 days. Your time and consideration are greatly appreciated.

In the meantime, if you have any questions or require further clarification regarding the project please call me at (301) 415-7293, or Jennifer Davis at (301) 415-5874.

Sincerely,

Phyllis Sobel for Charlotte Abrams

Charlotte E. Abrams, Section Chief
Environmental and LLW Section
Environmental and Performance
    Assessment Branch
Division of Waste Management
Office of Nuclear Material Safety
    and Safeguards

Docket: 70-3098
Enclosures: Location maps

cc: See attached list

Letter dated <u>September 24, 2001</u> to:

Gilbert Blue, Chairperson,
Catawba Indian Nation
PO Box 188
Catawba, SC 29704

Virginia Montoya
Pee Dee Indian Association
101 E. Tatum Avenue
McColl, SC 29570-057

Nancy Carnley
Ma Chis Lower Alabama Creek Indian Tribe
Route 1, 708 S. John Street
New Brockton, AL 36351

John Ross, Chief
United Keetowah Band of Cherokee Indians
PO Box 746
Tahlequah, OK 74465-0746

Tom Berryhill
Council Member
National Council of the Muscogee Creek
PO Box 158
Okmulgee, OK 74447

Andrew Skeeter, Chairman
Yuchi Tribal Organization, Inc.
PO Box 1990
Sapulpa, OK 74067

Julie Moss
United Keetowah Band
Tahlequah, OK 74464

R. Perry Beaver, Principal Chief
Muscogee Creek Nation
PO Box 580
Okmulgee, OK 74447

3

Letter dated September 24, 2001

Distribution:

cc: Arthur B. Gould, Jr.
      DOE Savannah River Indian Relations
         Officer
      PO Box A
      Aiken, SC 29802

      Mary Birch, P.E., C.H.P.
      Environment, Safety and Health Manager
      Duke COGEMA Stone & Webster
      PO Box 31847
      Charlotte, NC 28231-1847

      Edwin D. Pentecost
      Argonne National Laboratory
      9700 South Cass Avenue
      Argonne, IL 60439

      James Johnson
      U.S. Department of Energy
      MD-12
      1000 Independence Avenue, SW
      Washington, DC 20585

      Henry Potter
      Division of Waste Management
      Bureau of Land and Waste Management
      SC Dept of Health & Environmental Control
      2600 Bull Street
      Columbia, SC 29201

      File Center
      EPAB r/f
      NMSS r/f
      JPiccone
      MWeber
      ELeeds
      APersinko
      TJohnson
      JHull
      AFernandez
      R Martin
      TEssig
      DAyres/RII
      THarris
      DBrown
      EMcAlpine/RII

Dennis Ryan
DOE Savannah River Site
PO Box A
Aiken, SC 29802

Don Moniak
Blue Ridge Defense League
PO Box 3487
Aiken, SC 29802

Glenn Carroll
Georgians Against Nuclear Energy
PO Box 8574

Ruth Thomas
Environmentalists, Inc.
1339 Sinkler Road
Columbia, SC 29206

John T. Conway, Chairman
Defense Nuclear Facilities Safety Board
625 Indiana Avenue, Suite 700
Washington, DC 20004

UNITED STATES
NUCLEAR REGULATORY COMMISSION
WASHINGTON, D.C. 20555-0001

September 24, 2001

Mr. Roger Banks, Field Supervisor
U.S. Fish and Wildlife Service
Charleston Ecological Services Field Office
176 Croghan Spur Road, Suite 200
Charleston, SC 29404

SUBJECT: INFORMAL CONSULTATION UNDER SECTION 7 OF THE ENDANGERED
SPECIES ACT FOR THE MIXED OXIDE FUEL FABRICATION FACILITY

Dear Mr. Banks:

The U.S. Nuclear Regulatory Commission (NRC) published its Notice of Intent (NOI) to prepare an Environmental Impact Statement (EIS) for construction, operation and deactivation of a Mixed Oxide Fuel Fabrication Facility (MOX FFF) in the Federal Register (66 FR 13794) on March 7, 2001 (copy of NOI enclosed). The MOX FFF would be used to manufacture mixed oxide fuel from surplus weapons plutonium.

The proposed MOX FFF would be constructed in F Area at the Department of Energy's (DOE) Savannah River Site (SRS) in Aiken County, South Carolina (see three enclosed figures). The F Area of SRS is within the Savannah River watershed. NRC requests that the U.S. Fish and Wildlife Service provide information regarding federally listed threatened and endangered species (including candidate and proposed species) that may occur on or in the vicinity of F Area at SRS that should be considered in preparing the EIS.

The MOX FFF and associated infrastructure would occupy approximately 6.9 ha of the 16.6 ha project site. The remainder of the site will be landscaped. The F area is an upland plateau that is designated for Industrial use. A portion of the MOX FFF project site was previously used to store excavation material from another facility constructed in the F Area. About 68 percent of the MOX FFF site is composed of planted longleaf and slash pine forest. No wetlands occur within the area proposed for the MOX FFF (see enclosed land cover figure), but wetlands associated with Upper Three Runs Creek are adjacent to and downslope of the project site. Operational impacts would be minimized as airborne and aqueous releases would comply with applicable standards and permit requirements. Preliminary analyses presented in the environmental report for the MOX FFF prepared by the applicant DCS (a consortium formed by Duke Engineering & Services, COGEMA, Inc., and Stone and Webster) indicate that the impacts to ecological resources from facility construction and operation would be limited to the immediate project site vicinity.

R. Banks   2

I wish to thank you in advance for the information on the threatened, endangered, candidate, and proposed species that you believe should be addressed in the EIS. Please mail your response to me within 30 days of the date of this letter.

If you have any questions, please contact me at (301) 415-7293, or Jennifer Davis at (301) 415-5874.

Sincerely,

Phyllis Sobel for Charlotte Abrams

Charlotte E. Abrams, Section Chief
Environmental and LLW Section
Environmental and Performance
    Assessment Branch
Division of Waste Management
Office of Nuclear Material Safety
    and Safeguards

Docket: 70-3098

Enclosures: Notice of Intent
              F Area Figures
              Land Cover Figure

cc: See attached list

R. Banks   3

Letter to Roger Banks, U.S. Fish and Wildlife Service, SC
dan dated September 24, 2001

cc: Mary Birch, P.E., C.H.P.   Edna Foster
     Environment, Safety and Health Manager   120 Balsam Lane
     Duke COGEMA Stone & Webster   Highlands, NC 28741
     PO Box 31847
     Charlotte, NC 28231-1847

     Edwin D. Pentecost
     Argonne National Laboratory
     9700 South Cass Avenue
     Argonne, IL 60439

     James Johnson
      U.S. Department of Energy
     MD-12
     1000 Independence Avenue, SW
     Washington, DC 20585

     Henry Potter
     Division of Waste Management
     Bureau of Land and Waste Management
     SC Dept of Health & Environmental Control
     2600 Bull Street
     Columbia, SC 29201

     John T. Conway, Chairman
     Defense Nuclear Facilities Safety Board
     625 Indiana Avenue, Suite 700
     Washington, DC 20004

     Don Moniak
     Blue Ridge Defense League
     PO Box 3487
     Aiken, SC 29802

     Glenn Caroll
     Georgians Against Nuclear Energy
     PO Box 8574
     Atlanta, GA 30306

     Ruth Thomas
     Environmentalists, Inc.
     1339 Sinkler Road
     Columbia, SC 29206

**UNITED STATES
NUCLEAR REGULATORY COMMISSION**
WASHINGTON, D.C. 20555-0001

September 24, 2001

Mr. Jon Ambrose, Program Manager
Georgia Department of Natural Resources
Wildlife Resources Division
Nongame Wildlife & Natural Heritage Section
Georgia Natural Heritage Program
2117 US Hwy 278 SE
Social Circle, GA 30025

SUBJECT: STATE LISTED THREATENED, ENDANGERED AND RARE SPECIES IN THE
VICINITY OF THE MIXED OXIDE FUEL FABRICATION FACILITY

Dear Mr. Ambrose:

The U.S. Nuclear Regulatory Commission (NRC) published its Notice of Intent (NOI) to prepare
an Environmental Impact Statement (EIS) for construction, operation and deactivation of a
Mixed Oxide Fuel Fabrication Facility (MOX FFF) in the Federal Register (66 FR 13794) on
March 7, 2001 (copy of NOI enclosed). The MOX FFF would be used to manufacture mixed
oxide fuel from surplus weapons plutonium.

The proposed MOX FFF would be constructed in F Area at the Department of Energy's (DOE)
Savannah River Site (SRS) in Aiken County, South Carolina, (see three enclosed figures). The
SRS borders several counties of Georgia (i.e., Burke and Screven Counties) and the F Area is
within the Savannah River watershed. NRC requests that the Georgia Department of Natural
Resources provide information regarding state listed threatened, endangered, and rare species
that may occur on or in these areas that should be considered in preparing the EIS.

The MOX FFF and associated infrastructure would occupy approximately 6.9 ha of the 16.6 ha
project site. The remainder of the site will be landscaped. The F area is an upland plateau that
is designated for industrial use. A portion of the MOX FFF project site was previously used to
store excavation material from another facility constructed in the F Area. About 68 percent of
the MOX FFF site is composed of planted longleaf and slash pine forest. No wetlands occur
within the area proposed for the MOX FFF (see enclosed land cover figure), but wetlands
associated with the Upper Three Runs Creek are adjacent to and downslope of the project site.
Operational impacts would be minimized as airborne and aqueous releases would comply with
applicable standards and permit requirements. Preliminary analyses presented in the
environmental report for the MOX FFF prepared by the applicant DCS (a consortium formed by
Duke Engineering & Services, COGEMA, Inc., and Stone and Webster) indicate that the
impacts to ecological resources from facility construction and operation would be limited to the
immediate project site vicinity.

J. Ambrose   2

I wish to thank you in advance for the information on the threatened, endangered and rare species in Georgia that you believe should be addressed in the EIS. Please mail your response to me within 30 days of the date of this letter.

If you have any questions, please contact me at (301) 415-7293, or Jennifer Davis at (301) 415-5874.

Sincerely,

Phyllis Sobel for Charlotte Abrams

Charlotte E. Abrams, Section Chief
Environmental and LLW Section
Environmental and Performance
   Assessment Branch
Division of Waste Management
Office of Nuclear Material Safety
   and Safeguards

Docket: 70-3098

Enclosures: Notice of Intent
             F Area Figures
             Land Cover Figure

cc: See attached list

J. Ambrose   3

Letter to Jon Ambrose, Georgia Department of
  Natural Resources, dated September 24, 2001

cc: Mary Birch, P.E., C.H.P.   Edna Foster
     Environment, Safety and Health Manager   120 Balsam Lane
     Duke COGEMA Stone & Webster   Highlands, NC 28741
     PO Box 31847
     Charlotte, NC 28231-1847

     Edwin D. Pentecost
     Argonne National Laboratory
     9700 South Cass Avenue
     Argonne, IL 60439

     James Johnson
      U.S. Department of Energy
     MD-12
     1000 Independence Avenue, SW
     Washington, DC 20585

     Henry Potter
     Division of Waste Management
     Bureau of Land and Waste Management
     SC Dept of Health & Environmental Control
     2600 Bull Street
     Columbia, SC 29201

     John T. Conway, Chairman
     Defense Nuclear Facilities Safety Board
     625 Indiana Avenue, Suite 700
     Washington, DC 20004

     Don Moniak
     Blue Ridge Defense League
     PO Box 3487
     Aiken, SC 29802

     Glenn Caroll
     Georgians Against Nuclear Energy
     PO Box 8574
     Atlanta, GA 30306

     Ruth Thomas
     Environmentalists, Inc.
     1339 Sinkler Road
     Columbia, SC 29206

UNITED STATES
NUCLEAR REGULATORY COMMISSION
WASHINGTON, D.C. 20555-0001

September 24, 2001

Ms. Julie Holling
South Carolina Department of Natural Resources
Wildlife and Freshwater Fisheries Division
P.O. Box 167
Columbia, SC 29202

SUBJECT: STATE LISTED THREATENED, ENDANGERED AND RARE SPECIES IN THE VICINITY OF THE MIXED OXIDE FUEL FABRICATION FACILITY

Dear Ms Holling:

The U.S. Nuclear Regulatory Commission (NRC) published its Notice of Intent (NOI) to prepare an Environmental Impact Statement (EIS) for construction, operation and deactivation of a Mixed Oxide Fuel Fabrication Facility (MOX FFF) in the Federal Register (66 FR 13794) on March 7, 2001 (copy of NOI enclosed). The MOX FFF would be used to manufacture mixed oxide fuel from surplus weapons plutonium.

The proposed MOX FFF would be constructed in F Area at the Department of Energy's (DOE) Savannah River Site (SRS) in Aiken County, South Carolina, (see three enclosed figures). NRC requests that the South Carolina Department of Natural Resources provide information regarding state listed threatened, endangered, and rare species that may occur on or in the vicinity of F Area at SRS that should be considered in preparing the EIS.

The MOX FFF and associated infrastructure would occupy approximately 6.9 ha of the 16.6 ha project site. The remainder of the site will be landscaped. The F area is an upland plateau that is designated for industrial use. A portion of the MOX FFF project site was previously used to store excavation material from another project site facility constructed in the F Area. About 68 percent of the MOX FFF site is composed of planted longleaf and slash pine forest. No wetlands occur within the area proposed for the MOX FFF (see enclosed land cover figure), but wetlands associated with Upper Three Runs Creek are adjacent to and downslope of the project site. Operational impacts would be minimized as airborne and aqueous releases would comply with applicable standards and permit requirements. Preliminary analyses presented in the environmental report for the MOX FFF prepared by the applicant DCS (a consortium formed by Duke Engineering & Services, COGEMA, Inc., and Stone and Webster) indicate that the impacts to ecological resources from facility construction and operation would be limited to the immediate project site vicinity.

J. Holling   2

I wish to thank you in advance for the information on the threatened, endangered, and rare species in South Carolina that you believe should be addressed in the EIS. Please mail your response to me within 30 days of the date of this letter.

If you have any questions, please contact me at (301) 415-7293, or Jennifer Davis at (301) 415-5874.

Sincerely,

Phyllis Sobel for Charlotte Abrams

Charlotte E. Abrams, Section Chief
Environmental and LLW Section
Environmental and Performance
    Assessment Branch
Division of Waste Management
Office of Nuclear Material Safety
    and Safeguards

Docket: 70-3098

Enclosures: Notice of Intent
             F Area Figures
             Land Cover Figure

cc: See attached list

J. Holling   3

Letter to Julie Holling, South Carolina Department of
Natural Resources, dated September 24, 2001

cc: Mary Birch, P.E., C.H.P.   Edna Foster
     Environment, Safety and Health Manager   120 Balsam Lane
     Duke COGEMA Stone & Webster   Highlands, NC 28741
     PO Box 31847
     Charlotte, NC 28231-1847

     Edwin D. Pentecost
     Argonne National Laboratory
     9700 South Cass Avenue
     Argonne, IL 60439

     James Johnson
      U.S. Department of Energy
     MD-12
     1000 Independence Avenue, SW
     Washington, DC 20585

     Henry Potter
     Division of Waste Management
     Bureau of Land and Waste Management
     SC Dept of Health & Environmental Control
     2600 Bull Street
     Columbia, SC 29201

     John T. Conway, Chairman
     Defense Nuclear Facilities Safety Board
     625 Indiana Avenue, Suite 700
     Washington, DC 20004

     Don Moniak
     Blue Ridge Defense League
     PO Box 3487
     Aiken, SC 29802

     Glenn Caroll
     Georgians Against Nuclear Energy
     PO Box 8574
     Atlanta, GA 30306

     Ruth Thomas
     Environmentalists, Inc.
     1339 Sinkler Road
     Columbia, SC 29206

Enclosures to letter from Abrams not included.

Department of Energy
National Nuclear Security Administration
Washington, DC 20585

September 25, 2001

Dr. Roger Stroup
South Carolina State Historic Preservation Officer
South Carolina Department of Archives and History
8301 Parklane Road
Columbia, South Carolina 29223

Subject: Designation of Department of Energy as Lead Agency for Mitigation at
Proposed Location of the Mixed Oxide Fuel Fabrication Facility

Dear Dr. Stroup:

This letter is to notify you that the U.S. Department of Energy (DOE) and the U.S. Nuclear
Regulatory Commission (NRC) are designating DOE as the lead federal agency, pursuant to 36
C.F.R. 800.2 (a)(2), for mitigation of potential effects to Register-eligible site 38AK546. That
site is located at the proposed location of the Mixed Oxide Fuel Fabrication Facility (MOX FFF)
at DOE's Savannah River Site in South Carolina.

In May 2000, your office concurred with DOE's recommendation that archaeological site
38AK546/547 was eligible under Criterion (d), 36 C.F.R. 60.4, for inclusion in the National
Register of Historic Places. Following consultation, your office also concurred with the
Mitigation Plan for the site. The mitigation field work is expected to begin in October, 2001.

The MOX FFF, which could affect site 38AK546, is being funded by DOE and is part of DOE's
program for the disposition of surplus plutonium. The MOX FFF is also subject to NRC
licensing. Thus, the MOX FFF constitutes an undertaking by both DOE and NRC pursuant to 36
C.F.R. 800.16 (y) and sections 106 and 301 of the National Historic Preservation Act, 16 U.S.C.
470f and 470w(7).

Because the potential MOX FFF and affected archaeologic site are located at DOE's Savannah
River Site, they are also within the purview of DOE's 1990 "Programmatic Memorandum of
Agreement Among the Savannah River Operations Office, United States Department of Energy,
the South Carolina State Historic Preservation Officer and the Advisory Council on Historic
Preservation Concerning the Management of Archaeological Sites on the Savannah River Site,
Aiken, Allendale and Barnwell Counties, South Carolina." Under these circumstances, DOE and
NRC have determined that it is appropriate to identify DOE as the lead federal agency for

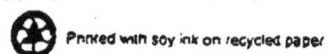 Printed with soy ink on recycled paper

mitigation of site 38AK546.

Please be advised that the NRC will analyze archaeologic impacts and related mitigation in its Environmental Impact Statement, pursuant to the National Environmental Policy Act (NEPA), for the proposed licensing of the MOX FFF. As part of the NEPA process, archaeological information has also been submitted to NRC in the Environmental Report prepared by Duke, Cogema, Stone and Webster (DCS), the MOX FFF license applicant under contract with DOE.

The lead agency approach outlined above has been discussed previously with your office by staff from DOE's Savannah River Site. Both agencies agree that any written correspondence from the SHPO to DOE should be sent to NRC as well. If you should have any additional questions regarding the MOX program or the identification of DOE as lead agency, please feel free to contact Mr. James Johnson at (202) 586-5960 for appropriate coordination.

Sincerely,

J. David Nulton, Director
Reactors Group
Office of Fissile Materials Disposition

Charlotte E. Abrams, Section Chief
Environmental and Low Level Waste Section
Division of Waste Management
U.S. Nuclear Regulatory Commission

cc:
Peter Hastings, DCS
Andrew Persinko, NRC
Allison Blackmon, SRS, ODNN
James V. Johnson, DOE, NN-61

OCT 2 4 2002

Mr. Chad Long
South Carolina State Historic Preservation Office
South Carolina Department of Archives and History
8301 Parklane Road
Columbia, SC 29223

Dear Mr. Long:

SUBJECT: Data Recovery Projects 38AK546 and 38AK757 at the Savannah River Site

I am writing to inform you that the Savannah River Archaeological Research Program (SRARP) has completed the data recovery projects conducted at 38AK546 and 38AK757. The excavations were conducted to mitigate impacts caused by construction of the Mixed Oxide Fuel Facility and the Pit Disassembly and Conversion Facility on the Department of Energy's Savannah River Site.

Investigations at 38AK546 were completed on 19 April 2002 and the excavations at 38AK757 ended on 15 September 2002. Exceeding the recommendations in the data recovery plans for each site, a total of 427 square meters was excavated at 38AK546 and 300 square meters were investigated at 38AK757. Except for monitoring ground disturbing activities associated with actual construction of these facilities, all fieldwork outlined in the data recovery plans has been completed. Staff members from the SRARP will monitor the removal of fill on the site areas when construction begins late next year.

Pending environmental documentation requires that DOE-SR have written concurrence from the SC SHPO that our field obligations have been met. At your convenience please send a written concurrence or if you have questions or need addition information before sending that concurrence, then call Dennis Ryan at (803) 725-8162. Thank you for your assistance in this matter.

Sincerely,

Original Signed by
A. B. Gould

A. B. Gould, Director
Environmental Quality
Management Division

EQMD:DR:orc

OE-03-003

bc: EQMD Read File   AMEST Read File
Dennis Ryan, EQMD   A. King, SRARP
A. M. Blackmon, ODNN

RECEIVED

02 NOV 25 AM 11: 47

MAIL CONTROL

022345

origEQmD

November 21, 2002

Mr. A.B. Gould, Director
Environmental Quality
Department of Energy, Savannah River Operations Office
P.O. Box A
Aiken, SC 29802

RE: Data Recovery Projects 38AK546 and 38AK757 at the Savannah River Site

Dear Mr. Gould:

I am writing to inform you that our office concurs with the Department of Energy's determination that field obligations have been met for data recovery investigations at 38AK546 and 38AK757. The excavations exceeded the requirements of the approved data recovery plans. We look forward to reviewing the results.

This letter was written to assist you with your obligations under Section 106 of the National Historic Preservation Act, as amended, and the regulations codified at 36 CFR Part 800. Please contact me at 803-896-6181 if you have any questions or comments regarding this matter.

Sincerely,

Chad C. Long
Staff Archaeologist
State Historic Preservation Office

cc: Mark Brooks

S. C. Department of Archives & History • 8301 Parklane Road • Columbia • South Carolina • 29223-4905 • (803) 896-6100 • www.state.sc.us/scdah

APPENDIX C:

TRANSPORTATION RISK ANALYSIS

APPENDIX C:

TRANSPORTATION RISK ANALYSIS

This appendix provides the detailed methodology, input parameters and assumptions, and results for the transportation risk analysis performed in support of this Mixed Oxide Fuel Fabrication Facility Environmental Impact Statement (MOX EIS). The analysis evaluates transportation of depleted uranium hexafluoride (UF_6) from the Portsmouth Gaseous Diffusion Plant in Portsmouth, Ohio, to the Global Nuclear Fuel-Americas, LLC Fuel Fabrication Facility in Wilmington, North Carolina; transportation of the uranium dioxide (UO_2) conversion product from Wilmington to the proposed MOX facility; transportation of plutonium metal from U.S. Department of Energy (DOE) storage sites; and transportation of the fresh MOX fuel from the proposed MOX facility to a surrogate nuclear power plant site.

C.1 Methodology

C.1.1 Overview

The transportation risk assessment considers human health risks from routine transport (normal, incident-free conditions) of hazardous materials and from potential accidents. In both cases, risks associated with the nature of the cargo itself, or "cargo-related" impacts, and those related to the transportation vehicle (regardless of type of cargo), or "vehicle-related" impacts, are considered.

C.1.1.1 Routine Transportation Risk

The radiological risk associated with routine transportation is cargo-related and results from the potential exposure of people to low levels of external radiation near a loaded shipment. It is assumed that there are no cargo-related risks posed by incident-free transport of hazardous chemicals. No direct chemical exposure to radioactive material will occur during routine transport because, as discussed in Section C.2.2, these materials will be in packages that are designed and maintained to ensure that they will contain and shield their contents during normal transport. Any leakage or unintended release would be considered under accident risks.

Vehicle-related risks during routine transportation are caused by potential exposure to increased vehicular emissions. These emissions include diesel exhaust, tire and brake particulate emissions, and fugitive dust raised from the roadbed by passing vehicles.

C.1.1.2 Accident Transportation Risk

The cargo-related radiological risk from transportation-related accidents lies in the potential release and dispersal of radioactive material into the environment during an accident and the subsequent exposure of people through multiple exposure pathways, such as exposure to contaminated soil, inhalation, or the ingestion of contaminated food. Cargo-related hazardous chemical accident impacts to human health during transportation come from immediate inhalation exposure resulting from container failure and chemical release during an accident.

Vehicle-related accident risks refer to the potential for transportation-related accidents that result in fatalities caused by physical trauma unrelated to the cargo.

C.1.2 Routine Risk Assessment Methodology

The RADTRAN 4 computer code (Neuhauser and Kanipe 1992) was used in the routine and accident cargo-related risk assessments to estimate the radiological impacts to collective populations. RADTRAN 4 was developed by Sandia National Laboratories to calculate population risks associated with the transportation of radioactive materials by truck, rail, air, ship, or barge. The code has been used extensively for transportation risk assessments since it was originally issued in the late 1970s as RADTRAN (RADTRAN 1) and has been reviewed and updated periodically. RADTRAN 1 was originally developed to facilitate the calculations presented in NUREG-0170 (NRC 1977b).

C.1.2.1 Collective Population Risk

The radiological risk associated with routine transportation results from the potential exposure of people to low-level external radiation in the vicinity of loaded shipments. Even under routine transportation, some radiological exposure could occur. Because the radiological consequences (dose) would occur as a direct result of normal operations, the probability of routine consequences is taken to be 1 in the RADTRAN 4 code. Therefore, the dose risk is equivalent to the estimated dose.

For routine transportation, the RADTRAN 4 computer code considers major groups of potentially exposed persons. The RADTRAN 4 calculations of risk for routine highway and rail transportation include exposures of the following population groups:

- *Persons along the Route (Off-Link Population).* Collective doses were calculated for all persons living or working within 0.8 km (0.5 mi) of each side of a transportation route. The total number of persons within the 1.6-km (1-mi) corridor was calculated separately for each route considered in the assessment.

- *Persons Sharing the Route (On-Link Population).* Collective doses were calculated for persons in all vehicles sharing the transportation route. This group includes persons traveling in the same or opposite directions as the shipment, as well as persons in vehicles passing the shipment.

- *Persons at Stops.* Collective doses were calculated for people who might be exposed while a shipment was stopped en route. For truck transportation, these stops include those for refueling, food, and rest.

- *Crew Members.* Collective doses were calculated for truck transportation crew members involved in the actual shipment of material. Workers involved in loading or unloading were not considered. The doses calculated for the first three population groups were added together to yield the collective dose to the public; the dose calculated for the fourth group represents the collective dose to workers.

The RADTRAN 4 calculations for routine dose generically compute the dose rate as a function of distance from a point source (Neuhauser and Kanipe 1995). Associated with the calculation of routine doses for each exposed population group are parameters such as the radiation field strength, the source-receptor distance, the duration of exposure, vehicular speed, stopping time, traffic density, and route characteristics (such as population density). The RADTRAN manual contains derivations of the equations used and descriptions of these parameters (Neuhauser and Kanipe 1995).

C.1.2.2 Maximally Exposed Individual Risk

In addition to the assessment of the routine collective population risk, the risk to a maximally exposed individual (MEI) was estimated. In RADTRAN 4, the MEI is assumed to be located 30 m (100 ft) from the transport route as the radioactive shipment passes by at a speed of 24 km/h (15 mph).

C.1.2.3 Vehicle-Related Risk

Vehicle-related health risks resulting from routine transportation are associated with the generation of air pollutants by transport vehicles during shipment and would be independent of the radioactive or chemical nature of the shipment. The health endpoint assessed under routine transportation conditions was the excess latent mortality from inhalation of vehicular emissions. These emissions consist of particulate matter in the form of diesel engine exhaust, tire and brake particulates, and fugitive dust raised from the roadway by the transport vehicle. Risk factors for pollutant inhalation in terms of latent mortality have been used in this analysis. Vehicle-related risks from routine transportation were calculated for each shipment by multiplying the total distance traveled by the appropriate risk factor.

C.1.3 Accident Assessment Methodology

As stated above, the radiological transportation accident risk assessment also uses the RADTRAN 4 code for estimating collective population risks. The hazardous chemical transportation accident risk assessment relies on the HGSYSTEM model (Post 1994a,b; Hanna et al. 1994). The model is a widely applied code recognized by the U.S. Environmental Agency (EPA) for use in chemical accident consequence predictions. The FIREPLUME model (Brown et al. 1997) was used to supplement the HGSYSTEM model in the analysis of fire scenarios involving depleted uranium releases. The HGSYSTEM and FIREPLUME models were used previously in assessing the hazardous chemical transportation impacts from transportation of depleted uranium materials (Biwer et al. 1997).

The risk analysis for potential accidents differs fundamentally from the risk analysis for routine transportation because occurrences of accidents are statistical in nature. The accident risk assessment is treated probabilistically in RADTRAN 4 for radiological risk and in the HGSYSTEM approach used to estimate the hazardous chemical component of risk. Accident risk is defined as the product of the accident consequence (dose or exposure) and the probability of the accident's occurring. In this respect, RADTRAN 4 and the HGSYSTEM approach both estimate the collective accident risk to populations by considering a spectrum of transportation-related accidents. The spectrum of accidents was designed to encompass a range of possible accidents, including low-probability accidents that have high consequences, and high-probability accidents that have low consequences (such as "fender benders"). For radiological risk, the results for collective accident risk can be directly compared with the results for routine collective risk because the latter results implicitly incorporate a probability of occurrence of 1 if the shipment takes place. Such is not the case for chemical materials, because routine transport would pose no exposure risk.

C.1.3.1 Radiological Accident Risk Assessment

The RADTRAN 4 calculation of collective accident risk uses models that quantify the range of potential accident severities and the responses of transported packages to accidents. The spectrum of accident severity is divided into several categories, each of which is assigned a conditional probability of occurrence — that is, the probability that if an accident does occur, it will be of a particular severity. Release fractions, defined as the fraction of the material in a package that could be released in an accident, are assigned to each accident severity category on the basis of the physical and chemical form of the material. The model takes into account the mode of transportation and the type of packaging through selection of the appropriate accident probabilities and release fractions, respectively. The accident rates, the definition of accident severity categories, and the release fractions used in this analysis are discussed further in Sections C.2 and C.3.

For accidents involving the release of radioactive material, RADTRAN 4 assumes that the material is dispersed in the environment according to standard Gaussian diffusion models. For the risk assessment, default data for atmospheric dispersion were used, representing an instantaneous ground-level release and a small-diameter source cloud (Neuhauser and Kanipe

1995). The calculation of the collective population dose following the release and dispersal of radioactive material includes the following exposure pathways:

- External exposure to the passing radioactive cloud,

- External exposure to contaminated ground,

- Internal exposure from inhalation of airborne contaminants, and

- Internal exposure from the ingestion of contaminated food.

For the ingestion pathway, state-average food transfer factors, which relate the amount of radioactive material ingested to the amount deposited on the ground, were calculated in accordance with the methods described by U.S. Nuclear Regulatory Commission (NRC) Regulatory Guide 1.109 (NRC 1977a) and were used as input to the RADTRAN code. Doses of radiation from the ingestion or inhalation of radionuclides were calculated by applying standard dose conversion factors (DOE 1988a,b).

C.1.3.2 Chemical Accident Risk Assessment

The risks from exposure to hazardous chemicals during transportation-related accidents can be either acute (resulting in immediate injury or fatality) or latent (resulting in cancer that would present itself after a latency period of several years). The acute health endpoint, potential irreversible adverse effects, was evaluated for the assessment of cargo-related population impacts from transportation accidents. Accidental releases during transport of the uranium compounds (UF_6 and UO_2) were evaluated quantitatively.

The acute effects evaluated were assumed to exhibit a threshold nonlinear relationship with exposure; that is, some low level of exposure can be tolerated without inducing a health effect. To estimate risks, chemical-specific concentrations were developed for potential irreversible adverse effects. All individuals exposed at these levels or higher following an accident were included in the transportation risk estimates. In addition to acute health effects, the cargo-related risk of excess cases of latent cancer from accidental chemical exposures could be evaluated. However, none of the chemicals that might be released in any of the accidents would be carcinogenic. As a result, no predictions for excess latent cancers from accidental chemical releases are presented in this report.

The primary exposure route of concern with respect to accidental release of hazardous chemicals would be inhalation. Although direct exposure to hazardous chemicals via other pathways, such as ingestion or absorption through the skin (dermal absorption), would also be possible, these routes would be expected to result in much lower exposure than the inhalation pathway doses for the uranium compounds. The likelihood of acute effects would be much less for the ingestion and dermal pathways than for inhalation.

The HGSYSTEM model (Version 3.0) (Hanna et al. 1994) has a built-in source-term algorithm that is used to compute the rate, quantity, and type of atmospheric release of a hazardous air pollutant, including pool evaporation from a spill of a volatile organic liquid. The model can be used to evaluate frequently encountered accidental releases from ruptured tanks, drums, and pipes. The model incorporates a chemical data library of physical and chemical properties (such as vapor pressure, boiling point, and molecular weight) for 30 compounds. Physical properties of the chemical released, along with container content input, such as the container geometry and rupture characteristics (e.g., hole size), are used by HGSYSTEM to compute chemical release rate and duration. The risk assessment for hazardous chemicals assumed that particulate releases would be of short duration as liquid and solid (as respirable fraction) aerosols.

The approach for hazardous chemicals incorporates the same accident severity categories and release fractions used by RADTRAN 4 for radiological accidents. The risks associated with the consequences estimated with the HGSYSTEM code were computed separately with a risk quantification spreadsheet program.

C.1.3.3 Vehicle-Related Accident Risk Assessment

The vehicle-related accident risk refers to the potential for transportation accidents that could result directly in fatalities not related to the nature of the cargo in the shipment. This risk represents fatalities from physical trauma. State-average rates for transportation fatalities are used in the assessment. Vehicle-related accident risks are calculated by multiplying the total distance traveled by the rates for transportation fatalities. In all cases, the vehicle-related accident risks are calculated on the basis of distances for round-trip shipment since the presence or absence of cargo would not be a factor in accident frequency.

C.2 Input Parameters and Assumptions

The principal input parameters and assumptions used in the transportation risk assessment are discussed in this section. Where appropriate, applicable government regulations are referenced. Transportation of hazardous chemical and radioactive materials is governed by U.S. Department of Transportation (DOT), NRC, and EPA regulations, and by the Hazardous Materials Transportation Act. These regulations may be found in the *Code of Federal Regulations* (CFR) at 49 CFR Parts 171-178, 49 CFR Parts 383-397, 10 CFR Part 71, and 40 CFR Parts 262 and 265, respectively. State organizations are also involved in regulating such transport within their borders. All transportation-related activities must be in accordance with applicable regulations of these agencies. However, the DOT and NRC have primary regulatory responsibility for shipment of radioactive materials. Those regulations most pertinent to this risk assessment can be found in 49 CFR 173 (*Shippers—General Requirements for Shipments and Packagings*), 49 CFR 397 (*Transportation of Hazardous Materials; Driving and Parking Rules*), and 10 CFR 71 (*Packaging and Transportation of Radioactive Material*).

C.2.1 Route Characteristics

The transportation route selected for a shipment determines the total potentially exposed population and the expected frequency of transportation-related accidents. For truck transportation, the route characteristics most important to the risk assessment include the total shipping distance between each origin-and-destination pair of sites and the population density along the route.

C.2.1.1 Route Selection

The DOT routing regulations concerning radioactive materials on public highways are prescribed in 49 CFR 397.101(*Requirements for Motor Carriers and Drivers*). The objectives of the regulations are to reduce the impacts of transporting radioactive materials, to establish consistent and uniform requirements for route selection, and to identify the role of state and local governments in routing radioactive materials. The regulations attempt to reduce potential hazards by prescribing that populous areas be avoided and that travel times be minimized. In addition, the regulations require that the carrier of radioactive materials ensure that the vehicle is operated on routes that minimize radiological risks, and that accident rates, transit times, population density and activity, time of day, and day of week are considered in determining risk. However, the final determination of the route is left to the discretion of the carrier, such as for shipments of depleted UF_6 and UO_2, unless the shipment contains a "highway route controlled quantity" (HRCQ) of radioactive material as defined in 49 CFR 173.403 (*Definitions*), such as the plutonium metal or the MOX fuel.

A vehicle transporting an HRCQ of radioactive materials is required to use the interstate highway system except when moving from origin to interstate or from interstate to destination, when making necessary repair or rest stops, or when emergency conditions make continued use of the interstate unsafe or impossible. Carriers are required to use interstate circumferential or bypass routes, if available, to avoid populous areas. Any state or Native American tribe may designate other "preferred highways" to replace or supplement the interstate system. Under its authority to regulate interstate transportation safety, the DOT can prohibit state and local bans and restrictions as "undue restraint of interstate commerce." State or local bans can be preempted if inconsistent with the HRCQ regulations. Shipments of TRU waste will follow designated Waste Isolation Pilot Plant (WIPP) routes to the WIPP repository.

For this analysis, representative shipment routes were identified using the WebTRAGIS (Version 1.10) routing model (Johnson and Michelhaugh 2000) for the truck shipments. The routes were selected to be reasonable and consistent with routing regulations and general practice, but they are considered only representative because the actual routes used would be chosen in the future and are often determined by the shipper. At the time of shipment, route selection would reflect current road conditions, including road repairs and traffic congestion.

The HIGHWAY data network in WebTRAGIS is a computerized road atlas that includes a complete description of the interstate highway system and of all U.S. highways. In addition, most principal state highways and many local and community highways are identified. The code is periodically updated to reflect current road conditions and has been compared with reported mileages and observations of commercial trucking firms.

Routes are calculated within the model by minimizing the total impedance between origin and destination. The impedance is basically defined as a function of distance and driving time along a particular segment of highway. The population densities along a route are derived from 2000 census data from the U.S. Bureau of the Census.

The WebTRAGIS database version used was Highway Data Network 2.1. Summary route information on the truck routes used in the analysis is provided in Table C.1.

C.2.1.2 Population Density

Three population density zones — rural, suburban, and urban — were used for the population risk assessment. The fractions of travel and average population density in each zone were determined with the WebTRAGIS routing model. Rural, suburban, and urban areas are characterized according to the following breakdown: rural population densities range from 0 to 54 persons/km^2 (0 to 139 persons/mi^2); suburban densities range from 55 to 1,284 persons/km^2 (140 to 3,326 persons/mi^2); and urban covers all population densities greater than 1,284 persons/km^2 (3,326 persons/mi^2). Use of these three population density zones is based on an aggregation of the 11 population density zones provided in the WebTRAGIS model output. For calculation purposes, information about population density was generated at the state level and used as RADTRAN input for all routes. Route average population densities and other route characteristics are given in Table C.1.

C.2.1.3 Accident and Fatality Rates

For calculating accident risks, vehicle accident involvement and fatality rates are taken from data provided in Saricks and Tompkins (1999). For each transport mode, accident rates are generically defined as the number of accident involvements (or fatalities) in a given year per unit of travel by that mode in the same year. Therefore, the rate is a fractional value — the accident-involvement count is the numerator, and vehicular activity (total traveled distance) is the denominator. Accident rates are derived from multiple-year averages that automatically account for such factors as heavy traffic and adverse weather conditions. For assessment purposes, the total number of expected accidents or fatalities is calculated by multiplying the total shipping distance for a specific case by the appropriate accident or fatality rate.

For truck transportation, the rates presented in Saricks and Tompkins (1999) are specifically for heavy combination trucks involved in interstate commerce. Heavy combination trucks are rigs composed of a separable tractor unit containing the engine and one to three freight trailers connected to each other and the tractor. Heavy combination trucks are typically used for shipping radioactive wastes. Truck accident rates are computed for each state on the basis of

Table C.1. Summary route data

Route		Total distance [km (mi)]	Fraction of travel			Average population density [persons/km² (persons/mi²)]		
Origin	Destination		Rural	Suburban	Urban	Rural	Suburban	Urban
Portsmouth, OH	Wilmington, NC	936 (581)	55.5	40.7	3.9	18.5 (47.8)	366.7 (949.8)	2,155 (5,582)
Wilmington, NC	PDCF	443 (275)	60.1	37.5	2.4	15.7 (40.7)	353.1 (914.6)	2,140 (5,543)
Pantex	PDCF	2,179 (1,354)	67.8	28.5	3.7	13.4 (34.6)	332.4 (861.0)	2,271 (5,882)
Hanford Site	PDCF	4,434 (2,755)	76.7	20.9	2.3	11.3 (29.2)	320.5 (830.1)	2,244 (5,811)
Proposed MOX facility	WIPP	2,442 (1,518)	70.7	26.7	2.6	13.2 (34.2)	315.6 (817.4)	2,173 (5,627)
Proposed MOX facility	Surrogate Nuclear Power Plant	2,147 (1,334)	57.1	37.4	5.5	18.5 (47.8)	342.1 (886.1)	2,366 (6,128)

statistics compiled by the DOT Office of Motor Carriers for 1994 to 1996. Saricks and Tompkins (1999) present accident involvement and fatality counts, estimated kilometers of travel by state, and the corresponding average accident involvement and fatality rates for the 3 years investigated. Fatalities (including of crew members) are deaths that are attributable to the accident and that occurred within 30 days of the accident.

The truck accident assessment presented in this EIS uses accident (fatality) rates for travel on interstate highways. The total accident risk for a case depends on the total distance traveled in various states and does not rely on national average accident statistics. However, for comparative purposes, the national average truck accident rate on interstate highways presented in Saricks and Tompkins (1999) is 3.15×10^{-7} accidents/truck-km (5.07×10^{-7} accidents/mi).

Note that the accident rates used in this assessment were computed using all interstate shipments, regardless of the cargo. Saricks and Kvitek (1994) point out that shippers and carriers of radioactive material generally have a higher-than-average awareness of transportation risk and prepare cargoes and drivers for such shipments accordingly. This preparation should have the twofold effect of reducing component and equipment failure and mitigating the contribution of human error to accident causation. However, these mitigating effects were not considered in the accident assessment.

C.2.2 Packaging

Shipment packaging for radioactive materials must be designed, constructed, and maintained to ensure that it will contain and shield the contents during normal transportation. For more highly radioactive material, the packaging must contain and shield the contents in severe accidents. The type of packaging used is determined by the radioactive hazard associated with the packaged material. The basic types of packaging required by the applicable regulations are designated as Type A, Type B, or industrial packaging (generally for low-specific-activity [LSA] material).

C.2.2.1 Depleted UF$_6$ and UO$_2$ Packaging

Depleted UF$_6$ and UO$_2$ shipments would use Type A and industrial packaging, respectively. These types of packaging must withstand the conditions of normal transportation without the loss or dispersal of the radioactive contents. "Normal" transportation refers to all transportation conditions except those resulting from accidents or sabotage. Approval of Type A packaging is obtained by demonstrating that the packaging can withstand specified testing conditions intended to simulate normal transportation. Type A packaging usually does not require special handling, packaging, or transportation equipment. The depleted UF$_6$ would be shipped in Model 30B cylinders (USEC 1999) with overpacks, and the depleted UO$_2$ would be shipped in 55-gal drums.

C.2.2.2 Plutonium Metal, MOX Fuel, and TRU Waste

The plutonium metal, MOX fuel, and TRU waste would be shipped in Type B packaging. In addition to meeting all the Type A standards, Type B packaging must also provide a high degree of assurance that the package integrity will be maintained even during severe accidents, with essentially no loss of the radioactive contents or serious impairment of the shielding capability. Type B packaging is required for shipping large quantities of radioactive material and must satisfy stringent testing criteria (as specified in 10 CFR 71). The testing criteria were developed to simulate conditions of severe hypothetical accidents, including impact, puncture, fire, and immersion in water. The most widely recognized Type B packagings are the massive casks used to transport highly radioactive spent nuclear fuel from nuclear power stations. Large-capacity cranes and mechanical lifting equipment are usually necessary for handling Type B packagings. Many Type B packagings are transported on trailers specifically designed for that purpose.

Plutonium metal as pits is expected to be shipped in DOE-approved FL containers, while piece parts might be shipped in DOE-approved USA/9975 containers (DOE 1999b). TRU waste would be transported to the WIPP in Type B containers referred to as the Transuranic Package Transporter-II (TRUPACT-II).

The MOX fresh fuel package is a Type B cylindrical container designed to carry three MOX fuel assemblies. MOX fuel does not require specific shielding material, and the containment shell provides a single containment boundary in accordance with 10 CFR 71.63(b)(1). The current design (DCS 2001b) specifies 4.46 m (175 in.) as the overall package length without the impact limiters. The impact limiters themselves are of a conventional polyurethane filled design and have an outer diameter of 1.5 m (60 in.). The outer diameter of the package containment shell is 0.74 m (29 in.). The package is designed to accommodate 3,200 kg (7,100 lb) of payload, including internal supports and the fuel assemblies. The package gross weight is 6,580 kg (14,500 lb).

C.2.3 Shipment Configurations and Number of Shipments

The anticipated shipment information for the proposed action is summarized in Table C.2. Table C.3 lists the radionuclide inventory for each shipment type. Depleted UF_6 shipments would consist of five overpacked 30B cylinders per truck, as depicted in Figure C.1. Each cylinder would contain about 2,277 kg (5,020 lb) of depleted UF_6. Depleted UO_2 shipments would consist of 24 55-gal drums in a commercial covered tractor trailer. Each drum would contain approximately 667 kg (1,470 lb) of depleted UO_2. For this analysis, sufficient quantities of UF_6 and UO_2 were assumed to be shipped so that a total of 34 MT(37.5 tons) of plutonium could be fabricated into MOX fuel assemblies for irradiation as reactor fuel (DCS 2002a). Thus, a total of 110 shipments of depleted UF_6 and 60 shipments of depleted UO_2 would be required.

As discussed in Section 4.4.1.1, it was assumed that 26.7 MT (29.4 tons) of plutonium would require transportation to the PDCF from Pantex and Hanford. On the basis of the information

Table C.2. Shipment information

Origin	Destination	Material	Package type	Amount per package [kg (lb)]	Packages per shipment	Number of shipments
Portsmouth, OH	Wilmington, NC	UF_6	30B cylinder	2,277 (5,020)	5	110
Wilmington, NC	MOX facility	UO_2	30-gal drum	667 (1,470)	24	60
Pantex	PDCF	Pu metal	Type B	62.3 (137)[a]	NA[b]	343
Hanford	PDCF	Pu metal	Type B	62.3 (137)[a]	NA	87
MOX facility	Surrogate nuclear power plant	MOX fuel	Type B	3 assemblies	1	598
WSB	WIPP	TRU waste	TRUPACT-II	2,590 (5,700)[a]	3	299–2,314

[a]Estimated amount per shipment.

[b]Not available, dependent on actual container used.

Table C.3. Single-shipment radionuclide inventories (Ci)[a]

Isotopes	UF_6	UO_2	Pu metal	MOX fuel[b]	TRU Waste[c,d] Volume Reduction	No Volume Reduction
U-234	0.474	0.868	NA[e]	NA	0.0231	0.00299
U-235	0.0445	0.0752	NA	0.00706	0.000530	6.87×10^{-5}
U-238	2.57	4.74	NA	0.438	5.43×10^{-6}	7.04×10^{-7}
Th-234	2.57	4.74	NA	NA	NA	NA
Pa-234m	2.57	4.74	NA	NA	NA	NA
Pu-236	NA	NA	NA	2.22	NA	NA
Pu-238	NA	NA	836	429	0.0822	0.0107
Pu-239	NA	NA	7,070	4,860	0.567	0.0735
Pu-240	NA	NA	1,730	1,080	0.110	0.0142
Pu-241	NA	NA	129,000	43,000	9.88	1.28
Pu-242	NA	NA	0.494	0.0956	3.76×10^{-5}	4.87×10^{-6}
Am-241	NA	NA	3,820	NA	3,650	474

[a]To convert from Ci to Bq, multiply by 3.7×10^{10}.

[b]Source: DCS (2001b).

[c]Source: DCS (2002b).

[d]Source: DCS (2004).

[e]NA = not applicable.

Figure C.1. Trailer carrying five UF$_6$ cylinders in overpacks (Photo courtesy of United States Enrichment Corporation [USEC 1999]).

presented in Didlake (1998), approximately 62.3 kg (137 lb) of plutonium would be in each shipment. The plutonium would be packaged in a suitable Type B container and shipped via the Safeguards Transporter (SGT) discussed later in this section.

Approximately 1,748 MOX fuel assemblies would be shipped to commercial reactor sites. Transport of the MOX fuel would be by SGT, one MOX fuel package per shipment. Figure C.2 shows a representative shipment configuration. With three assemblies per shipping cask, 598 shipments would be expected between the years 2007 and 2021 (DCS 2002a).

The SGT is a specially designed component of a tractor-trailer vehicle and is used by the Office of Secure Transportation of the DOE Albuquerque National Nuclear Security Administration (NNSA) Service Center for the transport of special nuclear materials, such as plutonium. Since 1975, more than 151 million km (94 million mi) of travel transporting DOE-owned cargo has been accumulated without an accident involving a fatality or a release of radioactive material. Although details of vehicle enhancements and some operational aspects are classified, key characteristics are as follows (DOE 1999b):

- Enhanced structural characteristics and a highly reliable tie-down system to protect the cargo from impact;

- Heightened thermal resistance to protect the cargo in case of fire;

- Established operational and emergency plans and procedures governing the shipment of nuclear materials;

- Couriers who are armed federal officers and who have received vigorous specialized training;

- An armored tractor component that provides courier protection against attack and contains advanced communications equipment;

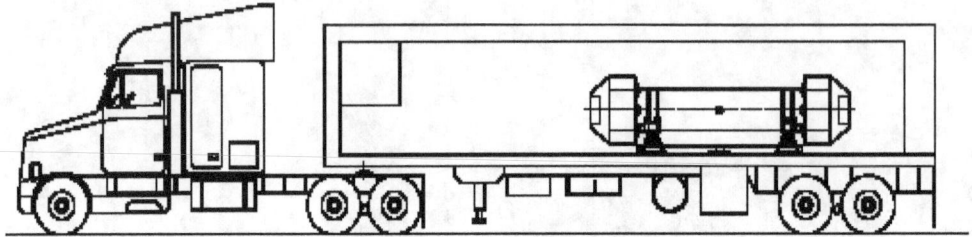

Figure C.2. MOX fresh fuel package loaded in SGT.

- Specially designed escort vehicles containing advanced communications and additional couriers;

- 24-hour-a-day, real-time communications to monitor the location and status of all SGT shipments; and

- Significantly more stringent maintenance standards than those for commercial transport equipment.

TRU waste was assumed to be fixed in cement, placed in standard waste boxes (SWBs), and shipped in TRUPACT-II containers from the WSB to the WIPP for disposal (DCS 2002b; 2004). Each TRUPACT-II contained 2 SWBs, and each truck shipment consisted of 3 TRUPACT-II containers. The number of TRU waste shipments could range from about 23 to 178 shipments per year (DCS 2004). The upper end of the range assumes that no volume reduction of the waste occurs, but the annual throughput in either case contains the same amount of americium. Thus, the total number of shipments over the 13-year operational life of the WSB would range from 299 to 2,314.

C.2.4 Accident Characteristics

Assessment of transportation accident risk takes into account the fraction of material in a package that would be released or spilled to the environment during an accident, commonly referred to as the release fraction. The release fraction is a function of the severity of the accident and the material packaging. For instance, a low-impact accident, such as a "fender-bender," would not be expected to cause any release of material. Conversely, a very severe accident would be expected to release nearly all of the material in a shipment into the environment. The method used to characterize accident severities and the corresponding release fractions for estimating both radioactive and chemical risks are described below.

C.2.4.1 Accident Severity Categories

A method to characterize the potential severity of transportation-related accidents has been described in the NRC NUREG-0170 report, *Final Environmental Statement on the*

Transportation of Radioactive Material by Air and Other Modes (NRC 1977b). The NRC method divides the spectrum of transportation accident severities into eight categories. Other studies have divided the same accident spectrum into six categories (Wilmot 1981), 20 categories (Fischer et al. 1987), or more (Sprung et al. 2000); however, these latter studies focused primarily on accidents involving shipments of spent nuclear fuel (SNF). In this analysis, the NUREG-0170 scheme was used for all shipments.

The NUREG-0170 scheme for accident classification is shown in Figure C.3 for truck transportation. Severity is described as a function of the magnitudes of the mechanical forces (impact) and thermal forces (fire) to which a package may be subjected during an accident. Because all accidents can be described in these terms, severity is independent of the specific accident sequence. In other words, any sequence of events that results in an accident in which a package is subjected to forces within a certain range of values is assigned to the accident severity category associated with that range. The scheme for accident severity is designed to take into account all credible transportation-related accidents, including those accidents with low probability but high consequences and those with high probability but low consequences.

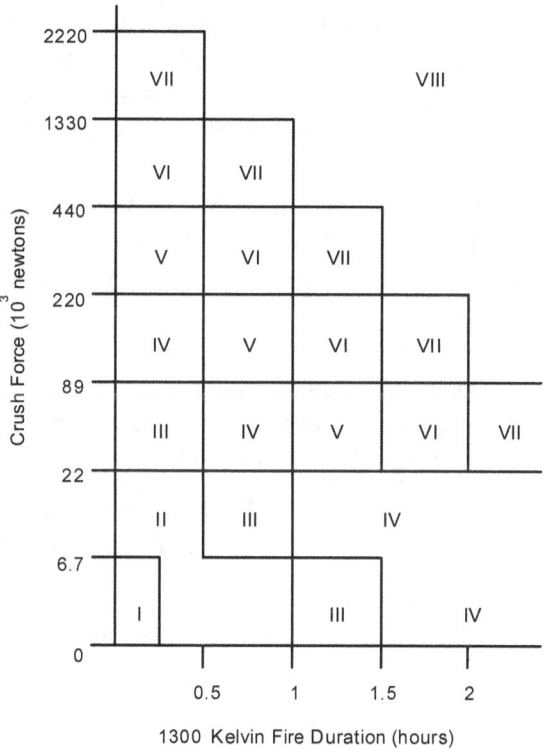

Figure C.3. Scheme for NUREG-0170 classification by accident severity category for truck accidents (*Source*: NRC 1977b).

Each severity category represents a set of accident scenarios defined by a combination of mechanical and thermal forces. A conditional probability of occurrence — that is, the probability that if an accident occurs, it is of a particular severity — is assigned to each category. The fractional occurrences for accidents by accident severity category and population density zone are shown in Table C.4 and are used for estimating both radioactive and chemical risks.

Category I accidents are the least severe but the most frequent; Category VIII accidents are very severe but very infrequent. To determine the expected frequency of an accident of a given severity, the conditional probability in the category is multiplied by the baseline accident rate. Each population density zone has a distinct distribution of accident severities related to differences in average vehicular velocity, traffic density, location (rural, suburban, or urban), and other factors.

C.2.4.2 Package Release Fractions

In NUREG-0170, radiological and chemical consequences are calculated by assigning package release fractions to each accident severity category. The release fraction is defined as the fraction of the material in a package that could be released from the package as the result of an accident of a given severity. Release fractions take into account all mechanisms necessary to create release of material from a damaged package to the environment. Release fractions vary according to the type of package and the physical form of the material.

Representative release fractions for accidents involving depleted UF_6 and UO_2 shipments were taken from NUREG-0170 (NRC 1977b). The recommendations in NUREG-0170 are based on best engineering judgments and have been shown to provide conservative estimates of

Table C.4. Fractional occurrences for truck accidents by severity category and population density zone

Severity category	Fractional occurrence	Fractional occurrence by population density zone		
		Rural	Suburban	Urban
Truck				
I	0.55	0.1	0.1	0.8
II	0.36	0.1	0.1	0.8
III	0.07	0.3	0.4	0.3
IV	0.016	0.3	0.4	0.3
V	0.0028	0.5	0.3	0.2
VI	0.0011	0.7	0.2	0.1
VII	8.5×10^{-5}	0.8	0.1	0.1
VIII	1.5×10^{-5}	0.9	0.05	0.05

Source: NRC (1977b).

material releases following accidents. The release fractions used are those reported in NUREG-0170 for both low-specific-activity (LSA) drums and NRC Type A packages. Release fractions for accidents of each severity category are given in Table C.5. As shown in that table, the amount of material released from the package ranges from zero for minor accidents to 100% for the most severe accidents. As shown in Table C.5, representative release fractions for accidents involving fresh MOX fuel were assumed to be the same as those developed for SNF in the NRC's study (Fischer et al. 1987), commonly referred to as the Modal Study, on the behavior of SNF in Type B containers under accident conditions. These values were derived on the basis of best engineering judgments. These values are expected to be conservative when applied to fresh MOX fuel because the fuel has not yet become embrittled through use.

Also important for the purposes of risk assessment are the fraction of the released material that can be entrained in an aerosol (part of an airborne contaminant plume) and the fraction of the aerosolized material that is also respirable (of a size that can be inhaled into the lungs). These fractions depend on the physical form of the material. Most solid materials are difficult to release in particulate form and are, therefore, relatively nondispersible. Conversely, liquid or gaseous materials are relatively easy to release if the container is breached in an accident.

Table C.5. Estimated release fractions for Type A and Type B packages under various accident severity categories

| Severity category | Release fraction[a] | | | |
| | NUREG-0170 | | | |
	Type A[b]	Type B[c]	Type B[d]	TRUPACT-II[e]
I	0	0	0	0
II	0.01	0	6×10^{-8}	0
III	0.1	0.01	2×10^{-7}	8×10^{-9}
IV	1	0.1	2×10^{-6}	2×10^{-7}
V	1	1	2×10^{-6}	8×10^{-5}
VI	1	1	2×10^{-5}	2×10^{-4}
VII	1	1	2×10^{-5}	2×10^{-4}
VIII	1	1	2×10^{-5}	2×10^{-4}

[a]Values are for total material release fraction (the fraction of material in a package released to the environment during an accident).

[b]*Source:* NRC (1977b), used for depleted UF_6 and UO_2 shipments.

[c]*Source:* NRC (1977b), used for Pu metal shipments.

[d]*Source:* Fischer et al. (1987), used for fresh MOX fuel shipments.

[e]*Source:* DOE (1997). Aerosolized and respirable fractions are both assumed to equal 1.0.

The aerosolized fraction for the UF_6 was taken to be 0.01 except in the case of higher severity accidents (Categories VI through VIII) involving fire, for which it was taken to be 0.33 (Policastro et al. 1997). The respirable fraction was taken to be 1 for all accidents. For UO_2, which was assumed to behave as a loose powder, the aerosolized fraction was set to 0.1, with a respirable fraction of 0.05 (Biwer et al. 1997). The aerosolized fraction and the respirable fraction were taken to be 1×10^{-6} and 0.05, respectively, for the Pu metal expected to behave as immobile material (Neuhauser and Kanipe 1992). For the MOX fuel, the aerosolized fraction was taken to be 1, and the respirable fraction taken to be 0.05 in accordance with spent fuel particulates as derived from NUREG-0170 in Neuhauser and Kanipe (1992). Release fractions used for the TRU waste shipments are given separately in Table C.5.

C.2.4.3 Atmospheric Conditions during Accidents

Hazardous material released to the atmosphere is transported by the wind. The amount of dispersion, or dilution, of the contaminant material in the air depends on the meteorologic conditions at the time of the accident. Because predicting the specific location of an off-site transportation-related accident and the exact meteorologic conditions at the time of the accident is impossible, generic atmospheric conditions were selected for the accident risk assessment. Neutral weather conditions were assumed. These conditions were represented by Pasquill atmospheric stability Class D with a wind speed of 4 m/s (9 mph). Because neutral meteorological conditions are the most frequently occurring atmospheric stability condition in the United States, these conditions are most likely to be present in the event of an accident involving a hazardous material shipment. Observations at National Weather Service surface meteorological stations at more than 300 U.S. locations indicate that on a yearly average, neutral conditions (represented by Pasquill Classes C and D) occur about half (50%) the time; stable conditions (Pasquill Classes E and F) occur about one-third (33%) of the time; and unstable conditions (Pasquill Classes A and B) occur about one-sixth (17%) of the time (Doty et al. 1976). The neutral category predominates in all seasons, but it is most prevalent (nearly 60% of the observations) during winter.

C.2.5 Radiological Risk Assessment Input Parameters and Assumptions

The dose (and, correspondingly, the risk) to populations during routine transportation of radioactive materials is directly proportional to the assumed external dose rate from the shipment. The actual dose rate from the shipment is a complex function of the composition and configuration of shielding and containment materials used in the packaging, the geometry of the loaded shipment, and the characteristics of the radioactive material itself.

Shipments of depleted UF_6 and UO_2 have been studied previously (Biwer et al. 1997) for the Depleted UF_6 Programmatic EIS (PEIS) (DOE 1999a). Representative shipment dose rates were developed using the MicroShield™ shielding code (Negin and Worku 1992). The input to MicroShield™ consisted of the activity of a material, the geometry and composition of the shipping package, and the amount of material in the package. Where multiple packages per shipment were assumed, a dose rate for the shipment was derived from the summation of the individual package dose rates, taking into consideration the configuration of the packages on the transport vehicle and the relative distances to a receptor.

Table C.6 lists the external dose rates developed for the Depleted UF$_6$ PEIS and used in this transportation analysis. The dose rates are presented in terms of the transport index (TI), which is the dose rate at 1 m (3 ft) from the lateral sides of the transport vehicle. The regulatory limit established in 49 CFR Part 173.441 (*Radiation Level Limitations*) and 10 CFR Part 71.47 (*External Radiation Standards for All Packages*) to protect the public is 0.1 mSv/h (10 mrem/h) at 2 m (6 ft) from the outer lateral sides of the transport vehicle. The estimated dose rate at a distance of 1 m (3 ft) from a truck shipment of depleted UO$_2$ identical to that considered for this analysis was 0.0076 mSv/h (0.76 mrem/h). Depleted UF$_6$ in larger, 14-ton cylinders in overcontainers was estimated to have external dose rates of 0.0023 mSv/h (0.23 mrem/h) and 0.0024 mSv/h (0.24 mrem/h) for truck (1 cylinder/tractor-trailer) and rail (4 cylinders/railcar) shipments, respectively. For this analysis, depleted UF$_6$ shipments, each involving five 30B cylinders, were assumed to have an external dose rate of 0.0024 mSv/h (0.24 mrem/h), which is more consistent with the line source geometry of the railcar shipments in the Depleted UF$_6$ PEIS (DOE 1999a). These estimated dose rates for the depleted uranium shipments are less than 5% of the allowed maximum value. A value of 0.040 mSv/h (4.0 mrem/h) was used for the WSB TRU waste shipments. This value represents the highest estimated dose rate for TRUPACT-II truck shipments estimated for any TRU waste generator site considered in the *Waste Isolation Pilot Plant Disposal Phase Final Supplemental Environmental Impact Statement* (DOE 1997). For MOX fuel shipments, preliminary analysis has estimated a conservative value of 0.0484 mSv/h (4.84 mrem/h) for the external dose rate at 1 m (DCS 2001a). The regulatory maximum of 0.10 mSv/h (10 mrem/h) at 2 m was assumed for the plutonium metal. This dose rate corresponds approximately to 0.14 mSv/h (14 mrem/h) at 1 m.

In addition to the specific parameters discussed previously, values for a number of general parameters must be specified within the RADTRAN code to calculate radiological risks. These general parameters define basic characteristics of the shipment and traffic and are specific to the mode of transportation. The user's manual for the RADTRAN code (Neuhauser and Kanipe 1992) contains derivations and descriptions of these parameters. The general RADTRAN input parameters used in the radiological transportation risk assessment are summarized in Table C.7.

C.2.6 Hazardous Chemical Risk Assessment Input Parameters and Assumptions

To estimate the consequences of chemical accidents, two potential health effects end points were evaluated: (1) adverse effects and (2) irreversible adverse effects. Potential adverse effects range from mild and transient effects — such as respiratory irritation, redness of the eyes, and skin rash — to more serious and potentially irreversible effects. Potential irreversible adverse effects are defined as effects that generally occur at higher concentrations and are permanent in nature — including death, impaired organ function (such as damaged central nervous system or lungs), and other effects that may impair everyday functions.

Table C.6. External dose rates and package sizes used in RADTRAN

Shipment	Dose rate at 1 m [mSv/h (mrem/h)]	Package size (m)
UF$_6$	0.0024 (0.24)	12[a]
UO$_2$	0.0076 (0.76)	6.0[a]
Pu metal	0.14 (14)	9
TRU waste	0.040 (4.0)	7.4
MOX fuel	0.0484 (4.84)	3.66[b]

[a]*Source:* Biwer et al. (1997).

[b]Active length of fuel assembly (DCS 2001a).

Table C.7. General RADTRAN input parameters[a]

Parameter	Truck[b]
Number of crew members	2
Distance from source to crew (m)	3.1
Average vehicular speed (km/h)[c]	
Rural	88.49
Suburban	40.25
Urban	24.16
Stop time (h/km)	0.011
Number of people exposed while stopped	50
Distance for exposure while stopped (m)	20
Number of people per vehicle sharing route	2
Population densities (persons/km^2)[d]	Route specific
One-way traffic count (vehicles/h)	
Rural	470
Suburban	780
Urban	2,800

[a]Accident conditional probabilities are listed by severity category in Table C.4; accident release fractions are given in Table C.5.

[b]*Source:* Biwer et al. (1997).

[c]Fraction of rural and suburban travel on freeways was set to 1 in RADTRAN. Thus, the rural speed was used for both urban and suburban zones.

[d]Route-specific population densities are listed in Table C.1.

For uranium compounds, an intake of 10 mg or more was assumed to cause potential adverse effects (McGuire 1991), and an intake of 30 mg or more was assumed to cause potential irreversible adverse effects. These intake levels are based on NRC guidance (NRC 1994). For hydrogen fluoride (HF), which is a by-product of UF_6 reacting with moisture in the air following an accidental release, potential adverse effects levels were assumed to occur at levels that correspond to Emergency Response Planning Guideline No. 1 (ERPG-1) or equivalent levels, and potential irreversible adverse effects levels were assumed to occur at levels that correspond to ERPG-2 or equivalent levels. The ERPG values have been generated by teams of toxicologists who review all published (as well as some unpublished) data for a given chemical (AIHA 1996). In addition to potential irreversible adverse effects, the number of fatalities from accidental chemical exposures was estimated to facilitate comparisons with radiological impacts. For exposures to uranium and HF, it was estimated that the number of fatalities occurring would be about 1% of the number of irreversible adverse effects (EPA 1993a; Policastro et al. 1997).

Application of the FIREPLUME code involves the choice of a number of parameters that affect the results. Input values were selected to represent reasonable conditions at a generic location without being too conservative. More details about the models and input parameters are presented in Post et al. (1994a,b) and Brown et al. (1997).

C.2.7 Routine Nonradiological Vehicle Emission Risks

Vehicle-related risks during incident-free transportation include incremental risks caused by potential exposure to airborne particulate matter from fugitive dust and vehicular exhaust emissions. The health end point assessed under routine transport conditions is the excess (additional) latent mortality caused by inhalation of vehicular emissions. These emissions are primarily in the form of diesel exhaust and fugitive dust (resuspended particulates from the roadway). Strong epidemiological evidence exists suggesting that increases in ambient air concentrations of PM_{10} (particulate matter with a mean aerodynamic diameter less than or equal to 10 µm) lead to increases in mortality (EPA 1996a,b). Currently, it is assumed that no threshold exists and that the dose-response functions for most health effects associated with PM_{10} exposure, including premature mortality, are linear over the concentration ranges investigated (EPA 1996a). Over both the short and long terms, fatalities (mortality) may result from life-shortening respiratory or cardiovascular diseases (EPA 1996a; Ostro and Chestnut 1998). The long-term fatalities also are assumed to include those from cancer.

The increased ambient air particulate concentrations caused by the transport vehicle, due to fugitive dust and diesel exhaust emissions, were related to such premature latent fatalities in the form of risk factors by Biwer and Butler (1999) for transportation risk assessments. Thus, in this assessment, a value of 8.36×10^{-10} latent fatalities/km for truck transport was used. This value is for heavy combination trucks (truck class VIIIB). The risk factor is for areas with an assumed population density of 1 person/km^2. One-way shipment risks are obtained by multiplying the appropriate risk factor by the average population density along the route and the route distance. The risks reported for routine vehicle risks in this analysis are for round-trip travel of the transport vehicle.

The vehicle risks reported here are estimates based on the best available data. However, as is true for the radiological risks, there is a large, not readily quantifiable, degree of uncertainty in the vehicle emission risk factors. For example, large uncertainties exist as to the extent of increased mortality with an incremental rise in particulate air concentrations and as to whether there are threshold air concentrations that are applicable. Also, estimates of the particulate air concentrations caused by transport vehicles are dependent on location, road conditions, vehicle conditions, and weather.

As discussed by Biwer and Butler (1999), there are large uncertainties in the human health risk factors used to develop the emission risks. In addition, because of the conservatism of the assumptions made to reconcile results with those presented in an EPA study (EPA 1993b), latent fatality risks estimated with the above risk factor may be considered to be near an upper bound. Use of this risk factor for truck class VIIIB will give estimated fatalities comparable to those from accident fatalities in some cases. In addition, the question as to what exactly constitutes a fatality as a direct consequence of increased PM_{10} levels from vehicle emissions is still an open question, but long-term fatalities have been associated with increased levels of PM_{10} (Biwer and Butler 1999).

C.3 Transportation Impacts

Single shipment transportation impacts are presented in Table C.8. Total collective population transportation impacts are presented in Section 4.4.1.3.

C.4 References for Appendix C

AIHA (American Industrial Hygiene Association) 1996. *The AIHA 1996 Emergency Response Planning Guidelines and Workplace Environmental Exposure Level Guides Handbook.* Fairfax, VA.

Biwer, B.M., and J.P. Butler 1999. "Vehicle Emission Unit Risk Factors for Transportation Risk Assessments." *Risk Analysis* 19:1157-1171.

Biwer, B.M., et al. 1997. *Transportation Impact Analyses in Support of the Depleted Uranium Hexafluoride Programmatic Environmental Impact Statement.* Attachment to intraoffice memorandum from Biwer to H.I. Avci (Argonne National Laboratory, Argonne, IL). May 21.

Brown, D., et al. 1997. *FIREPLUME: Modeling Plume Dispersion from Fires and Applications to UF_6 Cylinder Fires.* ANL/EAD/TM-69. Argonne National Laboratory, Argonne, IL.

DCS (Duke Cogema Stone & Webster) 2001a. *Responses to Request for Clarification of Additional Information for the Duke Cogema Stone & Webster (DCS) Mixed Oxide (MOX) Fuel Fabrication Facility (FFF) Environmental Report (ER).* Docket Number 070-03098. Charlotte, NC.

DCS 2001b. "Response to Clarification Request — *Responses to Request for Additional Information on the Duke Cogema Stone & Webster Oxide Fuel Fabrication Facility Environmental Report.*" Letter, with attachments, from P.S. Hastings (DCS, Charlotte, NC) to U.S. Nuclear Regulatory Commission (Washington, DC). Oct.

Table C.8. Single-shipment collective population transportation risks

Impact category	Depleted UF$_6$ From Portsmouth, OH to Wilmington, NC	Depleted UO$_2$ From Wilmington, NC to MOX facility	Pu metal From Pantex to PDCF	Pu metal From Hanford to PDCF	MOX From MOX facility to surrogate commercial reactor	TRU Waste From WSB to WIPP
Population impacts						
Cargo-related[a]						
Radiological impacts						
Dose risk[b]						
(person-rem)						
Routine crew	0.0055	0.0075	0.14	0.27	0.16	0.16
Routine public						
Off-link	4×10^{-4}	2.2×10^{-4}	0.026	0.035	0.0064	0.0063
On-link	9.8×10^{-4}	5.8×10^{-4}	0.072	0.12	0.016	0.020
Stops	0.0041	0.0029	0.33	0.67	0.057	0.10
Total	0.0054	0.0037	0.43	0.82	0.079	0.13
Accident[c]	0.0023	8.2×10^{-4}	8.8×10^{-5}	3.8×10^{-4}	0.027	0.0027–0.021
Latent cancer fatalities[d]						
Crew fatalities	3×10^{-6}	5×10^{-6}	9×10^{-5}	2×10^{-4}	9×10^{-5}	9×10^{-5}
Public fatalities	5×10^{-6}	3×10^{-6}	3×10^{-4}	5×10^{-4}	6×10^{-5}	8×10^{-5}– 9×10^{-5}
Chemical impacts						
Irreversible adverse effects[e]	1.2×10^{-9}	0	NA[f]	NA	NA	NA
Vehicle-related[g]						
Emission fatalities	4×10^{-4}	1×10^{-4}	7×10^{-4}	9×10^{-4}	0.001	6×10^{-4}
Accident fatalities	2.7×10^{-5}	2×10^{-5}	5.4×10^{-5}	1.1×10^{-4}	4.8×10^{-5}	5.8×10^{-5}

[a]Cargo-related impacts are impacts attributable to the radioactive or chemical nature of the waste material.

[b]To convert from person-rem to person-Sv, multiply by 0.01.

[c]Dose risk is a societal risk and is the product of accident probability and accident consequence.

[d]Latent cancer fatalities are calculated by multiplying dose by the FGR 13 health risk conversion factor of 0.06 fatal cancer per person-Sv (6×10^{-4} fatal cancer per person-rem) (Eckerman et al. 1999).

[e]Potential for irreversible adverse effects from chemical exposures. Exposure to HF or uranium compounds is estimated to result in fatality of approximately 1% or less of those persons experiencing irreversible adverse effects (Policastro et al. 1997).

[f]NA = not applicable.

[g]Vehicle-related impacts are impacts independent of the cargo in the shipment.

DCS 2002a. *Mixed Oxide Fuel Fabrication Facility Environmental Report, Revision 1 & 2.* Docket No. 070-03098. Charlotte, NC.

DCS 2002b. *Responses to the Request for Additional Information on the Environmental Report, Revisions 1 & 2.* DCS-NRC-000116, Docket No. 070-03098. Charlotte, NC. Oct. 29.

DCS 2004. *Mixed Oxide Fuel Fabrication Facility Environmental Report, Revision 5.* Docket Number 070-03098. Charlotte, NC. June 10.

Didlake, J. 1998. *Fissile Materials Disposition Program SST/SGT Transportation Estimation.* SAND98-8244. Sandia National Laboratories, Livermore, CA. June.

DOE (U.S. Department of Energy) 1988a. *External Dose Rate Conversion Factors for Calculation of Dose to the Public.* DOE/EH-0070. Office of Environment, Safety, and Health, Washington, DC.

DOE 1988b. *Internal Dose Conversion Factors for Calculation of Dose to the Public.* DOE/EH-0071. Office of Environment, Safety, and Health, Washington, DC.

DOE 1997. *Waste Isolation Pilot Plant Disposal Phase Final Supplemental Environmental Impact Statement, Vol. II Appendices.* DOE/EIS-0026-S-2. Carlsbad, NM. Sept.

DOE 1999a. *Programmatic Environmental Impact Statement for Alternative Strategies for the Long-Term Management and Use of Depleted Uranium Hexafluoride.* DOE/EIS-0269. Washington, DC.

DOE 1999b. *Surplus Plutonium Disposition Final Environmental Impact Statement.* DOE/EIS-0283. Office of Fissile Materials Disposition, Washington, DC.

Doty, S.R., et al. 1976. *A Climatological Analysis of Pasquill Stability Categories Based on STAR Summaries.* National Oceanic and Atmospheric Administration, National Climatic Center, Asheville, NC. April.

Eckerman, K.F., et al. 1999. *Cancer Risk Coefficients for Environmental Exposure to Radionuclides, Federal Guidance Report No. 13.* EPA 402-R-99-001. Prepared by Oak Ridge National Laboratory, Oak Ridge, TN, for U.S. Environmental Protection Agency, Office of Radiation and Indoor Air, Washington, DC. Sept.

EPA (U.S. Environmental Protection Agency) 1993a. *Hydrogen Fluoride Study, Report to Congress, Section 112(n)(6), Clean Air Act as Amended, Final Report.* EPA550-R-93-001. Chemical Emergency Preparedness and Prevention Office, Washington, DC. Sept.

EPA 1993b. *Motor Vehicle-Related Air Toxics Study.* EPA 420-R-93-005. Office of Mobile Sources, Ann Arbor, MI. April.

EPA 1996a. *Review of the National Ambient Air Quality Standards for Particular Matter: Policy Assessment of Scientific and Technical Information.* EPA-452/R-96-013. Office of Air Quality Planning and Standards, Research Triangle Park, NC. July.

EPA 1996b. *Air Quality Criteria for Particulate Matter.* EPA-600/P-95-001aF, Vols. 1-3. Office of Air Quality Planning and Standards, Research Triangle Park, NC. April.

Fischer, L.E., et al. 1987. *Shipping Container Response to Severe Highway and Railway Accident Conditions.* NUREG/CR-4829, UCID-20733. Prepared by Lawrence Livermore National Laboratory for U.S. Nuclear Regulatory Commission, Division of Reactor System Safety, Office of Nuclear Regulatory Research, Washington, DC.

Hanna, S.R., et al. 1994. *Technical Documentation of HGSYSTEM/UF6 Model.* Earth Technology Corporation, Concord, MA.

Johnson, P.E., and R.D. Michelhaugh 2000, *Transportation Routing Analysis Geographic Information System (WebTRAGIS) User's Manual.* ORNL/TM-2000/86. Prepared by Oak Ridge National Laboratory, Computational Physics and Engineering Division, Oak Ridge, TN, for U.S. Department of Energy, National Transportation Program, Albuquerque, NM. April.

McGuire, S.A. 1991. *Chemical Toxicity of Uranium Hexafluoride Compared to Acute Effects of Radiation, Final Report.* NUREG-1391. U.S. Nuclear Regulatory Commission, Office of Nuclear Regulatory Research, Washington, DC. Feb.

Negin, C.A. and G. Worku 1992. *MicroShield, Version 4, User's Manual.* Grove 92-2. Grove Engineering, Inc., Rockville, MD.

Neuhauser, K.S., and F.L. Kanipe 1992. *RADTRAN 4: Volume 3, User Guide.* SAND89-2370. Sandia National Laboratories, Albuquerque, NM.

Neuhauser, K.S., and F.L. Kanipe 1995. *RADTRAN 4: Volume 2, Technical Manual.* SAND89-2370. Sandia National Laboratories, Albuquerque, NM.

NRC (U.S. Nuclear Regulatory Commission) 1977a. *Calculation of Annual Dose to Man from Routine Releases of Reactor Effluents for the Purpose of Evaluating Compliance with 10 CFR Part 50, Appendix I, Rev. 1.* Regulatory Guide 1.109. Washington, DC.

NRC 1977b. *Final Environmental Statement on the Transportation of Radioactive Material by Air and Other Modes.* NUREG-0170. Washington, DC.

NRC 1994. "10 CFR Part 19, et al., Certification of Gaseous Diffusion Plants, Final Rule, discussion on Section 76.85, 'Assessment of Accidents.'" *Federal Register* 59(184):48954-48955, Sept. 23.

Ostro, B., and L. Chestnut 1998. "Assessing the Health Benefits of Reducing Particular Matter Air Pollution in the United States." *Environmental Research A* 76:94-106.

Policastro, A.J., et al. 1997. Unpublished information. Argonne National Laboratory, Argonne, IL.

Post, L. 1994a. *HGSYSTEM 3.0, User's Manual.* TNER.94.058. Shell Research Limited, Thornton Research Centre, Chester, United Kingdom.

Post, L. (ed.) 1994b. *HGSYSTEM 3.0, Technical Reference Manual.* TNER.94.059. Shell Research Limited, Thornton Research Centre, Chester, United Kingdom.

Saricks, C., and T. Kvitek 1994. *Longitudinal Review of State-Level Accident Statistics for Carriers of Interstate Freight.* ANL/ESD/TM-68. Argonne National Laboratory, Argonne, IL. July.

Saricks, C., and M. Tompkins 1999. *State-Level Accident Rates of Surface Freight Transportation: A Re-Examination.* ANL/ESD/TM-150. Argonne National Laboratory, Argonne, IL.

Sprung, J.L., et al. 2000. *Reexamination of Spent Fuel Shipment Risk Estimates.* NUREG/CR-6672, SAND2000-0234. Prepared by Sandia National Laboratories for the U.S. Nuclear Regulatory Commission, Office of Nuclear Material Safety and Safeguards, Washington, DC.

USEC (United States Enrichment Corporation) 1999. *The UF6 Manual, Good Handling Practices for Uranium Hexafluoride.* USEC-651, Rev. 8. Jan.

Wilmot, E.L. 1981. *Transportation Accident Scenarios for Commercial Spent Fuel.* SAND80-2124. Sandia National Laboratories, Albuquerque, NM.

APPENDIX D:

SOCIOECONOMICS

APPENDIX D:

SOCIOECONOMICS

This appendix (1) discusses the methods and briefly describes the data sources that were used to perform the socioeconomic analyses for this environmental impact statement (EIS) (Section D.1) and (2) presents fiscal data collected from each of the counties, cities, and school districts in the region of influence (as defined below) (Section D.2).

D.1 Impact Assessment Methods

The socioeconomic analysis for a Mixed Oxide Fuel Fabrication Facility (the proposed MOX facility), including its supporting facilities, the Pit Disassembly and Conversion Facility (PDCF) and the Waste Solidification Building (WSB), at the Savannah River Site (SRS) assessed impacts at two geographic scales. A regional economic area (REA) was used to assess impacts on employment and income for the various alternatives. An REA is a broad market area defined by the economic linkages among the regional industrial and service sectors and the communities within a region. In this case, the REA consists of 15 counties in South Carolina and Georgia (see Table D.1). A region of influence (ROI) that consists of the four counties in which the majority (90%) of the SRS employees live was used to assess impacts on population, housing, community services, and traffic (see Table D.1).

D.1.1 Impacts on Regional Employment and Income

The assessment of projected impacts of the proposed facilities on regional employment and income was based on the use of regional economic multipliers. These multipliers capture the indirect (off-site) effects of on-site activities associated with construction and operation.

To estimate employment impacts of the proposed MOX facility, the PDCF, and the WSB at the SRS, direct and indirect employment impacts associated with construction and operation were taken from data provided in the Surplus Plutonium Disposition (SPD) EIS (DOE 1999, Appendix F, Section 9.2). The indirect (off-site) employment impacts were estimated from these data by using the relationship between direct and indirect employment of the facilities in the REA at the SRS as estimated in the SPD EIS. By using direct (on-site) facility employment data taken from the project Environmental Report (ER)(DCS 2002) as the basis for calculation, the indirect employment impacts were estimated for the peak year of construction and for the first year of operations.

The impact of facility construction and operation on regional incomes was estimated by using facility employment impact estimates together with average regional income multipliers for the REA taken from Intelligent Multi-Resource Planning (IMPLAN) regional economic data (MIG, Inc., 2001). IMPLAN input-output economic accounts show the flow of commodities to

Table D.1. Jurisdictions included in the regional economic area and ROI at the SRS

Regional Economic Area

Georgia	**South Carolina**
Counties	Counties
Burke	Aiken
Columbia	Allendale
Glascock	Bamberg
Jefferson	Barnwell
Jenkins	Edgefield
Lincoln	
McDuffie	
Richmond	
Warren	
Wikes	

Region of Influence

Georgia	**South Carolina**
Counties	Counties
Columbia	Aiken
Richmond	Barnwell
Cities	Cities
Augusta	Aiken
Blythe	Jackson
Grovetown	New Ellenton
Harlem	North Augusta
Hephzibah	Wagener
School Districts	School Districts
Columbia County	Aiken County
Richmond County	Barnwell #19
	Barnwell #29
	Barnwell #45

industries from producers and institutional consumers. The accounts also show consumption activities by workers, owners of capital, and imports from outside the region. The IMPLAN model contains 528 sectors representing industries in agriculture, mining, construction, manufacturing, wholesale and retail trade, utilities, finance, insurance and real estate, and consumer and business services. The model also includes information for each sector on employee compensation; proprietary and property income; personal consumption expenditures; federal, state, and local expenditures; inventory and capital formation; imports; and exports.

Impacts on employment are described in terms of the total number of jobs created in the region in the peak year of construction and in the first year of operation. The relative impact of the increase in employment in the REA was calculated by comparing total facility construction employment over the period in which construction would occur with baseline REA employment forecasts over the same period. Impacts are expressed in terms of the percentage point difference in the average annual employment growth rate with and without facility construction. The forecasts were based on data from the U.S. Department of Commerce (U.S. Bureau of the Census 1992, 2002b).

D.1.2 Impacts on Population

An important consideration in assessing potential impacts of the proposed facilities was the number of workers, families, and children who might move into the ROI (in-migrate), either temporarily or permanently, with construction and operation of the proposed facilities. The capacity of regional labor markets to provide sufficient workers in the appropriate occupations required for facility construction and operation is closely related to the occupational profile of the REA and to occupational unemployment rates. To estimate the in-migration that would occur to satisfy direct labor requirements, the analysis developed estimates of available labor in each direct labor category on the basis of REA unemployment rates applied to each occupational category. In-migration associated with indirect labor requirements was derived from estimates of available labor in the REA economy as a whole able to satisfy the demand for labor by industry sectors in which facility spending would initially occur. The national average household size was used to calculate the number of additional family members who would accompany direct and indirect in-migrating workers.

Impacts on population are described in terms of the total number of in-migrants arriving in the region in the peak year of construction and in the first year of operation. The relative impact of the increase in population in the REA was calculated by comparing total facility construction in-migration over the period in which construction would occur with baseline REA population forecasts over the same period. Impacts are expressed in terms of the percentage point difference in the average annual population growth rate with and without project construction. The forecasts were based on data from the U.S. Census Bureau (U.S. Bureau of the Census 2002a).

D.1.3 Impacts on Local Housing Markets

The in-migration of workers that would occur during construction and operation would have the potential to substantially affect the housing market in the ROI. The analysis considered these impacts by estimating the increase in demand for rental housing units in the peak year of construction and for owner occupied housing in the first year of operation that would result from the in-migration of both direct and indirect workers into the ROI. The impacts on housing are described in terms of the number of rental units required in the peak year of construction and the number of owner occupied units required in the first year of operations. The relative impact on the existing housing in the ROI was estimated by comparing the calculated facility-related housing demand with the forecasted number of vacant rental housing units in the peak year of construction and the forecasted number of vacant owner occupied units in the first year of operations. The forecasts were based on data from the U.S. Census Bureau (U.S. Bureau of the Census 1994, 2002a).

D.1.4 Impacts on Community Services

In-migration associated with construction and operation of the facilities could increase demand for educational services and for other public services (e.g., police and fire protection, health services) in the ROI. Estimates of the total number of in-migrating workers and their families for facility construction and operation were used as a basis for calculating the potential increase in public service demands in the core ROI counties in which the majority of new workers would be expected to locate. Impacts of the facilities on county, city, and school district revenues and expenditures were also calculated on the basis of baseline data provided in the jurisdictions' annual comprehensive financial reports. Impacts were forecasted for the peak year of construction and in the first year of operations on the basis of per capita revenues and expenditures for each jurisdiction. The population forecasts were based on data from the U.S. Census Bureau (U.S. Bureau of the Census 2002a).

Impacts of facility-induced in-migration on community service employment were also calculated for the core ROI counties. The estimated numbers of in-migrating workers and families were used to calculate the numbers of new sworn police officers, firefighters, and general government employees required to maintain the existing levels of service for each community service. Calculations were based on the existing number of employees per 1,000 population for each community service. The analysis of the impact on educational employment estimated the number of teachers in each school district required to maintain existing teacher-student ratios across all student age groups. Impacts on health care employment were estimated by calculating the number of physicians in each county required to maintain the existing level of service. The estimated impacts are given in terms of the number of additional physicians and the number of additional staffed hospital beds required to maintain the existing levels of service (expressed in terms of number of doctors and number of staffed hospital beds per 1,000 population). Information on existing employment and levels of service was collected from the individual jurisdictions providing each service.

D.1.5 Impacts on Traffic

Impacts on traffic in the ROI are described in terms of the effects of the increase in traffic from the facilities on the "levels of service" of major road segments used to commute to and from the site by existing site employees. The analysis allocated trips made by construction workers to individual road segments on the basis of the residential distribution of existing site workers. The impact on the existing annual average number of daily trips was then calculated, and the impact on the level of service provided by each individual segment was estimated. Traffic information used in the analysis was collected from state and county transportation departments.

D.1.6 Impacts of Accidents

The impacts of accidents associated with a MOX facility on agriculture, water, and fisheries resources, and subsequently on the economies of communities surrounding SRS, were not estimated in the EIS because it is not expected that the impacts from an accident would be significant. In the case of the most serious accident, potential damage to crops under the plume in the event of an airborne release and the subsequent damage to water resources from the associated runoff would be small because the amount of radioactive material deposited per unit area would be relatively small. Dilution of runoff would occur fairly rapidly in the affected rivers and streams and would not cause any significant risk to the economies of the communities downstream of the location of the proposed facility. Any interdiction of crops as a result of the deposition of radioactive material would be a limited, one-time event, and if it were to occur at all, only would affect a small number of farm communities. Emergency response activities associated with a release from the facility would be handled by local emergency response and health authorities already prepared for accidents at SRS, with no resulting additional burden on local community financial resources.

D.2 Region of Influence Fiscal Data

Financial data for local governmental bodies and school districts in the ROI for the facilities are presented in Tables D.2 and D.3.

Table D.2. ROI local government financial data ($ millions)

Category	Columbia County, Georgia		
	Columbia County	Town of Grovetown	Town of Harlem
Revenues			
Taxes	23.4	0.7	0.9
Licenses and permits	0.3	0.0	0.0
Intergovernmental	1.6	1.0	0.0
Charges for services	1.1	0.4	0.2
Fines and forfeits	1.6	0.2	0.1
Miscellaneous	0.9	0.1	0.0
Total	28.9	2.5	1.3
Expenditures			
General government	8.1	0.7	0.2
Public safety	11.8	0.7	0.4
Highways and streets	3.3	0.3	0.2
Health, welfare and sanitation	0.9	0.4	0.1
Culture and recreation	2.5	0.0	0.0
Debt service	0.0	0.0	0.0
Intergovernmental	0.0	0.0	0.0
Other	0.9	0.0	0.0
Total	27.5	2.1	1.0
Revenues less expenditures	+1.4	+0.3	+0.3

Table D.2. Continued

	Richmond County, Georgia		
Category	City of Augusta/ Richmond County	City of Blythe	City of Hephzibah
Revenues			
Taxes	55.9	0.1	0.8
Licenses and permits	2.3	0.0	0.0
Intergovernmental	3.0	0.0	0.0
Charges for services	12.8	0.0	0.0
Fines and forfeits	9.0	0.0	0.0
Miscellaneous	3.0	0.1	0.1
Total	86.0	0.2	0.9
Expenditures			
General government	26.3	0.1	0.1
Public safety	34.2	0.1	0.4
Highways and streets	6.1	0.0	0.0
Health, welfare and sanitation	5.2	0.0	0.0
Culture and recreation	9.3	0.0	0.0
Debt service	2.0	0.0	0.0
Intergovernmental Other	2.4	0.0	0.0
Total	85.5	0.2	0.5
Revenues less expenditures	+0.5	0.0	+0.4

Table D.2. Continued

Aiken County, South Carolina

Category	Aiken County	City of Aiken	Town of Jackson
Revenues			
Taxes	16.0	6.4	0.2
Licenses and permits	0.6	4.6	0.1
Intergovernmental	7.7	1.5	0.0
Charges for services	2.0	3.7	0.2
Fines and forfeits	3.2	0.6	0.2
Miscellaneous	1.0	11.0	0.0
Total	30.5	27.8	0.7
Expenditures			
General government	12.1	1.6	0.5
Public safety	10.6	5.4	0.1
Highways and streets	3.7	1.9	0.0
Health, welfare and sanitation	1.7	2.4	0.1
Culture and recreation	2.4	2.3	0.0
Debt service	0.0	0.3	0.0
Intergovernmental	0.0	0.0	0.0
Other	0.0	11.8	0.3
Total	30.5	25.7	0.9
Revenues less expenditures	0.0	+2.1	-0.2

Table D.2. Continued

Aiken County, South Carolina

Category	Town of New Ellenton	City of North Augusta	Town of Wagener
Revenues			
Taxes	0.3	3.7	0.1
Licenses and permits	0.1	2.0	0.1
Intergovernmental	0.1	0.6	0.0
Charges for services	0.2	0.8	0.1
Fines and forfeits	0.1	0.5	0.0
Miscellaneous	0.0	0.3	0.1
Total	0.8	7.9	0.4
Expenditures			
General government	0.2	1.5	0.2
Public safety	0.4	3.4	0.1
Highways and streets	0.1	0.8	0.0
Health, welfare and sanitation	0.1	0.0	0.1
Culture and recreation	0.1	1.7	0.0
Debt service	0.0	0.0	0.0
Intergovernmental	0.0	0.0	0.0
Other	0.0	0.3	0.0
Total	0.9	7.7	0.4
Revenues less expenditures	-0.1	+0.2	0.0

Table D.2. Continued

	Barnwell County, South Carolina			
Category	Barnwell County	City of Barnwell	Town of Blackville	Town of Williston
Revenues				
Taxes	3.0	1.2	0.4	1.1
Licenses and permits	0.0	0.4	0.1	0.0
Intergovernmental	1.6	0.2	0.1	0.2
Charges for services	0.0	0.2	0.2	0.0
Fines and forfeits	0.0	0.1	0.2	0.0
Miscellaneous	4.3	0.0	0.0	0.0
Total	8.9	2.1	1.0	1.3
Expenditures				
General government	2.5	0.4	0.1	0.2
Public safety	2.0	0.9	0.5	0.6
Highways and streets	0.6	0.2	0.0	0.2
Health, welfare and sanitation	1.1	0.2	0.1	0.2
Culture and recreation	0.2	0.0	0.1	0.0
Debt service	0.2	0.0	0.0	0.0
Intergovernmental	0.0	0.0	0.0	0.0
Other	2.0	0.0	0.1	0.1
Total	8.6	1.7	0.9	1.3
Revenues less expenditures	+0.3	+0.4	+0.1	0.0

Sources: Columbia County, annual financial report, June 30, 2000; City of Grovetown Financial Report, December 31, 2000; City of Harlem Annual Financial Report, December 31, 2000; City of Augusta/Richmond County, Annual Financial Statements, December 31, 1999; City of Blythe, Annual Financial Report, December 31, 2000; City of Hephzibah, Financial Statements and Independent Auditors Report, June 30, 2000; Aiken County, Annual Financial Report, June 30, 2000; City of Aiken, Annual Report, June 30, 2000; Town of Jackson, Financial Statements, June 30, 2000; Town of New Ellenton, Financial Statements, June 30, 1999; City of North Augusta, Annual Financial Statements, December 31, 2000; Town of Wagener, Financial Statements, June 30, 1999; Barnwell County, Audited Financial Statements, June 30, 2000; City of Barnwell, Financial Statements, September 30, 2000; Town of Blackville, Audited General Purpose Financial Statements, June 30, 2000; Town of Williston, Financial Statements, June 30, 2000.

Table D.3. ROI school district financial data ($ millions)

Category	Georgia		South Carolina	
	Columbia County	Richmond County	Aiken County	Barnwell County[a,b]
Revenues				
Local sources	32.8	81.2	31.5	8.9
State sources	64.4	134.6	66.5	20.0
Federal sources	0.1	16.1	0.1	0.1
Other	2.2	0.0	0.0	0.0
Total	99.5	231.9	98.1	29.0
Expenditures				
Administration and instruction	65.5	161.1	65.0	17.7
Services	27.9	48.0	34.6	8.3
Debt service	0.0	0.0	0.0	1.0
Other	0.0	0.0	0.0	1.2
Total	93.4	209.1	99.8	28.2
Revenues less expenditures	+6.1	+22.8	-1.6	+0.8

[a]Includes Williston School District #19, #29, and #45.

[b]Revenue data estimated based on South Carolina Department of Education, 2001 School and District Report Cards, and Williston School District #29, Financial Statements, June 30, 2000.

Sources: Columbia County Board of Education, General Purpose Financial Statements, June 30, 2000; Georgia Department of Education, Local, State and Federal Revenue Report Fiscal Year 2001, available at http://dbl.doe.k12go.us:8001/ows-bin/owo/fin_pack_revenue.display.proc; Consolidated School District of Aiken County Financial Statements, June 30, 2000; South Carolina Department of Education, 2001 School and District Report Cards, available at http://www.unyscschools.com/reportcard/2001/; DCS 2002; Williston School District #29, Financial Statements, June 30, 2000.

D.3 References for Appendix D

DCS (Duke Cogema Stone & Webster) 2002. *Mixed Oxide Fabrication Facility Environmental Report, Revision 1 & 2.* Docket Number 070-03098. Charlotte, NC.

DOE (U.S. Department of Energy) 1999. *Surplus Plutonium Disposition Final Environmental Impact Statement.* DOE/EIS-0283. Office of Fissile Materials Disposition, Washington, DC.

MIG, Inc. 2001. "IMPLAN Data Files." MIG, Inc., Sillwater, MN.

U.S. Bureau of the Census 1992. *County Business Patterns, 1990.* Washington, DC. Available at http://www.census.gov/ftp/pub/epcd/cbp/view/cbpview.html.

U.S. Bureau of the Census 1994. *City and County Data Book, 1994.* Washington, DC.

U.S. Bureau of the Census 2002a. *U.S. Census American Fact Finder.* Washington, DC. Available at http://factfinder.census.gov/.

U.S. Bureau of the Census 2002b. *County Business Patterns, 2000.* Washington, DC. Available at http://www.census.gov/ftp/pub/epcd/cbp/view/cbpview.html.

APPENDIX E:

HUMAN HEALTH RISK

APPENDIX E:

HUMAN HEALTH RISK

This appendix provides detailed information concerning the input data and assumptions used in the chemical and radiological human health risk assessments performed for this Mixed Oxide (MOX) Fuel Fabrication Facility Environmental Impact Statement. For chemicals, only accidents are addressed in this appendix; the evaluation of health impacts from chemical exposures during normal operations is discussed in Sections 3.10, 4.2.2, and 4.3.1.

E.1 Chemical

Impacts from the accidental release of chemical materials were assessed for Savannah River Site (SRS) workers outside the restricted area of the facility ("SRS employees") and members of the public. Impacts to facility workers would be sensitive to the specific circumstances of each accident and are not estimated in this assessment.

About 30 MOX process chemicals were identified for use in the proposed MOX facility and support facilities. A chemical was eliminated from the analysis if it had a very low volatility (i.e., vapor pressure <1 Pa (7.5×10^{-3} mmHg), had a low toxicity (i.e., a temporary emergency exposure limit 1 [TEEL 1] $\geq$15 mg/m^3, was stored in small quantities (maximum container quantity <38 L [10 gal]), or was stored and used as a solid. Impacts of a chemical release with these characteristics would be expected to be minimal. Chemicals eliminated from evaporative spill analysis because of very low vapor pressures at ambient temperatures were (1) manganese nitrate, (2) oxalic acid, (3) silver nitrate, (4) uranyl nitrate, (5) sodium hydroxide, (6) aluminum nitrate, and (7) phosphoric acid. Chemicals eliminated because of low toxicity were (1) aluminum sulfate, (2) isopropanol, (3) sodium carbonate, (4) sodium sulfite, and (5) zirconium nitrate. Chemicals eliminated because they are solids were azodicarbonamide, sodium nitrite, and zinc stearate. All other material inventories were analyzed in detail. A spill of sulfuric acid at the PDCF was also eliminated from further analysis based on the assumption that it would contain a concentration of less than 30% sulfur trioxide (i.e., not fuming) and would therefore not pose a toxic inhalation hazard.

The quantity of material released to the atmosphere was determined on the basis of the available physical properties of the spilled chemical (e.g., vapor pressure, mass transfer coefficient), meteorological conditions (e.g., wind speed), and the chemical storage conditions (e.g., temperature, pressure) (see Table E.1). This quantity defined the source term, which was determined either by estimating chemical evaporation rates or pressurized release rates and the associated release durations. The evaporative source term was used as input to the National Oceanic and Atmospheric Administration (NOAA) Areal Locations of Hazardous Atmospheres (ALOHA) dispersion model (Reynolds 1992). Impacts from pressurized releases were simulated with the HGSYSTEM model.

Table E.1. Chemical inventory, spill quantity, concentrations, and mole fraction (MF) calculations[a]

Facility and Chemical name	Formula	MW$_{solute}$ (g/mole)	MW$_{sol}$ (g/mole)	Density		M$_{pure}$ [100%] (moles/L)	Concentrations		N (moles/ L x # H atoms)
				Pure Compound (kg/m^3)	Solution (kg/L)		M (moles/L)	(%)	
MOX (BAP and BRP)									
Dodecane	C$_{12}$H$_{26}$	170.34	170.3	750	0.75	4.4	4.4	100	44.28
Nitrogen tetroxide	N$_2$O$_4$	92.01	92	1,450	1.45	15.76	15.76	**100**	15.76
Hydrazine	H$_6$N$_2$O	32.05	50.1	1,030	1.01	32.14	11.25	35	67.5
Hydrazine–NaOH mixture	N$_2$H$_4$–NaOH	32.05	72.0	1,030	1.03	32.14	0.03	0.10	**0.16**
Hydrazine/hydroxylamine nitrate mixture	H$_4$N$_2$O$_4$ – N$_2$H$_4$	32.05	128.1	1,030	1.54	32.14	**0.15**	0.47	0.6
Hydroxylamine nitrate (HAN)	H$_4$N$_2$O$_4$	96.05	114.1	1,540	1.29	16.03	**1.90**	11.8	7.6
Hydrogen peroxide	H$_2$O$_2$	34.02	52	1,440	1.27	42.33	14.82	**35**	29.6
Tributyl phosphate (TBP)	C$_{12}$H$_{27}$O$_4$P	266.36	266.4	980	0.98	3.68	3.68	**100**	99.3
Nitric acid (13.6 M)	HNO$_3$	63.01	81	1,380	1.28	21.90	13.6	62.1	**13.6**
Nitric acid (2.1 M)	HNO$_3$	63.01	81	1,380	1.20	21.90	2.1	9.6	**2.1**
WSB									
Nitric acid (10.1 M)	HNO$_3$	63.01	81	1,276	1.18	15.74	10.1	**64**	10.1
PDCF[b]									
Chlorine (gas)	Cl$_2$	70.91	70.9	1,491	1.49	21.03	21.03	**100**	21.03

Table E.1. Continued

Facility and Chemical name	Inventory Process tank fill quantity (kg)	(gal)	Spill volume Solution (gal)	Solution (L)	Spill moles n_{solute} (moles)	$n_{solvent}$ (water) (moles)	$n_{solution}$ (moles)	MF	Spill mass m_{solute} (kg)	$m_{solvent}$ (x or H_2O) (kg)	$m_{sol.}$ (kg)
MOX (BAP and BRP)											
Dodecane	511	180	180	681	2,998.5	0	2,998.5	1	510.8	0.00	510.8
Nitrogen tetroxide	1,317	240	240	908	14,315.2	0	14,315.2	1	1,317	0.00	1,317
Hydrazine	491	126	126	478	5,371.4	3,491	8,862.8	0.6061	268.9	62.90	331.8
Hydrazine-NaOH mixture	1,497	384	384	1,455	46.6	46,725	46,771.3	0.0010	4.4	841.75	846.1
Hydrazine/hydroxylamine nitrate mixture	2,445	627	627	2,376	356.4	76,012	76,368.8	0.0047	45.7	1,369.36	1,415.0
Hydroxylamine nitrate (HAN)	1,166	200	200	758	1,440.2	1,270	2,709.7	0.5315	138.3	22.87	161.2
Hydrogen peroxide	300	55	55	208	3,088.6	2,008	5,096.2	0.6061	105.1	36.17	141.2
Tributyl phosphate (TBP)	467	126	126	478	1,757.0	0	1,757.0	1	468.0	0.00	468.0
Nitric acid (13.6 M)	841	161	161	610	8,298.6	3,145	11,443.8	0.7252	522.9	56.66	579.6
Nitric acid (2.1 M)	6,901	1,321	1,321	5,007	10,514	9,506	20,020	0.5252	662.5	171.24	833.8
WSB											
Nitric acid (10.1 M)	1,690	350	350	1,327	13,365	4,811	18,176	0.7353	842.1	86.67	928.8
PDCF[b]											
Chlorine (gas)	430	240	240	911	19,158	0	19,158	1	430	0.00	430

[a] In general, chemicals used, concentrations, and process tank fill quantities for the proposed MOX facility were obtained from DCS (2004b, Table 8-2a and DCS 2003a, 2004a); values for the PDCF were obtained from DOE (1999, Appendix E). Concentrations obtained from these sources are in bold italics; others are calculated values.

[b] Sulfuric acid was also listed for this facility with an annual usage of 470 kg. The concentration was not given, so quantitative spill modeling was not performed. If dilute, the solution would have low volatility and would present minimal hazards from accidental spills. However, if it was a concentration of 30% or more sulfur trioxide, sulfuric acid is highly water reactive and could present inhalation risks to facility workers or SRS employees if spilled.

Abbreviations: MOX: proposed MOX Fuel Fabrication Facility; BAP: Aqueous Polishing Area; BRP: Reagent Processing Building; WSB: Waste Solidification Building; PDCF: Pit Disassembly and Conversion Facility.

For modeling potential impacts to the general public at the SRS site boundary (approximately 8.2 km [5.1 mi] from the proposed MOX facility), the estimated source term was used as input to the ALOHA dispersion model. For modeling potential impacts to SRS workers (assumed to be located a minimum of 100 m [330 ft] from the proposed MOX facility), the ARCON96 model (Ramsdell and Simonen 1997) was used because this model accounts for near-field concentrations affected by low wind speeds, plume meander, and building wake effects. This model is also used to be consistent with U.S. Nuclear Regulatory Commission (NRC) guidance regarding control room habitability during a hazardous chemical release (NRC Regulatory Guide 1.78 (NRC 2001). ARCON96 was used for modeling impacts for all receptors (SRS workers and general public) for uranium dioxide powder releases, similar to the modeling done for accidental releases of other radionuclides.

Two of the MOX process chemicals, nitrogen tetroxide and chlorine, are stored as pressurized liquids. Impacts from accidental releases of these two compounds were estimated with the HGSYSTEM model (Post 1994a,b).

Evaporative releases can be considered as either the "puddle" or "direct" source release mode in ALOHA. To use the puddle option, physical properties of the spilled chemical must be known. These properties, such as vapor pressure and molecular weight, are required in estimating evaporation rates. Physical properties are included for approximately 800 pure chemicals in ALOHA's chemical library. Because only two of the 13 MOX chemicals are included in the library and because the effect of dilute solution adjustments to vapor pressure are not allowed in ALOHA, the direct source release option was used to assess impacts for 11 evaporative spill scenarios. A simple evaporation algorithm, similar to ALOHA and other source evaporation codes, such as ADAM (Raj and Morris 1987; Kawamura and MacKay 1987), was incorporated into a spreadsheet along with the necessary physical properties for each of the eight chemicals. A brief description of the spreadsheet algorithm and its limitations and assumptions are given below:

$$Q_{evap} = \frac{A_p * k_m * MW_m * P_{sat}}{R * T_p},$$ (E-1)

where

A_p = pool area (m^2),

k_m = mass transfer coefficient (m/s),

MW_m = molecular weight of chemical (g/mole),

P_{sat} = saturation vapor pressure of chemical (Pa),

R = Universal Gas Constant (= 8314.472), and

T_p = pool temperature (K).

The evaporation rate from spilled chemical pools is conservatively assumed to be constant, along with the pool temperature and saturation vapor pressure, for the entire release duration. The saturation vapor pressure is set equal to the partial pressure over the pool. The saturation vapor pressure or the partial pressures of the vapors emanating from the pool are a function of the pool temperature through use of chemical-specific Antoine or Harlacher coefficients for inorganic compounds, and through the use of the Clausius-Clapeyron equation for organic compounds (e.g., tributyl phosphate [TBP]). In addition to the assumption that the saturation vapor pressure is equal to the vapor pressure of the chemical at ambient release conditions, the pool temperature is assumed equal to the ambient temperature for the entire release duration. Two ambient cases were assessed, one representing the 95[h] percentile temperature during the day and the other the 95th percentile during the night (see discussion of the full set of assumed weather conditions below). In cases where temperature-specific data (e.g., Antione coefficients and equations) were not available, temperature-dependent P_{vap} adjustments from a reference level (e.g., STP) were made using the ratio of vapor pressures (reference level to compound value at specified temperature) for compounds with similar physical properties for which these pressures were known at two representative temperature levels.

Two of the chemical compounds in the inventory are binary mixtures. The vapor pressure of mixtures was estimated using the following equation (CCPS 1996):

$$P_{mixvap} = \frac{\sum_{i=1}^{n} MF_i * P_{vapi} * MW_i * e^{-kP_{vap}t}}{\sum_{i=1}^{n} MF_i * MW_i} \qquad (E-2)$$

where

MF_i = mole fraction of component i,

P_{vapi} = vapor pressure of component i,

k = $k_m A_p / n_T RT$,

n_T = total number of moles of mixture,

MW_i = molecular weight of component i, and

t = 1.

Raoult's Law was used to make additional adjustments to spill vapor pressures to account for dilute solutions(such a solution lowers the vapor pressure of the solvent below that of the solute in proportion to the mole fraction of the solute). Table E.1 gives the computed mole fractions used in the analysis, along with the assumed spill volumes and the given chemical inventories and concentrations.

The mass transfer coefficient (k_m), used in most evaporative release models, is computed by one of two main methods used in source emission models, as shown in Equations E-3 and E-4 below. Both values were calculated for each chemical in the analysis and the expression giving the largest mass transfer rate between the liquid and the vapor was used in estimating the chemical-specific evaporative rate:

$$k_m = \frac{N_{Sh}D_{ma}}{d_p} \tag{E-3}$$

$$k_{m_2} = 0.0048 \, u_{10}^{7/9} \, d_p^{-1/9} N_{Sc}^{-2/3} \tag{E-4}$$

where

D_{ma} = molecular diffusivity;

d_p = pool depth;

u_{10} = wind speed at 10-m level;

υ_m = kinematic viscosity of the chemical;

N_{Re} = Reynolds number,

= $u_{10}d_p/\upsilon_m$;

N_{Sc} = Schmidt number,

= υ_m/D_{ma};

and

N_{Sh} = Sherwood number,

= $0.664 N_{Sc}^{1/3} N_{Re}^{1/2}$ *for* $N_{Re} < 320{,}000$

= $0.037 N_{Sc}^{1/3} [N_{Re}^{0.8} - 15{,}200]$ *for* $N_{Re} \geq 320{,}000$.

Chemical-specific molecular diffusivities (i.e., of chemical in air) and kinematic viscosities were used in all cases where data were available. In the absence of data (about one-third of the cases), the molecular diffusivity of water or the kinematic viscosity of air were used as substitutes. This estimate was made to be conservative (i.e., use of Graham's Law to estimate molecular diffusivity would produce a value smaller than that of water).

Pressurized releases (i.e., nitrogen tetroxide and chlorine) were modeled with HGSYSTEM's SPILL, AEROPLUME, and HEGADAS modules. To estimate the effects of building aerodynamic influence, the WAKE module was also run, assuming winds perpendicular to the largest building width. The source term was generated from the SPILL module, which simulates the transient liquid release from a pressurized vessel. AEROPLUME is a multicomponent, two-phase thermodynamic aerosol jet model that simulates steady-state release rates from a rupture or a leaking pressurized vessel and the near-field vapor cloud development of the flashed vapor and aerosol components in expelled jet release. Upon formation of the flow field from the release point and establishment of a heavy aerosol laden cloud, the release is linked to the HEGADAS module to simulate dense vapor cloud dispersion and entrainment of ambient air as the cloud moves and disperses downwind. For the building-influenced case, the WAKE module uses the source term from the SPILL module and simulates the aerodynamics in the wake of structures and neutrally buoyant vapor cloud dispersion beyond the wake. In the near-field, WAKE also simulates the concentration field of a release that may get trapped with the cavity recirculation region close to the building. It can also account for air entrainment and escape of vapors initially captured in the cavity region in back of the building, and the transport and dispersion of contaminants in the far wake and beyond.

Site-specific data used are from a 60-m meteorological tower in the H-Area, relatively close to the proposed MOX location. Hourly wind speed and direction and related fluctuating parameters at the 60-m level were available for a 5-year period from 1992 through 1996. The data were preprocessed at the SRS Plant and sent to Argonne for use in the MOX environmental evaluation. The data were reported in Greenwich Mean Time (GMT) and were adjusted in the analysis for local time. Winds at the 60-m level were adjusted to 10 m with a power-law equation.

As mentioned previously, two sets of meteorological conditions, representative of daytime and nighttime conditions and producing conservative emissions and dispersion, were simulated for each evaporative release scenario. Although daytime releases would have more favorable dispersion conditions than nighttime releases, a larger release rate would occur because of higher ambient temperatures and higher near ground-level wind speeds. Both cases needed to be examined in order to determine the controlling, or "worst-case," site-specific weather conditions.

To be consistent with the ARCON96 model, the 95th percentile daytime and nighttime winds were computed from the 5 years of tower data. Wind speeds were adjusted from the measured 60-m level to the 10-m level by using the standard power-law wind profiles employed in most EPA models (e.g., ISC). The 95th percentile day and night winds are representative of winds that occurred over the measurement period. By definition, 95% of all measured day and night wind speeds at the site would cause more plume dispersion. Similar computations were performed to derive the 95th percentile temperatures, defined as ambient temperatures producing reasonable upper-bound evaporative emission rates. Because higher wind speeds also tend to increase pool evaporation, the 5th percentile wind speeds (i.e., the 5th percentile here is defined as representing the largest wind speeds measured in the 5-year period studies) were also computed. Each of the meteorological cases, including the 95th percentile concentration ARCON case used for estimating 100-m downwind involved worker exposures, is summarized in Table E.2. In addition to wind speed and temperature, the complete set of meteorological parameters used in the ALOHA simulations and the temperatures and wind speeds used in the evaporative spreadsheet calculation tool are summarized in the table. A fourth set of conditions, typical during sunrise or sunset (given in the table), was also run to see if the larger wind speed and neutral conditions would result in more conservative impacts. These conditions resulted in lower impacts and are not further discussed.

Surface roughness was assumed to be 50 cm, which is representative of a good portion of the SRS (Weber 2002). This roughness is large enough to switch the ALOHA computed dispersion coefficients to that representative of urban environments, which will enhance the horizontal and vertical spread of released contaminant as it is advected downwind.

The spill scenario assumed that a forklift punctured a liquid storage tank containing the chemical. Estimates are needed for three key parameters used in determining the evaporation rates (Equation E-1). These parameters are the ambient temperature (T_a), pool area, and vapor pressure. Varying stability conditions, temperatures, and wind speeds were modeled to determine worst-case emission and dispersion conditions. Unlimited mixing was assumed to be consistent with U.S. Environmental Protection Agency (EPA) models (e.g., TSCREEN, ISC) for these conservative nighttime dispersion conditions. The maximum mixing height value, set as a default in ALOHA, is 1,524 m (5,000 ft).

All of the tanks were assumed to be cylindrical in shape with the puncture hole assumed to be located near the tank bottom. Tank dimensions varied depending on the specific chemical inventories. The calculated spill quantities were conservatively assumed to be the full contents of each liquid storage container. The spilled liquid was assumed to spread out on a concrete surface, with a surface roughness of around 3 cm (1.2 in.), to a pool depth of 2.54 cm (1 in.). The final pool area and diameter were computed by assuming a circular pool with a uniform

Table E.2. Scenario meteorology[a,b]

Parameter	Day (95% temp/ 95% winds)	ARCON (95% conc., ARCON)	Night (95% temp/ 95% winds)	Sunrise/ Sunset (95% temp/ 5% winds)
T_a (K)	304.0	299.2	299.2	299.2
T_a (°F)	87.5	78.5	78.5	78.5
u_{10} (m/s)	1.3	2.2	1.3	4.7
Stability	D	F	F	D
Frequency	27%	n/a	11%	100%
z_i (m)	416	n/a	Unlimited	Unlimited
Cloud cover	7/10	Clear to 4/10	Clear to 4/10	Clear to 4/10
RH (%)	85%	65%	65%	65%
Insolation	Slight	Night	Night	Slight

[a] T_a = ambient temperature, u_{10} = wind speed at 10 m, z_i = mixing height, RH = relative humidity.

[b] z_o = surface roughness = 50 cm, season = summer.

Table E.3. Evaporative release modeling results

Chemical	MET	Maximum evaporation rate Q (kg/h)	Release duration[a] t_d (h)	SRS worker exposure @100 m (mg/m³)	TEEL1 Passive (km)	TEEL1 Dense (km)	TEEL2 Passive (km)	TEEL2 Dense (km)	TEEL3 Passive (km)	TEEL3 Dense (km)
Dodecane	Day	0.51	1,000		< 0.01	0.016	< 0.01	< 0.01	< 0.01	< 0.01
	ARCON	0.96	6.1	0.16	-[b]	-	-	-	-	-
	Night	0.31	2,200		0.064	0.054	0.012	0.011	< 0.01	< 0.01
Hydrazine	Day	11.5	28.8		0.41	0.84	0.13	0.26	0.05	0.1
	ARCON	17.4	19.1	2.9	-	-	-	-	-	-
	Night	8.8	37.5		0.93	1.3	0.26	0.4	0.1	0.15
Hydrazine/NaOH	Day	0.13	18,000		0.032	NA[c]	0.01	NA	< 0.010	NA
	ARCON	0.098	23,000	0.02	-	-	-	-	-	-
	Night	0.11	15,000		0.064	NA	0.020	NA	< 0.010	NA
Hydrazine/HAN	Day	30.0	69.8		0.381	NA	0.117	NA	0.045	NA
	ARCON	12.8	163.4	2.2	-	-	-	-	-	-
	Night	23.3	116.5		0.855	NA	0.233	NA	0.086	NA
Hydroxyl-amine nitrate (HAN)	Day	7.7	20.9		0.07	0.135	0.053	0.101	0.024	0.044
	ARCON	11.6	13.9	2.0	-	-	-	-	-	-
	Night	5.9	10.0		0.135	0.206	0.102	0.151	0.045	0.063
Hydrogen peroxide	Day	0.70	202.6		0.023	0.044	< 0.010	0.02	< 0.010	0.011
	ARCON	1.44	155.7	0.2	-	-	-	-	-	-
	Night	0.7	108.4		0.050	0.075	0.022	0.033	0.016	0.022
Tributyl phosphate (TBP)	Day	6.51	71.8		0.103	NA	0.079	NA	0.014	NA
	ARCON	9.81	47.7	1.7	-	-	-	-	-	-
	Night	6.6	26.1		0.233	NA	0.178	NA	0.031	NA
Nitric acid 13.6 M	Day	9.9	58.3		0.197	0.388	0.08	0.146	0.022	0.032
	ARCON	12.9	44.8	2.2	-	-	-	-	-	-
	Night	7.9	38.5		0.417	0.616	0.158	0.223	0.042	0.041
Nitric acid 2.1 M	Day	34.9	23.9		0.378	0.752	0.151	0.282	0.041	0.06
	ARCON	49.5	16.8	8.4	-	-	-	-	-	-
	Night	27.8	11.7		0.86	1.2	0.310	0.412	0.079	0.072
Nitric acid 10.1 M	Day	18.1	46.6		0.269	0.536	0.108	0.204	0.03	0.043
	ARCON	23.5	35.9	4.0	-	-	-	-	-	-
	Night	14.4	58.6		0.586	0.837	0.217	0.299	0.056	0.054

Table E.3. Continued

Chemical		TEEL 1 (mg/m³)	Health index concentration TEEL 2 (mg/m³)	TEEL 3 (mg/m³)	Downwind concentration at SRS boundary (8.2 km) (mg/m³)
Dodecane	Day	7.5	60	750	< 0.7
	ARCON				-
	Night				< 0.7
Hydrazine	Day	0.7	6.6	40	0.004
	ARCON				-
	Night				0.009
Hydrazine/NaOH	Day	0.6	6	40	NS[d]
	ARCON				-
	Night				NS
Hydrazine/HAN	Day	0.6	6	40	NS
	ARCON				-
	Night				NS
Hydroxyl-amine nitrate (HAN)	Day	15	26	125	NS
	ARCON				-
	Night				NS
Hydrogen peroxide	Day	12.5	60	125	NS
	ARCON				-
	Night				NS
Tributyl phosphate (TBP)	Day	6	10	300	NS
	ARCON				-
	Night				NS
Nitric acid 13.6 M	Day	2.5	15	200	NS
	ARCON				-
	Night				0.009
Nitric acid 2.1 M	Day	2.5	15	200	0.011
	ARCON				-
	Night				0.028
Nitric acid 10.1 M	Day	2.5	15	200	NS
	ARCON				-
	Night				0.015

[a]Reported duration is based on maximum spill volume and evaporation rate. However, the ALOHA model restricts the maximum release duration to one hour. At constant wind speed, the highest concentration would occur in this first hour.

[b]_ = not applicable.

[c]NA = not available.

[d]NS = not significant (less than 0.001 mg/m³).

depth along with the spill volume. The pool size for each of the spill scenarios ranged from 8 m² (hydrogen peroxide spill outside the MOX BRP building) to 435 m² (nitric acid spill at the WSB).

As previously mentioned, the vapor pressures, as well as other the physical properties required in estimating the evaporation rate from Equation E-1, were computed by using chemical-specific coefficients in Antione or equivalent equations, or (in the absence of temperature dependent data) obtained directly from published literature (e.g., Linde 1999; Perry and Green 1984; NIST 2001; DIPPR 1989). Adjustments for dilute solutions were accounted for by multiplying by the computed mole fraction, the ratio of the number of moles of a substance to the total amount of that substance in a mixture. The physical properties, including the mole fraction adjusted vapor pressures, and the computed chemical specific nondimensional numbers used in computing evaporation rates (e.g., Reynolds Number), are summarized in Table E.4.

Accident consequences for evaporative releases, expressed as the ambient concentration at specified downwind distances, are reported in Table E.3. These concentrations are compared with (TEEL) values, criteria levels for accidental exposures adopted by the DOE Subcommittee on Consequence Assessment and Protective Action (SCAPA) (Craig 2002). TEEL values are available for about 2,000 substances; they are derived by using a hierarchy of other available criteria values (Craig et al. 2000). If Emergency Response Planning Guidelines (ERPGs) developed by panels of toxicologists for the American Conference of Governmental Industrial Hygienists (ACGIH) are available, these are used for the TEEL values. If ERPGs are not available, TEELs usually are based on emergency planning and other guideline levels developed for the protection of workers (Craig 2002). TEEL values are developed for evaluation of different levels of effects, ranging from no or very slight adverse effects to life-threatening effects (see text box in Section 4.3.5.3 for definitions).

To assess impacts for SRS employees, concentrations greater than TEEL-3 levels at 100 m for any chemical were defined as high consequence, and levels less than TEEL-3 but greater than TEEL-2 were defined as moderate consequence. To assess impacts for the general public, SRS boundary concentrations greater than TEEL-2 levels for any chemical were defined as high consequence, and levels less than TEEL-2 but greater than TEEL-1 were defined as moderate consequence. In addition, the hazard distances (i.e., maximum distances from the release point to which chemical TEEL-1, TEEL-2, and TEEL-3 air concentrations could extend) were estimated with the ALOHA model and are listed in Table E.3.

The impacts to SRS workers, located 100 m (330 ft) from the spill, were estimated by multiplying the ARCON96 95th percentile chi/Q value (0.00061 s/m³) by the estimated evaporation rate, assuming the same wind speed that produces the ARCON96 95th percentile chi/Q (2.2 m/s) and the 95th percentile site-specific temperature (78.5°F) derived from 5 years of data from the meteorological tower in the H-area. For evaporative releases, there would be no worker exposures above the TEEL-2 level. However, spills of hydrazine, hydrazine/HAN mixtures, and nitric acid have the potential to expose SRS employees above the TEEL-1 levels. The resulting health impacts would be temporary and mild. The 100-m (330-ft) concentration

Table E.4. Physical property data

Chemical/ property[a]	Dodecane	Nitrogen tetroxide (N_2O_4)	Nitric acid (HNO_3)	Hydrazine (H_6N_2O)	HAN[b] ($H_4N_2O_4$)
MW	170.4	92.0	63.1	50.06	96.04
ρ_l (kg/L)	0.75	1.443	1.383	1.03	1.54
ρ_v (kg/m^3)	–[c]	3.2-998.9[d]	2.012	0.95	0.981
k_m (m/s)	–	NA[e]	2.67×10^{-4} to 5.76×10^{-4}	5.26×10^{-3}	6.17×10^{-3}
D_m (m^2/s)	7.15×10^{-6}	NA	1.19×10^{-5}	1.65×10^{-5}	1.63×10^{-5}
ν_k (m^2/s)	–	NA	5.84×10^{-4}	1.28×10^{-5}	6.65×10^{-6}
P_{Vap} (Pa) (78.9 °F)	2,039	2,038.5	4,540.8 to 6,269.9[f]	1,235.5	281.5
P_{Vap} (Pa) (87.5 °F)	2,720	2,701.9	5,800.2 to 8,008.9[f]	1,637.5	373.1
N_{Sc}	–	NA	49.7	0.909	0.923
N_{sh}	–	NA	271 to 458	593	829
N_{Re}	–	NA	12,307 to 35,254	423,749	534,226

Chemical/ property[a]	Hydrazine-HAN ($H_4N_2O_4$-N_2H_4)	Hydrazine-NaOH (N_2H_4-NaOH)	Tributyl phosphate ($C_{12}H_{27}O_4P$)	Hydrogen peroxide (H_2O_2)	Chlorine (Cl)
MW	128.09	93.99	266.36	34.02	70.91
ρ_l (kg/L)	1.54[g]	2.13	0.979	1.44	1.49
ρ_v (kg/m^3)	–	–	–	2.72	4.72 to 432.5[d]
k_m (m/s)	3.82×10^{-4}	5.26×10^{-3}	8.86×10^{-4}	5.07×10^{-3}	NA
D_m (m^2/s)	–	–	–	1.62×10^{-5}	NA
ν_k (m^2/s)	–	–	–	7.92×10^{-4}	NA
P_{Vap} (Pa) (78.9 °F)	289.7	2.2	134.8	1,912.2	8.02×10^5
P_{Vap} (Pa) (87.5 °F)	379.0	2.6	135.3	1,978.0	9.37×10^5

Table E.4. Continued

Chemical/ property[a]	Hydrazine-HAN ($H_4N_2O_4$-N_2H_4)	Hydrazine-NaOH (N_2H_4-NaOH)	Tributyl phosphate ($C_{12}H_{27}O_4P$)	Hydrogen peroxide (H_2O_2)	Chlorine (Cl)
N_{Sc}	0.625	0.625	0.625	48.9	NA
N_{sh}	1,439	1,090	524	177	NA
N_{Re}	945,897	1,251,896	424,029	5,307	NA

[a] ρ_l = liquid density, ρ_v = vapor density, k_m = mass transfer coefficient, D_m = molecular diffusivity, P_{vap} = vapor pressure, ν_k = kinematic viscosity, N_{Sc} = Schmidt number, N_{Sh} = Sherwood number, N_{Re} = Reynolds number.

[b] Hydroxylamine nitrate.

[c] – = not available.

[d] Aerosol vapor mixture density from jet release is initially very high; it is diluted over time to its vapor density at ambient conditions.

[e] NA = not applicable, modeled as a pressurized release.

[f] Nitric acid (1.21 N) [4,540.8 (78.9°F), 5,800.2 (89.5°F)]; Nitric acid (7.9 N) [5,764.2 (78.9°F), 7,362.9 (89.5°F)]; Nitric acid (13.6 N) [6,269.9 (78.9°F), 8,008.9 (89.5°F)].

[g] No published value available, set equal to the HAN published density at STP.

reference level for SRS employees is consistent with the SRS Emergency Response Plan (SRS 2001), which defines the facility boundary as follows:

> "Generally, the facility boundary is the fence line for a property, protected area or a limited area, depending upon the facility. When a physical boundary is unavailable, the distance of 100 meters from the point of release or edge of the spill is used. Area/facility-specific Emergency Preparedness Hazard Assessment Documents identify facility boundaries and should be referenced."

Since the wind speed and atmospheric stability generating the upper-bound impacts for nighttime conditions were 1.3 m/s with stable conditions (i.e., PG Class F), the plume transport time or the time it would take the release to reach the nearest SRS boundary (8.2 km downwind) would be almost 2 hours. Because ALOHA restricts the maximum release duration and plume transport time to one hour or less, ALOHA impact estimates at the SRS boundary could not be made for the low wind speed assumed in the simulations. Therefore, maximum impact estimates at the SRS boundary were made by using a formula for a ground-level release producing maximum ground-level concentrations (i.e., on the plume centerline at the surface), similar to that used in ALOHA. Ground-level centerline passive plume concentrations were estimated using the following formula, derived from the standard Gaussian equation:

$C(x,0,0) = Q/\pi u \sigma_y \sigma_z$. Dense gas estimates at the fence line were estimated by increasing the wind speed from 1.3 to 2 m/s to shorten the transport time to the fence line to less than one hour. The ALOHA-estimated concentration was then multiplied by 1.3 [chi/u(2) x u(1.3)] to arrive at the estimated SRS boundary concentration. The highest concentrations at this distance occurred subsequent to transition to a purely passive plume (i.e., no negative buoyancy influences from density effects). Estimates at 100 m using the above expression compared well (no more than a 1 to 2% difference) with the ALOHA estimate at the same location.

The ALOHA estimated hazard distances are also given in Table E.3 for evaporative plumes exhibiting dense vapor cloud dispersion. These plumes disperse downwind to a transition point at which ambient air entrainment into the cloud sufficiently dilutes concentrations so that the plume continues to disperse from that point downwind as a neutrally buoyant plume. The releases considered that initially behave as dense clouds produced the largest hazard distance. The largest potential health hazard was shown to extend 1.3 km (0.8 mi) downwind for an accidental spill of 478 L (126 gal) of 35% hydrazine.

Releases of two materials, nitrogen tetroxide and chlorine, were modeled as pressurized releases. The analysis showed that these pressurized releases would potentially produce very large exposures to SRS workers at a distance of 100 m (330 ft) because the concentrated dense gas plume could extend to this distance for a short time. The concentrations within the jet plume would approach 10,000 and 1,500 mg/m^3 at 100 m (330 ft) for nitrogen tetroxide and chlorine, respectively. The TEEL-2 hazard distance for accidental releases of both substances could extend to 4 km (2.5 mi) from the release location. The high concentrations close to the source are primarily due to the release of a pressurized, two-phased vapor-aerosol, which forms a dense vapor cloud. It should be noted that building influences on the heavy vapor cloud are not accounted for in the AEROPLUME and HEGADAS simulations. Such influences on passive releases are accounted for in the WAKE model, but not the combination of building aerodynamics and density effects. The estimated 100-m (330-ft) exposure calculated with the WAKE model approached 1,600 mg/m^3 and 500 mg/m^3 for nitrogen tetroxide and chlorine, respectively. The actual concentrations would likely fall between the two modeled results for each chemical.

E.2 Radiological

Risks from radioactive materials were assessed for workers involved in facility operations ("facility workers") at the proposed MOX facility, the PDCF, and the WSB; other SRS workers outside the restricted area of the facility site ("SRS employees"); and members of the public.

E.2.1 Normal Operations

E.2.1.1 Facility Workers

For facility workers, external radiation from the direct handling of radioactive materials and/or the close working distances to radiation sources would be the primary exposure pathway. Radiation exposures through inhalation and incidental ingestion of contaminated particulates would be possible but for the average worker would be expected to be very small compared with exposures to external radiation.

Operations that could result in potential airborne radiological emissions would be conducted under fume hoods or in gloveboxes. Even if airborne releases from the gloveboxes did occur, the use of high-efficiency particulate air (HEPA) filters and protective air circulation systems would reduce the airborne pollutants in the working place to a minimal level. Exposures from inhalation could also be prevented by implementation of as-low-as-reasonably-achievable (ALARA) practices, such as requiring workers to wear respirators while performing activities with potential for generating airborne emissions. Potential exposure from incidental ingestion of particulate matter could be reduced by workers' wearing gloves and exercising good working practices.

For the proposed MOX facility, radiation exposure was estimated on the basis of exposures received during operation of a similar facility, the MELOX plant in Marcoule, France. External dose rates at the MELOX plant were extrapolated on the basis of the plutonium composition of the MELOX MOX fuel (8.5%) and proposed facility MOX fuel (5%) (DCS 2001b). Scaling was done by using the ratios of the photon and neutron intensities for the two concentrations. An annual collective external dose of 0.10 person-Sv (10 person-rem) was estimated for the processing area. An additional annual external dose of 0.02 person-Sv (2 person-rem) was assumed for the aqueous polishing area because no data were available (DCS 2001b). Thus, an annual external exposure of 0.12 person-Sv (12 person-rem) was estimated for facility workers.

Facility workers may also receive an internal dose. At the MELOX plant, from 1996 through July 2001, 41 individuals had received an internal radiation exposure: 30 had received <10% of the annual limit on intake (ALI), 10 ranging from 10% to 33.3% ALI, and 1 ranging from 33.3% to 100% ALI. With an intake of 100% ALI, an individual receives a dose of 0.05 Sv (5 rem). Because design and management measures at the MELOX plant are similar to those planned for the proposed facility, a MOX facility worker MEI may receive a dose of 0.017 Sv (1.7 rem), corresponding to a 33% ALI, in a year. The total dose of 0.13 person-Sv (13 person-rem) over this 5-year period results in an average internal dose of less than 0.03 person-Sv (3 person-rem) per year (assuming the full 50-year dose commitment in the year of exposure) (DCS 2001b). Thus, the annual collective facility worker exposure is estimated to be 0.15 person-Sv (15 person-rem), the sum of the estimated external and internal exposures.

For the PDCF and WSB, no historical operational experience is available to provide a reasonable estimate of the worker exposures. Because these two facilities would be owned and operated by the DOE, individual facility worker exposure would be maintained below 0.005 Sv/yr (0.5 rem/yr), the SRS site guideline, which is below the DOE administrative limit of 0.02 Sv/yr (2 rem/yr) (DOE 1994). However, using best practices under the ALARA principle, the average individual dose should be kept close to or lower than the average SRS radiological worker dose of 0.00048 Sv/yr (0.048 rem/yr) (DOE undated).

The information on radiation sources, worker activities, and number of required workers is subject to a large degree of uncertainty, as are the estimated collective and MEI worker doses. However, the radiation dose to the individual worker would be monitored and maintained below the NRC annual occupational total effective dose limit of 0.05 Sv (5 rem) (*Code of Federal Regulations*, Title 10, Part 20 [10 CFR 20]).

E.2.1.2 SRS Employees

Inhalation of contaminated particulates and external exposure to the plume of routine airborne releases from the plant and to soil contaminated by deposition of those airborne releases were considered for SRS employees. Because they would be located farther from the radiation sources handled in the three facilities than would facility workers, those SRS employees would not be exposed to direct external radiation from those sources. However, secondary external radiation would be possible from the deposited radionuclides on ground surfaces and from airborne radionuclides when the emission plume from the stack of the facilities passed the locations of the SRS employees.

The GENII computer code (Napier et al. 1988) was used to estimate radiological impacts to the SRS employees on the basis of emissions data shown in Table E.5. GENII has been used for the same application in several previous environmental impact statement projects, such as the *Final Waste Management Programmatic Environmental Impact Statement for Managing Treatment, Storage, and Disposal of Radioactive and Hazardous Waste* (WM PEIS) (DOE 1997). The GENII code uses either site-specific or representative meteorological data (joint frequency data) selected to estimate the air concentrations at downwind locations. The code implements the internal dosimetry models recommended by the International Commission on Radiological Protection (ICRP) in Publication 26 (ICRP 1977) and Publication 30 (ICRP 1979). The GENII code considers the transport of radioactive material in air, soil, water, and food sources to the human body.

The SRS employee population distribution used to estimate the SRS employee dose is given in Table E.6. This distribution is centered at the proposed MOX facility and involves a total population of 13,295 site workers. A stack height of 37 m (121 ft) (as specified in Section 3.1.1 of DCS 2002a) was used as the release height for normal emissions from the proposed MOX facility. WSB emissions were included in the proposed MOX facility estimates (DCS 2002a,b). An estimated stack height of 35 m (115 ft) was used as the release height for emissions from the PDCF (LANL 1998). Five years of weather information in the form of joint frequency data (1992-1996 average [as shown in Table E.7]) was used for the air dispersion calculations. On an annual basis, the total time of external exposure to the plume and contaminated soil for all SRS employees was assumed to be 0.5 year (NRC 1977). Resuspension of contaminated soil

Table E.5. Estimated annual radiological releases from the facilities during normal operations

	Airborne releases (μCi/yr)[a]	
Isotope	Proposed MOX facility and WSB[b]	PDCF[c]
Plutonium-236	1.3×10^{-8}	9.3×10^{-11}
Plutonium-238	8.5	0.065
Plutonium-239	91	0.69
Plutonium-240	23	0.18
Plutonium-241	101	0.69
Plutonium-242	6.1×10^{-3}	4.8×10^{-5}
Americium-241	48	0.37
Uranium-234	5.1×10^{-3}	NA[d]
Uranium-235	2.1×10^{-4}	NA
Uranium-238	0.012	NA
Tritium	NA	1.1×10^{9}

[a]To convert from microcuries (μCi) to becquerels (Bq), multiply by 3.7×10^4 (or 37,000).

[b]*Source:* DCS (2002a).

[c]*Source:* DOE (1999).

[d]NA = not applicable.

was not considered, and the soil was assumed to be previously uncontaminated. Ingestion of contaminated foodstuffs was not considered because food is not grown on-site and consumed.

The maximally exposed individual (MEI) for the SRS employees was assumed to be within the SRS boundary (but outside the facility site) at a location that would have the maximum air concentration and would thus yield the largest radiation dose. On an annual basis, the total time of annual external exposure to the plume and contaminated soil for the MEI was assumed to be 0.7 year. For the inhalation pathway, an exposure time of 1 year was assumed (NRC 1977).

E.2.1.3 Members of the Public

The GENII code was used to assess radiation exposures of members of the public outside the SRS boundaries. The exposure pathways analyzed included inhalation of contaminated particulates, external radiation from deposited radionuclides and from airborne radionuclides, and ingestion of contaminated food products (plants, meat, and dairy products). Plants grown in the area where the emission plume passed could become contaminated by deposition of

**Table E.6. SRS employee population distribution
centered at the proposed MOX facility on the SRS**

Direction	Population by distance (mi[a])						Total
	0 to 1	1 to 2	2 to 3	3 to 4	4 to 5	5 to 10	
S	1,191	0	225	171	0	397	1,984
SSW	592	0	0	0	0	7	600
SW	0	0	0	0	0	0	0
WSW	0	0	0	0	0	0	0
W	0	0	1,728	110	0	0	1,839
WNW	0	0	0	0	0	0	0
NW	0	0	0	0	2,408	897	3,305
NNW	0	0	0	0	0	0	0
N	0	0	0	0	0	0	0
NNE	0	0	0	0	0	0	0
NE	0	0	0	0	0	0	0
ENE	0	0	18	0	0	5	23
E	0	438	1,863	0	0	0	2,300
ESE	0	722	754	0	0	0	1,476
SE	70	101	26	0	0	25	221
SSE	282	0	0	1,164	0	100	1,547
Total	2,135	1,260	4,614	1,446	2,408	1,432	13,295

[a]To convert from miles to kilometers, multiply by 1.61.

Source: Birch (2001), Attachment A.10.

radionuclides on the leaves or ground surfaces. Radionuclides deposited on leaves could subsequently translocate to the edible portions of the plants, and those deposited on ground surfaces could subsequently be absorbed by plant roots. Livestock and their products could become contaminated if the livestock ate the contaminated surface soil and plants.

The off-site population distribution out to 80 km (50 mi), centered at F-Area, for the SRS area used in the assessment is given in Table E.8. The annual time of external exposure to the plume and contaminated soil for the general public off-site was assumed to be 0.5 year (NRC 1977). No credit for shielding was given for inhalation exposure. Ingestion parameters are provided in Table E.9. Food production data for the area surrounding the SRS are provided in Table E.10.

For the public, the location of the MEI was considered to be at the SRS boundary as a conservative assumption. Table E.11 lists the distance from the proposed MOX facility to the SRS boundary for the 16 compass directions from which the MEI was determined. Because of the close proximity of the PDCF and WSB to the proposed MOX facility, the same MEI receptor locations were used for these facilities. The annual external exposure to the plume and contaminated soil for the public off-site MEI was assumed to be 0.7 year (NRC 1977). No credit for shielding was given for inhalation exposure. Ingestion parameters are provided in Table E.9.

Table E.7. Joint frequency distribution used for calculation of receptor dose from facility air emissions

Wind speed (m/s)	Stability class	Wind direction															
		S	SSW	SW	WSW	W	WNW	NW	NNW	N	NNE	NE	ENE	E	ESE	SE	SSE
0.89	A	0.25	0.20	0.24	0.24	0.21	0.18	0.15	0.18	0.17	0.17	0.21	0.22	0.18	0.18	0.16	0.21
	B	0	0.03	0.03	0.03	0.01	0.00	0.00	0.01	0.01	0.01	0.03	0.03	0.00	0.03	0.03	0.02
	C	0.02	0.01	0.01	0.02	0.01	0.01	0.02	0.03	0.03	0.01	0.01	0.01	0.01	0.02	0.01	0.01
	D	0.01	0.02	0.00	0.02	0.02	0.01	0.01	0.02	0.02	0.02	0.02	0.01	0.01	0.01	0.00	0.03
	E	0.00	0.00	0.00	0.00	0.00	0.00	0.00	0.00	0.01	0.01	0.00	0	0	0.00	0.00	0.00
	F	0.00	0.00	0.00	0.00	0.00	0.00	0.00	0.00	0.00	0.00	0.00	0.00	0	0	0.00	0.00
	G	0.00	0.00	0.00	0.00	0.00	0.00	0.00	0.00	0.00	0.00	0.00	0.00	0.00	0	0.00	0.00
2.46	A	0.88	0.73	0.92	1.04	1.06	0.79	0.70	0.55	0.74	0.78	1.12	1.37	1.19	0.82	0.56	0.57
	B	0.24	0.36	0.43	0.44	0.35	0.25	0.19	0.21	0.26	0.24	0.34	0.38	0.29	0.25	0.16	0.16
	C	0.15	0.39	0.73	0.50	0.39	0.24	0.24	0.29	0.33	0.36	0.43	0.49	0.34	0.28	0.23	0.18
	D	0.09	0.25	0.59	0.34	0.31	0.27	0.34	0.37	0.42	0.39	0.38	0.33	0.30	0.22	0.26	0.21
	E	0.01	0.09	0.28	0.11	0.08	0.16	0.17	0.18	0.26	0.22	0.19	0.20	0.13	0.13	0.11	0.13
	F	0.01	0.02	0.02	0.01	0.00	0.03	0.02	0.03	0.03	0.03	0.02	0.05	0.00	0.01	0.02	0.04
	G	0.00	0.00	0.00	0.00	0.00	0.00	0.00	0.00	0.00	0.00	0.00	0.00	0.00	0.00	0.00	0.00
4.47	A	1.03	0.66	0.53	0.50	0.44	0.30	0.26	0.20	0.37	0.43	0.60	0.70	0.71	0.48	0.24	0.36
	B	0.21	0.57	0.65	0.67	0.32	0.23	0.16	0.19	0.31	0.33	0.55	0.75	0.55	0.36	0.16	0.18
	C	0.16	0.69	1.49	0.86	0.67	0.44	0.42	0.42	0.52	0.58	0.74	0.78	0.78	0.57	0.27	0.14
	D	0.12	0.52	1.64	0.95	0.81	0.70	0.84	1.12	1.48	1.05	1.26	1.27	1.01	0.88	0.50	0.20
	E	0.06	0.64	1.08	0.81	0.62	0.62	0.82	0.98	1.20	1.10	1.06	1.12	0.63	0.47	0.42	0.24
	F	0.02	0.22	0.19	0.07	0.10	0.16	0.18	0.17	0.22	0.16	0.21	0.27	0.07	0.06	0.05	0.06
	G	0.00	0.02	0.01	0.00	0.00	0.01	0.01	0.01	0.02	0.01	0.01	0.02	0.00	0.00	0.00	0.00
6.93	A	0.21	0.18	0.03	0.03	0.01	0.02	0.02	0.01	0.02	0.04	0.05	0.10	0.09	0.11	0.03	0.09
	B	0.02	0.17	0.12	0.04	0.04	0.03	0.05	0.04	0.04	0.09	0.18	0.31	0.46	0.34	0.09	0.03
	C	0.00	0.18	0.46	0.21	0.08	0.09	0.16	0.22	0.20	0.29	0.41	0.46	0.73	0.62	0.13	0.01
	D	0.00	0.09	0.19	0.08	0.05	0.06	0.13	0.46	0.43	0.24	0.24	0.12	0.13	0.11	0.07	0.00
	E	0.00	0.09	0.06	0.09	0.07	0.05	0.05	0.09	0.13	0.10	0.19	0.07	0.02	0.02	0.01	0.00
	F	0.00	0.04	0.02	0.03	0.01	0.03	0.02	0.01	0.01	0.01	0.03	0.02	0.01	0.00	0.00	0.00
	G	0.00	0.00	0.00	0.00	0.00	0.00	0.00	0.00	0.00	0.00	0.00	0.00	0.00	0.00	0.00	0.00
9.61	A	0.01	0.00	0.00	0.00	0.00	0.00	0.00	0.00	0.00	0.00	0.00	0.01	0.02	0.02	0.00	0.01
	B	0.00	0.01	0.00	0.00	0.00	0.00	0.00	0.00	0.00	0.00	0.02	0.03	0.08	0.06	0.01	0.00
	C	0.00	0.01	0.00	0	0.01	0.00	0.01	0.04	0.04	0.05	0.05	0.08	0.18	0.10	0.02	0.01
	D	0.00	0.00	0.00	0.00	0.00	0.00	0.00	0.03	0.02	0.02	0.01	0.00	0.02	0.00	0.00	0.00
	E	0.00	0.00	0.00	0.00	0.00	0.00	0.00	0.00	0.00	0.00	0.00	0.00	0.01	0.00	0.00	0.00
	F	0.00	0.00	0.00	0.00	0.00	0.00	0.00	0.00	0.00	0.00	0.00	0.00	0.00	0.00	0.00	0.00
	G	0.00	0.00	0.00	0.00	0.00	0.00	0.00	0.00	0.00	0.00	0.00	0.00	0.00	0.00	0.00	0.00

Table E.7. Continued

Wind speed (m/s)	Stability class	Wind direction															
		S	SSW	SW	WSW	W	WNW	NW	NNW	N	NNE	NE	ENE	E	ESE	SE	SSE
11.2	A	0	0	0	0	0	0	0	0	0	0	0	0	0	0	0	0
	B	0	0	0	0	0	0	0	0	0	0	0	0	0	0	0	0
	C	0	0	0	0	0	0	0	0	0	0	0	0	0	0	0	0
	D	0	0	0	0	0	0	0	0	0	0	0	0	0	0	0	0
	E	0	0	0	0	0	0	0	0	0	0	0	0	0	0	0	0
	F	0	0	0	0	0	0	0	0	0	0	0	0	0	0	0	0
	G	0	0	0	0	0	0	0	0	0	0	0	0	0	0	0	0

Source: DCS (2002a).

Table E.8. Projected off-site population distribution at the SRS for the public for the year 2030

| Direction | Population by distance (miles[a]) | | | | | | Total |
	0 to 5	5 to 10	10 to 20	20 to 30	30 to 40	40 to 50	
S	0	0	920	2696	11,367	6,013	20,996
SSW	0	15	1,317	3,692	8,115	4,376	17,515
SW	0	186	1,978	7,732	3,535	4,579	18,010
WSW	0	171	2,572	7,553	4,368	10,385	25,049
W	0	407	10,186	17,766	15,109	11,753	55,221
WNW	0	2,331	8,556	219,212	54,849	24,980	309,928
NW	0	1,861	25,692	137,243	15,851	5,567	186,214
NNW	0	1,978	33,320	18,925	11,627	5,648	71,498
N	0	3,500	36,210	15,530	11,294	17,670	84,204
NNE	0	397	3,010	3,515	6,925	28,857	42,704
NE	0	14	2,609	4,611	8,850	19,325	35,409
ENE	0	0	5,535	7,865	8,764	53,785	75,949
E	0	2	8,061	8,590	18,423	9,310	44,386
ESE	0	14	3,658	4,352	5,466	488	13,978
SE	0	0	951	7,673	7,409	17,619	33,652
SSE	0	0	615	1,154	1767	4,234	7,770
Total	0	10,876	145,190	468,109	193,719	224,589	1,042,483

[a]To convert from miles to kilometers, multiply by 1.61.

Source: DCS (2002a).

E.2.2 Accidents

For the proposed MOX facility, four accident events were considered for detailed analysis, as discussed in Section 4.3.5.1. In each case, the amount of material released to the atmosphere was determined by multiplying the amount of material present (material at risk [MAR]) by the fraction of material involved in the event (damage ratio), fraction of material released that is airborne and respirable, and the fraction of material transported through a confinement mechanism (leak path factor). The values used for these parameters and the initial amount of plutonium material assumed to be present for each accident considered are given in Table E.12. Table E.13 lists the activity by radionuclide estimated to be released to the environment for each hypothetical accident.

Accident events considered for the PDCF and the WSB were discussed in Section 4.3.5.1. Six accident events were considered for the PDCF as taken from DOE (1999). Three accident events for the WSB were considered (DCS 2002a,b; Bowling 2002; DCS 2003b). Table E.13 lists the activity by radionuclide estimated to be released to the environment for each hypothetical accident.

Table E.9. Ingestion parameters used in GENII
for calculation of radiological exposure of the public
for normal and accidental air emissions

Parameter	Value Maximally exposed individual	Value Population
Terrestrial food		
Consumption rate (kg/yr)[a]		
Leafy vegetables	43	21
Root vegetables	92	66
Fruit	120	60
Grain	64	67
Crop yield (kg/m^2)[b]		
Leafy vegetables	1.5	1.5
Root vegetables	4	4
Fruit	2	2
Grain	0.8	0.8
Hold time between harvest and storage (days)[b]		
Leafy vegetables	1	14
Root vegetables	5	14
Fruit	5	14
Grain	180	180
Animal products		
Consumption rate (kg/yr)		
Beef[a]	81	43
Milk[a]	230	120
Poultry[b]	18	8.5
Eggs[b]	30	20
Holdup time (days)[b]		
Beef	15	34
Milk	1	3
Poultry	1	34
Eggs	1	18
Production rate (kg/yr)	NA[c]	-[d]
Diet fraction for animal food sources[b]		
Stored feed		
Beef	0.25	0.25
Milk	0.25	0.25
Poultry	1	1
Eggs	1	1
Fresh forage		
Beef	0.75	0.75
Milk	0.75	0.75

Table E.9. Continued

Parameter	Value Maximally exposed individual	Value Population
Growing time for animal food sources (days)[b]		
Stored feed		
Beef	90	90
Milk	45	45
Poultry	90	90
Eggs	90	90
Fresh forage		
Beef	45	45
Milk	30	30
Yield of animal food sources (kg/m^3)[b]		
Stored feed		
Beef	0.8	0.8
Milk	2	2
Poultry	0.8	0.8
Eggs	0.8	0.8
Fresh forage		
Beef	2	2
Milk	1.5	1.5
Storage time for animal food sources (days)[b]		
Stored feed		
Beef	180	180
Milk	100	100
Poultry	180	180
Eggs	180	180
Fresh forage		
Beef	100	100
Milk	0	0

[a]*Source*: Arnett and Mamatey (2001).

[b]GENII default values.

[c]NA = not applicable.

[d]See Section E.1.3 and Table E.8.

Table E.10. Food production data used in GENII for calculation of radiological ingestion exposure of the public for normal and accidental air emissions

Product/ direction	Production (kg/yr) by distance (mi[a])					
	0 to 5	5 to 10	10 to 20	20 to 30	30 to 40	40 to 50
Leafy vegetables						
S	0	0	0	0	0	1.0×10^5
SSW	0	0	0	0	0	1.0×10^5
SW	0	3.4×10^5	0	0	0	1.1×10^3
WSW	0	3.7×10^2	3.3×10^1	0	1.6×10^3	8.8×10^3
W	0	1.3×10^3	1.3×10^2	0	2.8×10^3	4.1×10^3
WNW	0	1.4×10^3	3.4×10^3	0	0	0
NW	0	1.4×10^3	6.3×10^3	4.7×10^3	0	0
NNW	0	1.3×10^3	6.9×10^3	8.7×10^3	8.6	2.4×10^3
N	0	1.1×10^3	6.9×10^3	1.2×10^4	1.1×10^4	4.8×10^4
NNE	0	5.9×10^2	6.9×10^3	1.2×10^4	3.1×10^5	9.6×10^5
NE	0	4.6×10^1	6.0×10^3	3.1×10^4	2.5×10^5	7.7×10^5
ENE	0	0	7.6	3.2×10^4	1.6×10^5	2.1×10^5
E	0	0	0	0	2.3×10^4	1.3×10^5
ESE	0	0	0	0	0	1.0×10^5
SE	0	0	0	0	0	1.0×10^5
SSE	0	0	0	0	0	1.0×10^5
Root vegetables						
S	0	0	1.8×10^6	3.1×10^6	4.1×10^6	6.3×10^6
SSW	0	3.1×10^3	2.1×10^6	3.4×10^6	4.3×10^6	6.7×10^6
SW	0	9.7×10^7	2.2×10^6	3.6×10^6	4.8×10^6	5.8×10^6
WSW	0	1.1×10^5	2.1×10^6	3.6×10^6	5.3×10^6	8.0×10^6
W	0	1.8×10^5	2.3×10^5	1.3×10^6	3.4×10^6	4.4×10^6
WNW	0	1.9×10^5	5.0×10^5	1.1×10^5	5.4×10^4	3.2×10^5
NW	0	2.0×10^5	8.8×10^5	8.2×10^5	4.0×10^5	1.4×10^5
NNW	0	1.9×10^5	9.6×10^5	1.3×10^6	7.3×10^5	1.2×10^6
N	0	1.5×10^5	9.6×10^5	1.6×10^6	1.7×10^6	2.4×10^6
NNE	0	8.1×10^4	9.6×10^5	1.6×10^6	2.5×10^6	3.8×10^6
NE	0	6.3×10^3	1.2×10^6	2.6×10^6	4.2×10^6	5.1×10^6
ENE	0	0	3.4×10^6	6.3×10^6	7.8×10^6	9.9×10^6
E	0	0	3.6×10^6	6.3×10^6	7.9×10^6	1.0×10^7
ESE	0	0	3.3×10^6	6.6×10^6	8.4×10^6	5.3×10^6
SE	0	0	6.4×10^7	6.8×10^6	8.8×10^6	9.2×10^6
SSE	0	0	3.8×10^7	3.0×10^7	6.7×10^6	7.8×10^6
Fruit						
S	0	0	3.9×10^5	1.1×10^6	1.7×10^6	2.5×10^6
SSW	0	6.9×10^2	4.5×10^5	8.7×10^5	1.4×10^6	2.3×10^6
SW	0	3.3×10^7	4.8×10^5	7.9×10^5	1.2×10^6	1.2×10^6
WSW	0	4.4×10^4	4.7×10^5	7.9×10^5	1.0×10^6	8.8×10^5
W	0	1.1×10^5	4.5×10^4	2.7×10^5	4.4×10^5	3.9×10^5
WNW	0	1.2×10^5	2.8×10^5	1.1×10^3	2.3×10^2	1.3×10^3
NW	0	1.2×10^5	5.3×10^5	2.8×10^6	6.6×10^6	2.2×10^6
NNW	0	1.1×10^5	5.8×10^5	2.8×10^6	1.2×10^7	1.4×10^7
N	0	9.0×10^4	5.8×10^5	9.7×10^5	5.1×10^6	4.8×10^6
NNE	0	4.9×10^4	5.8×10^5	9.7×10^5	1.0×10^6	7.4×10^5
NE	0	3.9×10^3	5.3×10^5	8.9×10^5	1.0×10^6	7.5×10^5
ENE	0	0	2.5×10^5	4.9×10^5	8.5×10^5	1.1×10^6

Table E.10. Continued

Product/ direction	Production (kg/yr) by distance (mi[a])					
	0 to 5	5 to 10	10 to 20	20 to 30	30 to 40	40 to 50
E	0	0	2.6×10^5	3.4×10^5	1.6×10^5	7.0×10^5
ESE	0	0	2.4×10^5	4.0×10^5	1.8×10^5	5.6×10^4
SE	0	0	4.3×10^6	3.1×10^5	3.7×10^5	3.1×10^5
SSE	0	0	2.6×10^6	2.0×10^6	1.1×10^6	1.0×10^6
Grains						
S	0	0	2.6×10^6	7.4×10^6	1.1×10^7	1.5×10^7
SSW	0	4.5×10^3	2.9×10^6	6.0×10^6	1.1×10^7	1.4×10^7
SW	0	1.1×10^8	3.1×10^6	5.1×10^6	8.2×10^6	1.0×10^7
WSW	0	1.4×10^5	3.0×10^6	5.1×10^6	8.1×10^6	1.5×10^7
W	0	2.1×10^5	6.4×10^5	2.2×10^6	6.1×10^6	7.9×10^6
WNW	0	2.2×10^5	7.6×10^5	7.2×10^5	2.6×10^5	6.5×10^5
NW	0	2.2×10^5	1.0×10^6	1.2×10^6	7.5×10^5	3.3×10^5
NNW	0	2.1×10^5	1.1×10^6	1.6×10^6	1.3×10^6	2.0×10^6
N	0	1.7×10^5	1.1×10^6	1.8×10^6	2.3×10^6	4.1×10^6
NNE	0	9.3×10^4	1.1×10^6	1.8×10^6	2.7×10^6	3.6×10^6
NE	0	7.3×10^3	1.3×10^6	3.6×10^6	6.1×10^6	6.9×10^6
ENE	0	0	4.0×10^6	8.7×10^6	1.4×10^7	1.8×10^7
E	0	0	4.2×10^6	9.0×10^6	1.6×10^7	1.9×10^7
ESE	0	0	3.9×10^6	8.9×10^6	1.6×10^7	1.2×10^7
SE	0	0	8.2×10^7	1.1×10^7	1.5×10^7	1.7×10^7
SSE	0	0	5.2×10^7	5.2×10^7	1.3×10^7	1.6×10^7
Beef						
S	0	0	1.2×10^5	4.6×10^5	7.3×10^5	9.9×10^5
SSW	0	2.2×10^2	1.5×10^5	3.4×10^5	6.9×10^5	9.3×10^5
SW	0	6.0×10^4	1.5×10^5	2.5×10^5	4.6×10^5	6.1×10^5
WSW	0	1.0×10^4	1.5×10^5	2.5×10^5	4.1×10^5	7.9×10^5
W	0	2.1×10^4	4.0×10^4	1.2×10^5	3.4×10^5	5.1×10^5
WNW	0	2.2×10^4	7.0×10^4	5.0×10^4	9.5×10^4	1.8×10^5
NW	0	2.3×10^4	1.1×10^5	1.4×10^5	1.6×10^5	2.1×10^5
NNW	0	2.2×10^4	1.1×10^5	1.8×10^5	2.3×10^5	3.5×10^5
N	0	1.7×10^4	1.1×10^5	1.9×10^5	3.1×10^5	6.5×10^5
NNE	0	9.6×10^3	1.1×10^5	1.9×10^5	2.5×10^5	2.9×10^5
NE	0	7.5×10^2	1.0×10^5	2.6×10^5	4.3×10^5	5.0×10^5
ENE	0	0	2.4×10^4	2.2×10^5	8.2×10^5	1.1×10^6
E	0	0	2.6×10^4	1.4×10^5	5.2×10^5	8.8×10^5
ESE	0	0	2.4×10^4	8.2×10^4	3.4×10^5	4.5×10^5
SE	0	0	4.8×10^5	6.4×10^4	2.0×10^5	5.2×10^5
SSE	0	0	3.6×10^5	5.8×10^5	4.3×10^5	6.7×10^5
Poultry						
S	0	0	0	0	0	5.4×10^4
SSW	0	0	0	0	0	6.7×10^4
SW	0	4.7×10^7	0	0	0	4.5×10^1
WSW	0	5.1×10^4	4.5×10^3	0	6.1×10^1	3.5×10^2
W	0	1.7×10^5	1.8×10^4	0	1.1×10^2	1.6×10^2
WNW	0	1.9×10^5	4.6×10^5	0	0	5.1×10^3
NW	0	1.9×10^5	8.6×10^5	6.4×10^5	0	3.0×10^5
NNW	0	1.8×10^5	9.4×10^5	1.2×10^6	1.2×10^3	5.4×10^5
N	0	1.5×10^5	9.4×10^5	1.6×10^6	1.7×10^6	3.6×10^6
NNE	0	8.0×10^4	9.4×10^5	1.6×10^6	1.3×10^6	5.4×10^3

Table E.10. Continued

Product/ direction	Production (kg/yr) by distance (mi[a])					
	0 to 5	5 to 10	10 to 20	20 to 30	30 to 40	40 to 50
NE	0	6.3×10^3	8.2×10^5	1.2×10^6	9.7×10^5	0
ENE	0	0	1.1×10^3	0	0	0
E	0	0	0	0	0	1.0×10^5
ESE	0	0	0	0	0	1.0×10^5
SE	0	0	0	0	0	1.0×10^5
SSE	0	0	0	0	0	1.0×10^5
Milk						
S	0	0	5.5×10^5	6.2×10^5	6.5×10^5	7.6×10^5
SSW	0	9.7×10^2	6.4×10^5	2.9×10^6	7.9×10^6	8.1×10^6
SW	0	3.2×10^6	6.7×10^5	1.1×10^6	3.8×10^6	2.9×10^6
WSW	0	2.2×10^4	6.6×10^5	1.1×10^6	2.0×10^6	4.4×10^6
W	0	1.2×10^4	4.9×10^4	3.8×10^5	1.8×10^6	3.5×10^6
WNW	0	1.3×10^4	3.1×10^4	0	4.7×10^4	1.2×10^6
NW	0	1.3×10^4	5.8×10^4	4.4×10^5	1.1×10^6	7.9×10^5
NNW	0	1.2×10^4	6.4×10^4	4.3×10^5	2.0×10^6	3.3×10^6
N	0	9.9×10^3	6.4×10^4	1.1×10^5	1.9×10^6	7.4×10^6
NNE	0	5.4×10^3	6.4×10^4	1.1×10^5	3.9×10^5	9.7×10^6
NE	0	4.2×10^2	5.5×10^4	6.9×10^5	1.7×10^6	1.8×10^6
ENE	0	0	7.0×10^1	1.1×10^6	4.6×10^6	5.6×10^6
E	0	0	0	9.6×10^5	4.2×10^6	5.7×10^6
ESE	0	0	0	3.2×10^5	2.6×10^6	1.6×10^6
SE	0	0	2.4×10^4	1.2×10^4	4.2×10^4	1.2×10^5
SSE	0	0	2.0×10^5	3.2×10^5	3.5×10^5	3.9×10^5
Eggs						
S	0	0	6.3×10^2	0	0	8.3×10^4
SSW	0	0	0	0	0	1.0×10^5
SW	0	6.2×10^5	0	0	0	9.1×10^1
WSW	0	0	0	0	1.2×10^2	7.0×10^2
W	0	0	0	0	2.2×10^2	3.3×10^2
WNW	0	0	0	0	0	1.0×10^5
NW	0	0	0	1.2×10^5	3.2×10^5	1.1×10^5
NNW	0	0	0	1.0×10^5	5.9×10^5	6.4×10^5
N	0	0	0	0	1.7×10^5	2.9×10^1
NNE	0	0	0	0	0	1.0×10^5
NE	0	0	4.1×10^3	4.0×10^3	1.6×10^2	1.2×10^2
ENE	0	0	4.3×10^4	5.5×10^4	5.0×10^2	6.3×10^2
E	0	0	4.5×10^4	5.6×10^4	7.1×10^1	4.0×10^2
ESE	0	0	4.2×10^4	5.8×10^4	1.2×10^2	0
SE	0	0	6.3×10^5	1.2×10^3	0	0
SSE	0	0	3.1×10^5	0	0	0

[a]To convert from miles to kilograms, multiply by 1.61.

Source: DCS (2002a).

E.2.2.1 SRS Employees

SRS employees downwind of an accident might be exposed to airborne radioactive contamination. Exposure would result primarily from external radiation from the radioactive contamination in the passing plume (cloudshine) released from the accident location and inhalation of the airborne contaminants. Short-term exposure to external radiation from ground-deposited radionuclides (groundshine) might also occur.

The GENII computer code (Napier et al. 1988) was also used to assess the radiological impacts to the sitewide population of SRS employees for each accident considered. The SRS employee population distribution used for the accident analysis is given in Table E.6, and the joint-frequency weather data are given in Table E.7. A ground-level release (1-m [3.3-ft] release height) was assumed for all accidents. To provide a conservative estimate for the impacts, 95% meteorology (meteorological conditions that produce impacts that are not exceeded 95% of the time) was used. Employees were assumed to be unshielded during passage of the contaminant plume from an accident. Both the inhalation and external exposure pathways were considered. Further external exposure to ground contamination for a period of 5.6 hours (8 hours with a shielding factor of 0.7) after the accident was also considered. Resuspension of contaminated soil was not considered, and the soil was assumed to be previously uncontaminated. Ingestion of contaminated foodstuffs was not considered because food is not grown on-site and consumed. Accident impacts to the SRS employee population are presented in Section 4.3.5.2 (see Table 4.13).

Table E.11. Centerline distance to site boundary from the proposed MOX facility stack for the primary 16 compass directions

Direction	Distance (m)
S	20,480
SSW	17,700
SW	12,130
WSW	15,000
W	9,490
WNW	9,930
NW	9,070
NNW	9,720
N	10,680
NNE	13,060
NE	16,520
ENE	19,040
E	19,150
ESE	20,030
SE	21,130
SSE	20,580

Table E.12. Source terms for detailed accident analyses

Hypothetical accident event	Quantity of plutonium at risk (kg)	Damage ratio	Respirable release fraction	Leak path factor
Internal fire	62 (polished)	1	0.0006	0.0001
Load handling	254 (polished)	1	0.0006	0.0001
Explosion	75 (unpolished)	1	0.01	0.0001
Criticality	41.5 (unpolished)	1	0.0005[a]	0.0001[b]

[a]For particulate matter, respirable release fraction = 1 for gases.

[b]For particulate matter, leak path factor = 1 for gases.

Sources: DCS (2002a, 2004a); Brown (2001).

Table E.13. Radionuclide quantities (Ci)^a released to the atmosphere for each accident type

Isotope	Proposed MOX facility				WSB		
	Internal fire	Load handling	Explosion	Criticality	Loss of confinement	Fire	Earthquake
Pu-238	2.2×10^{-5}	9.2×10^{-5}	4.5×10^{-4}	6.0×10^{-13}	1.2×10^{-5}	2.4×10^{-4}	2.5×10^{-4}
Pu-239	1.9×10^{-4}	7.7×10^{-4}	3.8×10^{-3}	5.0×10^{-12}	8.0×10^{-5}	1.6×10^{-3}	1.6×10^{-3}
Pu-240	4.6×10^{-5}	1.9×10^{-4}	9.2×10^{-4}	1.3×10^{-12}	3.0×10^{-5}	5.7×10^{-4}	6.0×10^{-4}
Pu-241	3.4×10^{-3}	1.4×10^{-2}	6.8×10^{-2}	9.0×10^{-11}	1.4×10^{-3}	2.8×10^{-2}	3.0×10^{-2}
Pu-242	1.3×10^{-8}	5.3×10^{-8}	2.6×10^{-7}	3.5×10^{-16}	NA^b	NA	NA
Am-241	NA	NA	2.0×10^{-3}	2.1×10^{-12}	NA	NA	NA
U-234	NA	NA	NA	NA	5.2×10^{-4}	1.0×10^{-2}	1.1×10^{-2}
U-235	NA	NA	NA	NA	9.2×10^{-6}	1.8×10^{-4}	1.9×10^{-4}
U-238	NA	NA	NA	NA	1.4×10^{-4}	2.7×10^{-3}	2.9×10^{-3}
Kr-83m	NA	NA	NA	1.1×10^{2}	NA	NA	NA
Kr-85m	NA	NA	NA	7.1×10^{1}	NA	NA	NA
Kr-85	NA	NA	NA	8.4×10^{-4}	NA	NA	NA
Kr-87	NA	NA	NA	4.3×10^{2}	NA	NA	NA
Kr-88	NA	NA	NA	2.3×10^{2}	NA	NA	NA
Kr-89	NA	NA	NA	1.3×10^{4}	NA	NA	NA
Xe-131m	NA	NA	NA	1.0×10^{-1}	NA	NA	NA
Xe-133m	NA	NA	NA	2.2	NA	NA	NA
Xe-133	NA	NA	NA	2.7×10^{1}	NA	NA	NA
Xe-135m	NA	NA	NA	3.3×10^{3}	NA	NA	NA
Xe-135	NA	NA	NA	4.1×10^{2}	NA	NA	NA
Xe-137	NA	NA	NA	4.9×10^{4}	NA	NA	NA
Xe-138	NA	NA	NA	1.1×10^{4}	NA	NA	NA
Te-134	NA	NA	NA	NA	NA	NA	NA
I-131	NA	NA	NA	2.8	NA	NA	NA
I-132	NA	NA	NA	2.9×10^{2}	NA	NA	NA
I-133	NA	NA	NA	4.1×10^{1}	NA	NA	NA
I-134	NA	NA	NA	1.1×10^{3}	NA	NA	NA
I-135	NA	NA	NA	1.1×10^{2}	NA	NA	NA
H-3	NA	NA	NA	NA	NA	NA	NA

Table E.13. Continued

			PDCF			
Isotope	Criticality	Earthquake	Explosion	Fire	Leak/spill	Tritium release
Pu-238	NA	2.62×10^{-6}	2.15×10^{-5}	8.06×10^{-8}	2.62×10^{-8}	NA
Pu-239	3.5×10^{-8}	2.22×10^{-5}	1.82×10^{-4}	6.82×10^{-7}	2.21×10^{-7}	NA
Pu-240	3.3×10^{-8}	5.43×10^{-6}	4.45×10^{-5}	1.67×10^{-7}	5.40×10^{-8}	NA
Pu-241	8.3×10^{-7}	4.04×10^{-4}	3.31×10^{-3}	1.24×10^{-5}	4.00×10^{-6}	NA
Pu-242	6.1×10^{-11}	1.55×10^{-9}	1.27×10^{-8}	4.76×10^{-11}	1.53×10^{-11}	NA
Am-241	2.0×10^{-7}	1.20×10^{-5}	9.82×10^{-5}	3.68×10^{-7}	1.18×10^{-7}	NA
U-234	NA	NA	NA	NA	NA	NA
U-235	NA	NA	NA	NA	NA	NA
Kr-83m	1.3	NA	NA	NA	NA	NA
Kr-85m	3.0	NA	NA	NA	NA	NA
Kr-85	NA	NA	NA	NA	NA	NA
Kr-87	1.9×10^{1}	NA	NA	NA	NA	NA
Kr-88	5.5×10^{1}	NA	NA	NA	NA	NA
Kr-89	NA	NA	NA	NA	NA	NA
Xe-131m	NA	NA	NA	NA	NA	NA
Xe-133m	NA	NA	NA	NA	NA	NA
Xe-133	4.5×10^{-1}	NA	NA	NA	NA	NA
Xe-135m	2.8×10^{1}	NA	NA	NA	NA	NA
Xe-135	8.0	NA	NA	NA	NA	NA
Xe-137	NA	NA	NA	NA	NA	NA
Xe-138	3.5×10^{2}	NA	NA	NA	NA	NA
Te-134	2.1×10^{1}	NA	NA	NA	NA	NA
I-131	4.3×10^{-2}	NA	NA	NA	NA	NA
I-132	3.5×10^{-1}	NA	NA	NA	NA	NA
I-133	7.5×10^{-1}	NA	NA	NA	NA	NA
I-134	9.5	NA	NA	NA	NA	NA
I-135	2.5	NA	NA	NA	NA	NA
H-3	NA	NA	NA	NA	NA	1.90×10^{5}

[a]To convert from curies (Ci) to becquerels (Bq), multiply by 3.7×10^{10}.

[b]NA = not applicable.

Radiological impacts to an MEI of the SRS employee population were assessed by assuming that the MEI was located outside the facility boundary, 100 m (330 ft) from the accident location. Inhalation exposure and external exposure from the passing radioactive cloud were evaluated. The ARCON96 computer code (Ramsdell and Simonen 1997) was used to estimate contaminant air concentrations at the MEI receptor location following an accidental release. ARCON96 was designed to model air dispersion in the vicinity of buildings. The code uses hourly meteorological data in order to estimate relative air concentrations of atmospheric releases. Ten years, 1987 to 1996, of hourly meteorological data and a building area of 6,580 m^2 (70,825 ft^2) (DCS 2001a) were used as input to the code. The 95th percentile relative concentration, the air concentration that is more than what might be expected 95% of the time, in any given direction for the 0- to 2-hour averaging period was conservatively used to estimate impacts. This 95h percentile relative concentration was calculated to be 6.1 x 10^{-4} s/m^3.

An inhalation rate of 3.47 x 10^{-4} m^3/s (NRC 1972), which includes consideration of an 8-hour shift, was then used in conjunction with inhalation dose conversion factors from Federal Guidance Report (FGR) 11 (Eckerman et al. 1988) to estimate inhalation exposure. The most conservative (largest) dose conversion factor among the clearance classes for each radionuclide was used. For external exposure, the external dose conversion factors from FGR 12 (Eckerman and Ryman 1993) were used. Estimated impacts to the SRS employee MEI are presented in Table 4.13 (Chapter 4) of this EIS. With the exception of the criticality accidents, inhalation exposure was the dominant impact. External exposure to cloudshine from the passing radioactive cloud after the criticality accident accounted for approximately 93% of the estimated dose to the MEI.

E.2.2.2 Members of the Public

Radiation exposures to members of the off-site public were assessed for hypothetical accidental releases. Impacts from a short-term exposure and one-year exposures (with and without ingestion) were evaluated for each accident. Exposure pathways evaluated for short-term exposures were inhalation, cloudshine, and groundshine. For 1-year exposures with ingestion, ingestion of contaminated crops was considered in addition to the short-term exposure pathways.

The GENII computer code (Napier et al. 1988) was used to assess the radiological impacts to the collective off-site population (members of the public) for each accident considered. The off-site population distribution used for the accident analysis is given in Table E.8, and the joint-frequency weather data are given in Table E.7. A ground-level release (1-m [3.3-ft] release height) was assumed for all accidents. To provide a conservative estimate for the impacts, 95% meteorology (weather conditions that produce impacts that are not exceeded 95% of the time) was used. For the short-term exposure, no credit was given for shielding for the inhalation and external exposures to the passing airborne plume. Exposure to groundshine was evaluated for 8 hours, but a shielding factor of 0.5 (NRC 1977) was used.

For the 1-year exposure periods, the length of time of external exposure to contaminated soil was 0.5 year (NRC 1977), and no credit was given for shielding for the inhalation exposure and external exposure to the passing airborne plume. For the 1-year exposure period with ingestion, ingestion parameters are provided in Table E.9. Food production data for the area surrounding the SRS are provided in Table E.10. The estimated impacts for each accident in the short term and after 1 year of exposure are presented in Table 4.14 (Chapter 4). No mitigative actions were assumed.

Accident impacts to an MEI member of the public were determined using the GENII code for both short-term and 1-year exposures following an accidental release. Potential MEIs were assumed to live at the site boundary, one at each of the 16 compass directions, as given in Table E.11. Exposure pathways considered in the analysis included inhalation, external exposure from the passing plume and contaminated soil, and, in the case for 1-year exposure with ingestion, ingestion of contaminated foodstuffs. The same release height and meteorology conditions as used for the population accident impacts were used for the MEI analysis. The amount of time of external exposure to contaminated soil was 8 hours (with a 0.7 shielding factor) and 0.7 year (NRC 1977) for the short-term and 1-year exposure periods, respectively. No credit for shielding was given for the inhalation and external exposures to the passing airborne plume. As a conservative assumption, potential MEIs were assumed to consume locally grown food for the 1-year exposure period with ingestion. Ingestion parameters are provided in Table E.9. The estimated impacts for each accident are given in Table 4.15 (Chapter 4) for the short-term and 1-year exposure periods. No mitigative actions were assumed.

E.3 References for Appendix E

Arnett, M.W., and A.R. Mamatey (eds.) 2001. *Savannah River Site Environmental Data for 2000.* WSRC-TR-2000-00328. Westinghouse Savannah River Company, Aiken, SC.

Birch, M.L. 2001. Letter with attachments from Birch (ES&H Manager, Duke Cogema Stone & Webster, Charlotte, NC) to J.B. Davis (Office of Nuclear Material Safety and Safeguards, U.S. Nuclear Regulatory Commission, Washington, DC) relative to "Docket No. 070-03098, Duke Cogema Stone & Webster Mixed Oxide (MOX) Fuel Fabrication Facility NRC-ANL Site Visit Reference." June 19.

Bowling, T.J. 2002. E-mail transmittal from Bowling (DCS, Charlotte, NC) to E.D. Pentecost (Argonne National Laboratory, Argonne, IL) regarding "ANL Question on WSB Accident Analyses." Nov. 15.

Brown, D. 2001. E-mail correspondence from Brown (U.S. Nuclear Regulatory Commission, Washington, DC) to B. Biwer (Argonne National Laboratory, Argonne, IL). Nov. 15.

CCPS 1996. *Guidelines for Use of Vapor Cloud Dispersion Models.* Second Edition. American Institute of Chemical Engineering, New York, NY.

Craig, D.K. 2002. *Revision 19 of ERPGs and TEELs for Chemicals of Concern.* WSMS-SAE-02-0171, Westinghouse Safety Management Solutions, Inc., Aiken, SC. Dec.

Craig, D.K., et al. 2000. "Derivation of Temporary Emergency Exposure Limits." *Journal of Applied Toxicology* 20:11-20.

DCS (Duke Cogema Stone & Webster) 2001a. *Responses to Request for Additional Information for Duke Cogema Stone & Webster (DCS) Mixed Oxide (MOX) Fuel Fabrication Facility (FFF) Environmental Report (ER).* Report with letter and attachments on CD-ROM, submitted by P.S. Hastings (DCS, Charlotte, NC) to U.S. Nuclear Regulatory Commission (Washington, DC). July 12.

DCS 2001b. *Responses to Request for Clarification of Additional Information for Duke Cogema Stone & Webster (DCS) Mixed Oxide (MOX) Fuel Fabrication Facility (FFF) Environmental Report (ER).* Docket Number 070-03098. Charlotte, NC.

DCS 2002a. *Mixed Oxide Fuel Fabrication Facility Environmental Report, Revision 1 & 2.* Docket Number 070-03098. Charlotte, NC.

DCS 2002b. *Responses to the Request for Additional information on the Environmental Report, Revisions 1 & 2.* DCS-NRC-000116, Docket No. 070-03098. Charlotte, NC. Oct. 29.

DCS 2003a. *Mixed Oxide Fuel Fabrication Facility Environmental Report, Revision 3.* Docket No. 070-03098. Charlotte, NC. June.

DCS 2003b. *Mixed Oxide Fuel Fabrication Facility Environmental Report, Revision 4.* Docket No. 070-03098. Charlotte, NC. Aug.

DCS 2004a. *Mixed Oxide Fuel Fabrication Facility Environmental Report, Revision 5.* Docket Number 070-03098. Charlotte, NC. June 10.

DCS 2004b. *Mixed Oxide Fuel Fabrication Facility Construction Authorization Request, Revision 6/10/04.* Docket Number 070-03098. Charlotte, NC.

DIPPR (Design Institute for Physical Property Data) 1989. *TDS Numerica.* America Institute of Chemical Engineers, New York, NY.

DOE (U.S. Department of Energy) undated. *DOE Occupational Radiation Exposure 2000 Report.* Assistant Secretary for Environment, Safety and Health, Office of Safety and Health, Washington, DC.

DOE 1994. *Final Defense Waste Processing Facility Supplemental Impact Statement.* DOE/EIS 0082-S. Savannah River Operations Office, Aiken, SC.

DOE 1997. *Final Waste Management Programmatic Environmental Impact Statement for Managing, Treatment, Storage, and Disposal of Radioactive and Hazardous Waste.* DOE/EIS-0200-F. Office of Environmental Management, Washington, DC. May.

DOE 1999. *Surplus Plutonium Disposition Final Environmental Statement.* DOE/EIS-0283. Office of Fissile Materials Disposition, Washington, DC. Nov.

Eckerman, K.F., and J.C. Ryman 1993. *External Exposure to Radionuclides in Air, Water, and Soil, Federal Guidance Report No. 12.* EPA 402-R-93-081. U.S. Environmental Protection Agency, Office of Radiation and Indoor Air, Washington, DC. Sept.

Eckerman, K.F., et al. 1988. *Limiting Values of Radionuclide Intake and Air Concentration and Dose Conversion Factors for Inhalation, Submersion, and Ingestion, Federal Guidance Report No. 11.* EPA 520/1-88-020. U.S. Environmental Protection Agency, Office of Radiation Programs, Washington, DC. Sept.

ICRP (International Commission on Radiological Protection) 1977. *Recommendations of the International Commission on Radiological Protection (adopted January 17, 1977).* ICRP Publication 26. Pergamon Press, Oxford, United Kingdom.

ICRP 1979. *Limit for Intakes of Radionuclides by Workers.* ICRP Publication 30, Part 1 (and subsequent parts and supplements), Vol. 2, Nos. 3-4 through Vol. 8, No. 4. Pergamon Press, Oxford, United Kingdom.

Kawamura, P., and D. MacKay 1987. "The Evaporation of Volatile Liquids." *Journal of Hazardous Material* 15:343-364.

LANL (Los Alamos National Laboratory) 1998. *Pit Disassembly and Conversion Facility Environmental Impact Statement Data Report—Savannah River Site, Final Report.* LA-UR-97-2910. Los Alamos, NM. June.

Linde, D.R. (ed.) 1999. *CRC Handbook of Physics and Chemistry.* CRC Press, Boca Raton, FL.

Napier, B.A., et al. 1988. *GENII — The Hanford Environmental Radiation Dosimetry Software System.* PNL-6584. Prepared by Pacific Northwest Laboratory, Richland, WA, for U.S. Department of Energy, Washington, DC. Dec.

NIST (National Institute of Standards and Technology) 2001. "NIST Standard Reference Database Number 69." Bethesda, MD. July. Available at http://webbook.nist.gov/chemistry/name-ser.html.

NRC (U.S. Nuclear Regulatory Commission) 1977. Calculation of Annual Dose to Man from Routine Releases of Reactor Effluents for the Purpose of Evaluating Compliance with 10 CFR Part 50, Appendix I, Rev. 1, Regulatory Guide 1.109. Washington, DC.

NRC 2001. Regulatory Guide 1.78: Evaluating the Habitability of a Nuclear Power Plant Control Room during a Postulated Hazardous Chemical Release. Revision 1. Office of Nuclear Regulatory Research. Dec.

Perry, R.H., and D.W. Green 1984. *Perry's Chemical Engineering Handbook.* Sixth Edition. McGraw-Hill.

Post, L. 1994a. *HGSYSTEM 3.0, User's Manual.* TNER.94.058. Shell Research Limited, Thorton Research Centre, Chester, United Kingdom.

Post, L. (ed.) 1994b. *HGSYSTEM 3.0, Technical Reference Manual.* TNER.94.059. Shell Research Limited, Thorton Research Centre, Chester, United Kingdom.

Raj, P.K., and J.A. Morris 1987. *A User's Manual for ADAM (Air Force Dispersion Assessment Model).* AFGL-TR-88-0003 (II). Prepared by Technology & Management Systems, Inc., Burlington, MA, for U.S. Air Force, Air Force Geophysical Laboratory, Hanscom Air Force Base, MA. Dec.

Ramsdell, J.V., and C.A. Simonen 1997. *Atmospheric Relative Concentrations in Building Wakes.* NUREG/CR-6331, PNNL-10521, Rev. 1. Prepared by Pacific Northwest National Laboratory, Richland, WA, for Division of Reactor Program Management, Office of Nuclear Reactor Regulation, U.S. Nuclear Regulatory Commission, Washington, DC. May.

Reynolds, R.M. 1992. *ALOHA (Areal Locations of Hazardous Atmospheres) 5.0, Theoretical Description.* Technical Memorandum NOS ORCA-65. National Oceanic and Atmospheric Administration, Washington, DC.

SRS (Savannah River Site) 2001. *SRS Emergency Plan.* WSRC-SCD-7, Rev. 16. Savannah River Site, Aiken, SC. July 31.

Weber, A. 2002. Personal communication from Weber (Savannah River Site, Aiken, SC) to M. Lazaro (Argonne National Laboratory, Argonne, IL). Dec.

www.ingramcontent.com/pod-product-compliance
Lightning Source LLC
Chambersburg PA
CBHW080227180526
45167CB00006B/2238